Advances in Marine Mechanical and Structural Engineering

Advances in Marine Mechanical and Structural Engineering

Editors

Kun Liu
Bin Liu
Chenfeng Li

MDPI

Basel • Beijing • Wuhan • Barcelona • Belgrade • Novi Sad • Cluj • Manchester

Editors

Kun Liu
Jiangsu University of Science and Technology
Zhenjiang
China

Bin Liu
Wuhan University of Technology
Wuhan
China

Chenfeng Li
Harbin Engineering University
Harbin
China

Editorial Office
MDPI AG
Grosspeteranlage 5
4052 Basel, Switzerland

This is a reprint of articles from the Special Issue published online in the open access journal *Journal of Marine Science and Engineering* (ISSN 2077-1312) (available at: https://www.mdpi.com/journal/jmse/special_issues/NV6O509QDS).

For citation purposes, cite each article independently as indicated on the article page online and as indicated below:

Lastname, A.A.; Lastname, B.B. Article Title. *Journal Name* **Year**, *Volume Number*, Page Range.

ISBN 978-3-7258-1739-9 (Hbk)
ISBN 978-3-7258-1740-5 (PDF)
doi.org/10.3390/books978-3-7258-1740-5

Contents

Journal of
Marine Science and Engineering

Editorial

Advances in Marine Mechanical and Structural Engineering

Kun Liu [1], Bin Liu [2,*] and Chenfeng Li [3]

1 School of Naval Architecture and Ocean Engineering, Jiangsu University of Science and Technology, Zhenjiang 212100, China; kunliu@just.edu.cn
2 Green & Smart River-Sea-Going Ship, Cruise and Yacht Research Centre, Wuhan University of Technology, Wuhan 430063, China
3 College of Shipbuilding Engineering, Harbin Engineering University, Harbin 150001, China; lichenfeng@hrbeu.edu.cn
* Correspondence: liubin8502@whut.edu.cn

In the design of modern ship and offshore structures, one of the key issues is the accurate prediction of strength with regard to various new materials and structures used in the structural design stage and under extreme sea environment and accidental states. The advancements in marine, mechanical and structural engineering are used in the mechanical analyses of advanced materials, such as alloys and composite materials, and strength analyses of novel structures, such as sandwich structures and superstructures, in order to render structures lightweight, safe and economical throughout their lifetimes.

The present Special Issue contains 14 articles, with 4 papers addressing the impact of ship and offshore structures, 2 papers being related to the mechanical behavior of novel structures, 3 papers covering various aspects of the strength assessment of ship and offshore structures, 1 paper being dedicated to vibration control in novel marine equipment, 2 papers regarding the soil–structure interaction analysis of offshore structures, 1 paper dealing with the load identification of ships and 1 paper focusing on the displacement analysis of ship shafting.

The definition of material failure criteria is an issue in the impact analysis of structures. Wang et al. [1] analyzed the deformation characteristics of a common cruciform structure, which is a part of ship hull structures, under planar collision and quasi-static loading. The applicability of the EPS, BWH and RTCL failure criteria in the simulation of compressive structures was investigated via the finite element simulation of quasi-static tests and falling weight impact tests. The effects of mesh size on the deformation and impact force of cruciform structures under plane loading were comparatively analyzed. The results showed that under plane loading, a cruciform structure undergoes axial compression deformation first, followed by buckling and wrinkling deformation. It is worth noting that the RTCL failure criterion is effective in modeling the failure of compressive structures in simulations with structures with different compressive deformations.

Citation: Liu, K.; Liu, B.; Li, C. Advances in Marine Mechanical and Structural Engineering. *J. Mar. Sci. Eng.* **2024**, *12*, 1114. https://doi.org/10.3390/jmse12071114

Received: 30 May 2024
Accepted: 19 June 2024
Published: 2 July 2024

The risk of collisions between passing vessels and wind turbines is increasing, representing a serious threat to the safety of personnel and equipment. Zong et al. [2] established a fluid–structure coupling simulation method based on Star-CCM+ and ABAQUS. It was used to investigate the dynamic response of floating wind turbines following bow and side impacts from vessels. A sensitivity analysis was performed on parameters such as collision speed, collision angle, wind speed, and wave height. The findings indicated that the amplitude of pitching and heaving motions of a turbine exceed those observed under conditions devoid of collision loads, with the amplitude of motion intensifying with an increase in these parameters.

A comprehensive understanding of the dynamic behavior of materials and structures under impact loads is paramount for structural design and maintenance. Tian et al. [3] carried out quasi-static and dynamic tensile tests using Q235 steel to obtain its material properties. The acquired data were utilized to fit the parameters using the Johnson–Cook model, a widely accepted constitutive model employed in high-strain-rate applications. The tensile test was numerically simulated based on the acquired experimental parameters. The good agreement between the load–displacement curves of the tests and simulations provided robust validation of the accuracy of the dynamic mechanical parameters of Q235 steel.

The phenomenon of repeated impacts on engineering structures is very common, especially in ocean engineering. Guo et al. [4] established a theoretical model based on the rigid–plastic assumption in order to analyze the plastic mechanical behavior of metal foam sandwich beams suffering from repeated impacts. The theoretical predictions agreed well with the results of impact tests and numerical simulations, indicating that the theoretical model was accurate and reliable. The results showed that the dimensionless permanent deflection was sensitive to the core strength ratio and the face thickness ratio, and as the core strength ratio or the face thickness ratio increased, the dimensionless permanent deflection decreased gradually in an exponential form.

A honeycomb structure is one of the most representative advanced structures in recent investigations. Wang et al. [5] analyzed the effect of nodal configuration on the mechanical properties of honeycombs. Several quasi-static compression tests were performed, which revealed that nodal reinforcement can inhibit nodal aberrations during ligament winding, thus facilitating the "rotational" mechanism and improving the negative Poisson's ratio properties of the honeycomb. Experimental and finite-element analyses showed that nodal reinforcement mainly played a role in the stage in which stress increased, and the role of nodal filling was more significant than that of nodal thickening. Li et al. [6] analyzed the mechanical performance of an improved star-shaped honeycomb. A quasi-static compression test was carried out to investigate its deformation mode and mechanical properties. To further utilize the excellent performance of the structure and obtain a better negative Poisson's ratio effect and broader application, a three-dimensional improved star-shaped honeycomb structure was proposed and a finite element simulation was carried out. The effects of the structural design, materials and dimensions on the mechanical properties, such as the energy absorption and negative Poisson's ratio, of the structures were explored.

Global ship analysis involving structural, motion and vibration analyses are crucial to examine the structural safety of hulls and motion response in ship design and construction. Lim et al. [7] performed finite element analysis, and weight distribution was employed to make the weight lighter and adjust the center of gravity. An effective and accurate algorithm and program was proposed to tune the weight, longitudinal shear force, bending moment and center of gravity of the ship finite element model to the required target value. The accuracy of the newly developed algorithm was analyzed and compared by applying it to the shuttle tanker finite element model under the ballast and full load conditions.

Ship bows and stems are often subjected to wave slamming loads. Stiffened plates with curvatures in both longitudinal and transversal directions are the basic components of these structures. Guo et al. [8] investigated the lateral ultimate strength of doubly curved stiffened plates using the non-linear finite element method. An empirical formula was derived and verified for the prediction of the lateral ultimate strength of the doubly curved stiffened plates, having good agreement with the numerical calculations.

A bolted ball joint structure has been utilized in offshore floating platforms and deep-sea fish cages, and Du et al. [9] presented an innovative method to improve the sealing performance of bolted ball joints. The approach involved creating sealed surfaces within the contact gaps between sleeves and connecting components by adding circular grooves and sealing washers to both ends of the sleeve.

The analysis of the interaction between pipes and sandy soil is of importance in the design of submarine oil and gas pipelines. Wang et al. [10] carried out experiments to study pipe–soil interaction, specifically measuring the lateral soil resistance of flexible pipes at varying burial depths. To simulate the mechanical behavior of pipe–soil interaction, the coupled Eulerian–Lagrangian method was employed for numerical simulations. The research findings indicated that lateral soil resistance is influenced by the uplift height and accumulation width of the soil ahead of the pipe. The ultimate soil resistance exhibited an increasing trend with an increasing burial depth.

Submarine cable burial machines are used for trenching operations on the seabed. Yu et al. [11] proposed a novel mechanical burial machine design employing a chain-type structure based on a combination of theoretical considerations and practical requirements. Through theoretical analysis, simulation and experimental studies, the cutting process of the mechanical burial machine was investigated in detail, with special attention given to the seabed conditions in the Northeast Asia region.

A pendulum-tuned mass damper is a widely used vibration-damping device capable of transferring and dissipating structural vibration energy. Du et al. [12] investigated the application and design of a pendulum-tuned mass damper in a buoyancy platform and analyzed its vibration reduction performance. Finite-volume simulations of the structure were conducted, providing insights into its motion under flow field effects and validating the damper's effectiveness in mitigating the buoyancy platform's flow-induced vibrations.

The identification of longitudinal bending moments is a critical component in the health monitoring of ship structures. Xu et al. [13] examined the effect of the failure of measurement points on the accuracy of bending moment identification and presented a solution method. The impact of failure point position and quantity on strain fitting accuracy and bending moment identification was investigated by performing a four-point bending experiment in typical failure scenarios. Further numerical analysis was conducted to identify potential sources of errors in the measurement process.

Deviations between the design and actual shafting occur due to limitations in ship construction accuracy. It is challenging to accurately obtain the relationship between the actual shafting load and displacement relationship based on the shafting design. To address the issue of incomplete actual shafting data, Deng et al. [14] proposed a transfer learning-based method for the accurate calculation of bearing displacement values. By combining simulated data from the shafting design with measured data generated during the adjustment process of the actual shafting, higher accuracy could be achieved when calculating bearing displacement values.

In summary, the articles presented in this Special Issue cover broad research topics related to advancements in marine mechanical and structural engineering, guiding readers through the best approaches to analysis.

Author Contributions: Writing—original draft preparation, B.L.; writing—review and editing, K.L. and C.L. All authors have read and agreed to the published version of the manuscript.

Conflicts of Interest: The authors declare no conflicts of interest.

References

1. Wang, X.; Liu, K.; Xu, M.; Liu, H. Plastic Response and Failure of the Cruciform Structure under In-Plane Load. *J. Mar. Sci. Eng.* **2023**, *11*, 1478. [CrossRef]
2. Zong, S.; Liu, K.; Zhang, Y.; Yan, X.; Wang, Y. The Dynamic Response of a Floating Wind Turbine under Collision Load Considering the Coupling of Wind-Wave-Mooring Loads. *J. Mar. Sci. Eng.* **2023**, *11*, 1741. [CrossRef]
3. Tian, X.; Pei, J.; Rong, J. Dynamic Mechanical Properties and Damage Parameters of Marine Pipelines Based on Johnson–Cook Model. *J. Mar. Sci. Eng.* **2023**, *11*, 1666. [CrossRef]
4. Guo, K.; Mu, M.; Cai, W.; Xu, B.; Zhu, L. Theoretical Analysis of Plastic Behavior of Sandwich Beam with Metal Foam under Repeated Impacts. *J. Mar. Sci. Eng.* **2023**, *11*, 1974. [CrossRef]
5. Wang, Z.; Chen, G.; Cao, X.; Chen, W.; Li, C.B.; Li, X. Study on the Effect of Nodal Configuration on the Mechanical Properties of Hexa-Ligamentous Chiral Honeycombs. *J. Mar. Sci. Eng.* **2023**, *11*, 1692. [CrossRef]
6. Li, Q.; Cao, X.; Wu, X.; Chen, W.; Li, C.; Li, X. Investigation of the Energy Absorption Characteristics and Negative Poisson's Ratio Effect of an Improved Star-Shaped Honeycomb. *J. Mar. Sci. Eng.* **2023**, *11*, 1799. [CrossRef]
7. Lim, C.; Han, I.-S.; Kang, J.-Y.; Ban, I.-J.; Lee, B.; Park, J.S.; Shin, S.-C. Shear Force and Bending Moment Tuning Algorithm of Shuttle Tanker Model for Global Structural Analysis. *J. Mar. Sci. Eng.* **2023**, *11*, 1900. [CrossRef]
8. Guo, G.; Cui, J.; Wang, D. A Study on the Lateral Ultimate Strength and Collapse Modes of Doubly Curved Stiffened Plates. *J. Mar. Sci. Eng.* **2023**, *11*, 2315. [CrossRef]
9. Du, W.; Gu, J.; Sheng, G.; Guo, G.; Zhao, Y.; Liu, Z. Enhancing the Sealing Performance of Bolted Ball Joints by Gaskets: Numerical Simulation and Experiment. *J. Mar. Sci. Eng.* **2023**, *11*, 2050. [CrossRef]

10. Wang, C.; Liu, L.; Zhang, Y.; Lou, M. A Method for Predicting the Load Interaction between Reinforced Thermoplastic Pipe and Sandy Soil Based on Model Testing. *J. Mar. Sci. Eng.* **2023**, *11*, 2353. [CrossRef]
11. Yu, Z.; Jin, Z.; Wang, K.; Zhang, C.; Chen, J. Design and Study of Mechanical Cutting Mechanism for Submarine Cable Burial Machine. *J. Mar. Sci. Eng.* **2023**, *11*, 2371. [CrossRef]
12. Du, L.; Cui, Y.; Ma, Y.; Liu, J.; Liu, Z. Analysis of Flow-Induced Vibration Control in a Pontoon Carrier Based on a Pendulum-Tuned Mass Damper. *J. Mar. Sci. Eng.* **2023**, *11*, 1963. [CrossRef]
13. Xu, G.; Gan, J.; Li, J.; Liu, H.; Wu, W. A Strain Fitting Strategy to Eliminate the Impact of Measuring Points Failure in Longitudinal Bending Moment Identification. *J. Mar. Sci. Eng.* **2023**, *11*, 2282. [CrossRef]
14. Deng, Y.; Li, Y.; Zhu, H.; Fan, S. Displacement Values Calculation Method for Ship Multi-Support Shafting Based on Transfer Learning. *J. Mar. Sci. Eng.* **2024**, *12*, 36. [CrossRef]

Journal of
Marine Science
and Engineering

MDPI

Article

Plastic Response and Failure of the Cruciform Structure under In-Plane Load

Xiufei Wang [1], Kun Liu [1,*], Mingcai Xu [2] and Hewei Liu [1]

[1] School of Naval Architecture and Ocean Engineering, Jiangsu University of Science and Technology, Zhenjiang 212003, China; 182010037@stu.just.edu.cn (X.W.); 209010003@stu.just.edu.cn (H.L.)
[2] School of Ship and Ocean Engineering, Huazhong University of Science and Technology, Wuhan 430074, China; xumc@163.com
* Correspondence: kunliu@just.edu.cn; Tel.: +86-135-1169-2085; Fax: +86-0511-8444-6543

Abstract: In this study, a common cruciform structure in ship hulls was designed and experimented with in order to analyze its deformation characteristics under planar collision and quasi-static loading. The mechanical parameters of the materials were determined by performing tensile tests on the plates used in the specimens. The applicability of the EPS, BWH, and RTCL failure criteria in the simulation of compressive structures was investigated by finite element simulation of quasi-static tests and falling weight impact tests. The effects of mesh size on the deformation and impact force of the cruciform structure under plane loading were comparatively analyzed. The results show that under plane loading, the cruciform structure undergoes axial compression deformation first, followed by buckling and wrinkling deformation. Compared with the quasi-static test, the drop hammer impact test showed higher deformation concentration and smaller wrinkle height. Under the same axial deformation condition, the structural resistance of the drop hammer impact test was about 13% higher than that of the quasi-static test. It is worth noting that the RTCL failure criterion is effective in modeling the failure of compressive structures in simulations with structures with different compressive deformations.

Keywords: cruciform structure; model experiment; numerical simulation; ship collision and grounding; ductile failure criteria

Citation: Wang, X.; Liu, K.; Xu, M.; Liu, H. Plastic Response and Failure of the Cruciform Structure under In-Plane Load. *J. Mar. Sci. Eng.* **2023**, *11*, 1478. https://doi.org/10.3390/jmse11071478

Academic Editor: Vincenzo Crupi

Received: 2 July 2023
Revised: 20 July 2023
Accepted: 21 July 2023
Published: 24 July 2023

1. Introduction

The hull structural damage caused by collision and grounding accidents varies depending on the situation. Minor collision accidents usually only result in slight plastic deformation of the structure, which can be restored through repairs. However, severe accidents can cause extensive structural damage, resulting in cabin leakage, oil spills, and even causing the vessel to break due to insufficient remaining strength, leading to major disasters. According to statistics from the ITOPF organization on the causes of large oil spills from 1970 to 2022, 62% of accidents were caused by ship collisions and groundings [1]. In recent decades, to investigate the structural resistance of ships, many scholars have studied the structural response in collision and grounding accidents, using methods such as model testing, nonlinear finite element analysis, simplified analytical methods, and empirical formulae. They have investigated the response and damage modes of typical structures in ship structures subjected to impact loading, which has significant implications for ship structure design.

Experimental methods are the most reliable means to evaluate the resistance and collision response of ship structures. However, due to their high cost in terms of time and manpower, the practical operation of experimental methods is difficult. Simplified analytical methods are used for the preliminary design of structure collision prevention.

Stiffened plates are the most common structures in ship structures, and assessing their resistance performance is particularly important for the design of ship structures. Due to the demanding experimental methods that are time-consuming and labor-intensive, many scholars have verified other research methods through impact tests on stiffened plates. Cho and Lee [2] conducted collision tests on stiffened plates and proposed a simplified analytical method to predict the extent of damage to stiffened plates under in-plane loads. Liu K and Liu B [3,4], Yang L and Wang D [5], Alsos and Amdahl [6,7], and others have compared the influence of key parameters of material settings, including dynamic behaviors of materials and different failure models, on simulation results through experiments, optimizing numerical simulation methods and making simulation results more credible. Zhang M [8] conducted experimental and numerical simulation studies on a scaled ship side-shell quasi-statically punched at the mid-span by a raked bow indenter to investigate its damage characteristics from deformation to a large opening. For strong supporting structures commonly found in double-layer hull structures, such as the web girder and other supporting structures, Wang G [9], Zhang M [10], and Villavicencio R [11] proposed simplified analytical methods. The cruciform structure is widely used in ship structures as an energy absorption structure. Chen B [12] investigated the structural response of a double-hull structure under in-plane loading by changing the size of the indenter. Urban [13] conducted axial compression tests on aluminum L-shaped and cruciform structures and obtained types of damage modes. The supporting structure mainly produced relatively uniform wrinkling deformation, while the cruciform structure suffered damage at the axial weld seam. Zhou [14] investigated the energy absorption capacity of a similar cruciform structure by changing the axial weld seam under planer load. Haris S and Amdahl J [15,16] derived the resistance calculation formula for cruciform structures under overall compression and studied the buckling deformation process of the cruciform structures under planar loads. The proposed analytical method is used to estimate the energy absorbed by the structure in collision and grounding accidents. Hayduk and Wierzbicki [17] studied the deformation of cruciform structures, T-shaped structures, and L-shaped structures under planar loads and derived analytical calculation formulae for average resistance. Yang and Caldwell [18] conducted experimental research on the energy absorption and average compressive strength of cruciform structures and T-shaped structures under compression. Wang G [9] studied the variation of structural resistance of the cruciform structure under axial load in the longitudinal direction. The deformation of the cruciform structure differs when there are constraints around it compared to when it is alone. When the loading surface is large, the load-bearing condition of the cruciform structure is the same as that of the structure analyzed by Urban and Zhou [13,14]. When the loading surface is small, i.e., when the ship is subjected to minor structural impacts or grounded on small reefs, the outer side of the cruciform structure is not free and can be regarded as bearing local loads, as shown in Figure 1. The aforementioned studies conducted tests on designed structures under quasi-static or impact loads. Quasi-static tests are more suitable for accidents involving static compression and low-speed collisions, while collision tests can study the response of structures in high-speed collision scenarios. The impact of dynamic effects on structural response has been relatively less studied in experiments. This article will compare the structural response of a cross-shaped structure under two types of loads, quasi-static and impact, to investigate their differences.

In recent years, with the rapid development of computer performance, the finite element method has been widely used to simulate the structural response in ship collision and grounding scenarios. One of the most vital issues in the ship collision finite element analysis is the prediction of structural damage, including the location and damage size of the structure, which will have essential effects on the subsequent overall structural deformation mode and energy dissipation. The widely used failure criteria include the EPS failure criterion, BWH failure criterion, RTCL failure criterion, and so on. The EPS failure criterion has been widely applied because of its simplicity and convenience. However, due to the existence of strain gradients, the definition of failure strain is related to the element size [19–24]. References show that different failure strains correspond to different element sizes by comparing simulation curves with real experimental curves obtained by material tensile tests. Therefore, using the failure strain corresponding to the model mesh size can reduce the large deviations between simulation results and test data effectively. The BWH failure criterion proposed by Alsos [6] uses the necking of the element to determine its failure. Storheim et al. [25,26] improved the BWH failure criterion by considering the subsequent strength of the material after necking. Yang L [5] proposed the m-BWH failure criterion and validated its accuracy at high strain rates through various plane stress tests. Alsos and Amdahl [6] studied the mesh sensitivity of the BWH and RTCL failure criteria in the FEA of quasi-static penetration experiments on stiffened plates. Compared with the BWH failure criterion, the RTCL failure criterion is more sensitive to mesh size. Marinatos and Samuelides [27] obtained the same rule as Alsos [6] through finite element simulation of ship structures. In recent years, many scholars have also carried out research on new failure criteria, and these failure criteria are usually proposed by combining various classical failure criteria to bypass the inadequacy of a single failure criterion. Lu et al. [28,29] developed the MSSRT failure criterion and simulated the generation and extension of cracks in the structure in the test of a double-hulled hull. Costas [30] proposed the large-size shell unit regularization model for damage and validated it.

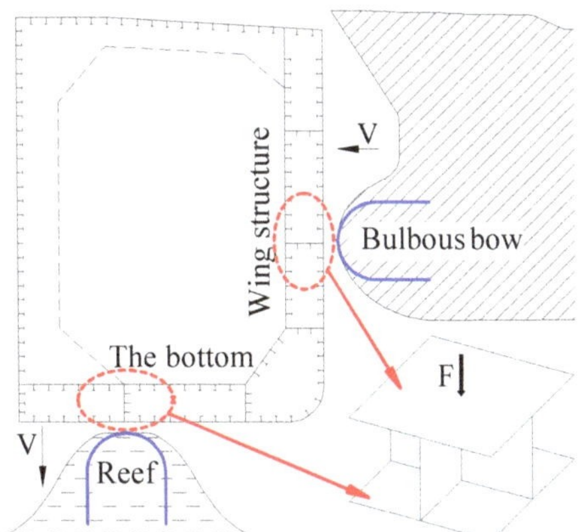

Figure 1. Collision and grounding scenario of the ship.

This paper investigates the structural response of the commonly found cruciform structure in typical double-hull ship structures under local quasi-static and impact loads. Tensile tests of materials were carried out to obtain the stress–strain relationship, and the corresponding finite element simulations were conducted to obtain the relationship between the mesh size and failure strain. The applicability of three failure criteria, EPS, BWH, and RTCL, was compared in the simulation of this type of compression structure, and the influence of mesh sizes on the deformation mode and structural resistance of the cruciform structure was studied. Additionally, the paper discusses the influence of friction coefficients and collision positions on the structural response.

2. Experiments of the Cruciform Structure

2.1. Specimen

To investigate the dynamic and quasi-static responses of a cruciform structure subjected to in-plane loads, reduced-scale test specimens were designed and manufactured. This structure is a simplified version of the typical double-hull structure, as shown in Figure 2. The cruciform structure comprises four 4 mm thick steel plates welded together. A front panel is also welded onto the cruciform structure, similar to the hull shell. Additionally, four support plates with lightning holes are welded around the cruciform structure, forming a frame that is in contact with the front panel. To prevent the frame from collapsing when subjected to in-plane loads, four 10# U-bars are welded around the top of the frame, with the sides of the square front panel welded onto the sides of the U-bars. The plate thicknesses of the main portion of the specimen are shown in Figure 3. The entire specimen is constructed from mild steel, and silver-gray paint is sprayed on it. Finally, a grid with a length of 20 mm is drawn on both the cruciform structure and front panel to observe the extent of deformation and degree of damage.

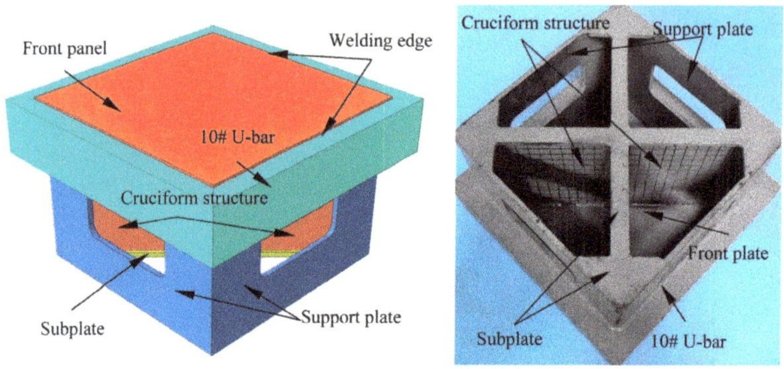

Figure 2. Specimen and its components.

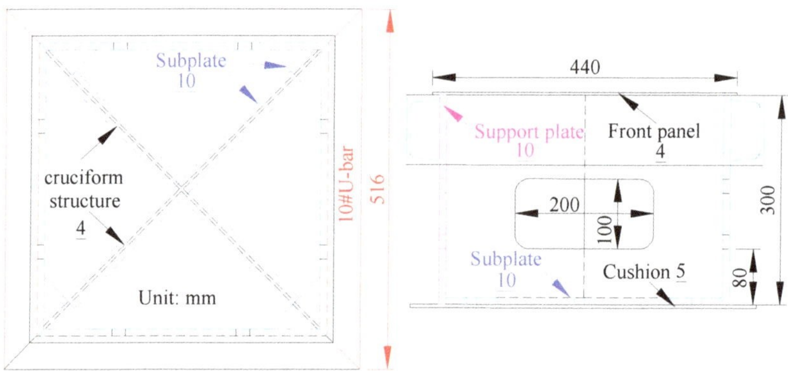

Figure 3. Main dimensions of the specimen.

2.2. Quasi-Static Indentation Test Set-Up

The tests were conducted at the Mechanics Lab at Jiangsu University of Science and Technology. The experiment utilized an electro-hydraulic servo universal testing machine (YNS1000), as depicted in Figure 4a, which is capable of imposing a 1000 kN load on the specimen being tested, with a maximum stroke of 100 mm of the indenter. A force transducer was installed at the end of the indenter to measure the applied force, while the dimensions of the indenter are displayed in Figure 4b. The indenter's contact point with the specimen is hemispherical, and it is composed of GCr15 material, which has high hardness. The testing system connected to the machine can record the displacement of the indenter and the contact force.

To begin the test, the specimen is positioned on the test bed and the center of the grid on the front panel is aligned with the axis of the indenter. The indenter is lowered using the testing machine and the displacement and load data are set to zero once the indenter makes contact with the front panel. Then, the indenter is moved downward at a constant speed of 10 mm/min while acquiring displacement and force data at a frequency of 20 Hz. The load profile was monitored in real time during the test. Due to the height of the specimen, the effective downward distance of the indenter is 65 mm, and the test is stopped when the displacement of the hammer head reaches the lowest limit.

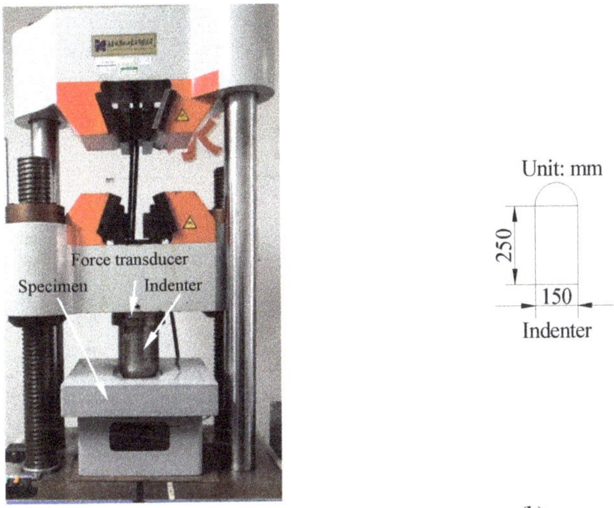

(a) (b)

Figure 4. (**a**) Electro-hydraulic servo universal testing machine, (**b**) dimensions of the indenter.

2.3. Falling Weight Impact Test

As shown in Figure 5a, the Falling Weight Impact Tester can be utilized to conduct impact tests and acquire the dynamic response and deformation characteristics of the cruciform structure. The indenter has a maximum lifting height of 3.7 m, enabling a maximum impact speed of 8.5 m/s, which can simulate most ship collision scenarios. The falling weight has a maximum mass of 1.35 tons and a designed maximum gravitational potential energy of 50 kilojoules. The acceleration sensor in Figure 5b is fixed on the main body of the falling weight to measure acceleration data, while a laser rangefinder is placed directly above the measuring point to record the displacement of the indenter during the experiment. A hemispherical indenter with the same size as used in the quasi-static test is utilized to compare the deformation of the cruciform structure under the two scenarios. The indenter is lifted to 0.5 m and 3 m, and the impact velocities are 3.13 m/s and 7.67 m/s, respectively. The mass of the falling weight is a maximum value of 1.35 tons. Both sensors in the test system start recording at the same time before the falling weight is released, with an acquisition frequency of 20 kHz.

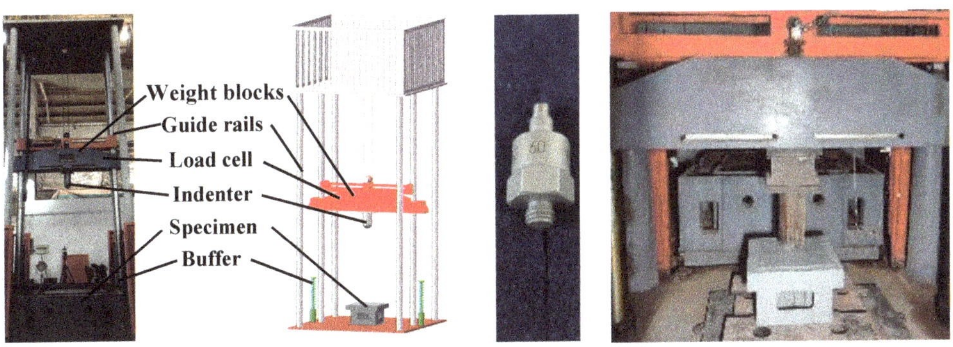

(a) (b) (c)

Figure 5. (**a**) Schematic diagram of the Falling Weight Impact Tester and (**b**) accelerometer, (**c**) their use during the experiment.

2.4. Experimental Results and Comparison

2.4.1. Quasi-Static Test Results

The deformation of the front panel and cruciform structure is displayed in Figure 6a. The peeled paint on the front panel reveals that the contact area between the indenter and the front panel forms a circle with a diameter of 10 cm. The significant deformation is concentrated in the circular area at the center of the indenter with a diameter of 20 cm. The front panel exhibits four cracks, with lengths of 40 mm, 45 mm, 43 mm, and 54 mm, all originating from the edge of the weld.

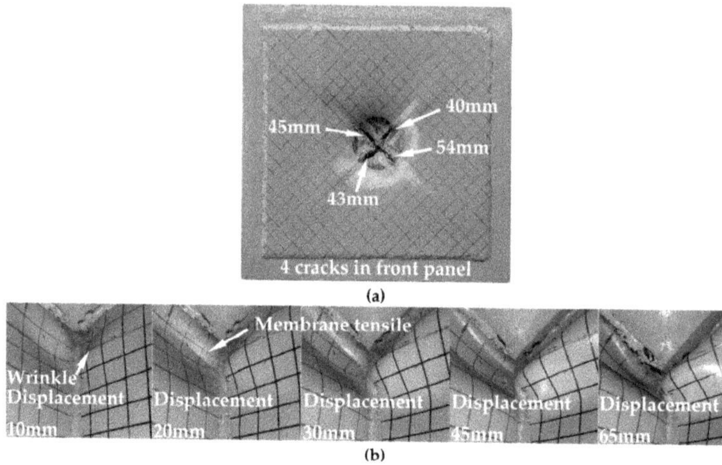

Figure 6. (**a**) Final deformation of the front panel, (**b**) deformation process of the cruciform structure.

Figure 6b shows the deformation process of the cruciform structure in the quasi-static test with indenter displacements of 10 mm, 20 mm, 30 mm, 45 mm, and 65 mm. It can be observed that as the indenter gradually presses down, the cruciform structure starts to bend at the joint of the weld with the front panel. At a distance of about 5 cm from the front panel, the cruciform structure bulges perpendicular to the web girder, and part of this pleat will produce membrane tensile deformation. With the increase in the indenter displacement, the fold on the web girder gradually forms. When the first fold is formed, the length of the fold is about 18 cm, and a crack is produced at the middle connection of the cruciform structure. The deformation direction of the two web girders is the same.

Figure 7 shows the force–indentation relationship in the test. The curve can be divided into three stages according to the sudden change in slope. The initial stage before point "A" is characterized by a rapid rise in reaction force, indicating an increase in the contact area and deformation degree. The sudden decrease in the curve slope at point A indicates the onset of buckling in the cruciform structure. There are two main reasons for the smooth increase in structural resistance in the "II" stage between points A and B in Figure 7. One of them is the increase in the contact area between the hammer head and the specimen and the increase in the deformed structure, which leads to the increase in the structural resistance, and on the other hand, it is due to the membrane tensile deformation of both the upper panel and the cross structure and the gradual increase in the component of the structural resistance along the direction of the load. At point B of the curve, the structural resistance begins to decrease because the front panel of the structure is torn, and the structural resistance gradually decreases with the increase in the crack. The "III" stage of the curve shows a wavy decrease, which is consistent with the characteristics of the gradual tearing of the plate structure.

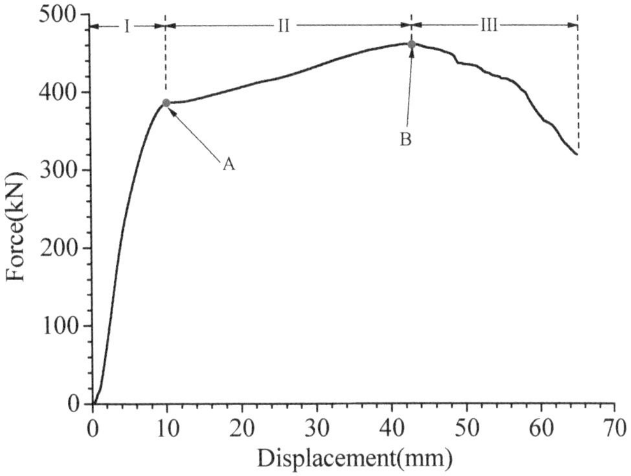

Figure 7. Force–indentation curve of the quasi-static test.

2.4.2. Falling Weight Impact Test Results

Figure 8 illustrates the deformation of the front panel and cruciform structure under two test conditions. It is evident from the figure that both test specimens experienced significant plastic deformation under the impact of the indenter, which caused severe paint peeling on the structure. In the 0.5 m drop test, the paint wear was concentrated in the contact area between the indenter and the specimen, and the cruciform structure showed obvious folding deformation. The length of the fold in the long side direction of the web girder was four grids, with a length of approximately 80 mm and a height of approximately 30 mm in the axial direction. The structure did not tear. In the 3.0 m drop test, a circular large deformation area was formed on the front panel because the entire indenter had fallen into it. Four cracks were produced along the direction of the web beam, and the length of a single crack was about 120 mm, as per the grid drawn on the front panel. The square front panel was divided into four right-angled triangle areas by the cruciform structure, and the deformation results showed that the triangle produced a fold parallel to the long side. This occurred because the web girder strongly supported the front panel in the axis direction, resulting in a straight fold in the triangle area. The four folds formed a concave deformation in a square area, as shown in the red box in Figure 8. From the damage and deformation of the cruciform structure, it was evident that the web girder forming the cruciform structure had a bulge in the normal direction, and the middle connecting part of the cruciform structure was also torn.

Figure 9 shows the force–indentation curve recorded in both test conditions. The load rose rapidly at the initial stage of the collision, primarily due to compression deformation along the central axis direction since the cruciform structure had not buckled yet. As the indenter moved downward, the cruciform structure buckled and produced fold deformation, causing the structural resistance fluctuation to rise until the rebound of the indenter. In both cases, the load at point A in the curve decreased significantly when the cruciform structure buckled, producing a large degree of fold deformation, and the contact force fluctuated. Since the structure did not tear when the indenter fell from a height of 0.5 m, the load fluctuation rose, and substantial unloading did not occur, decreasing to 0 after the indenter rebounded. However, when the indenter fell from a height of 3 m, the front panel of the specimen and the junction of the web girders had multiple tears, and the force–indentation curve exhibited more loading and unloading at the rear end.

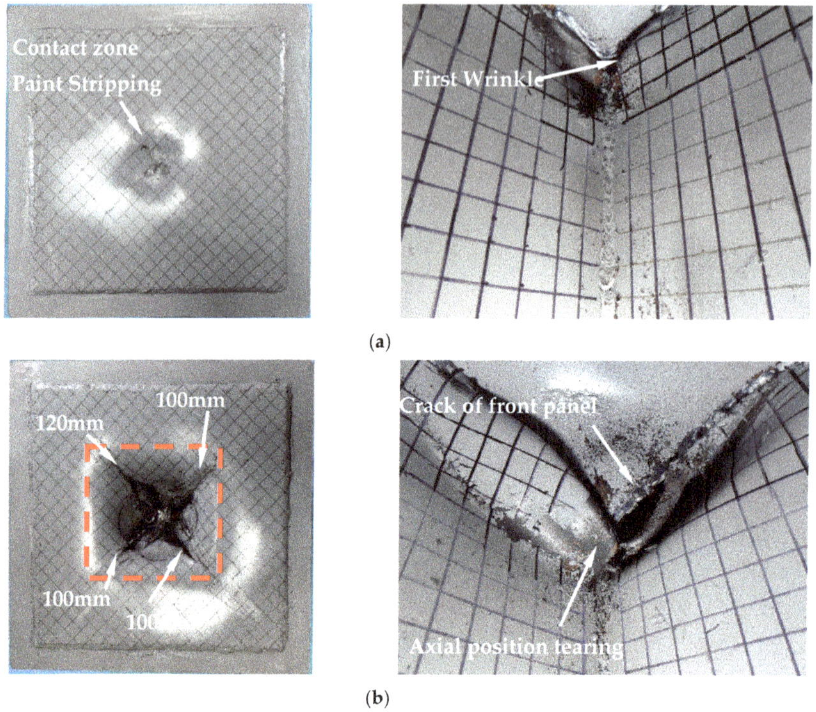

Figure 8. Deformation and damage of the specimen after impact, (**a**) 0.5 m falling impact condition, (**b**) 3 m falling impact condition.

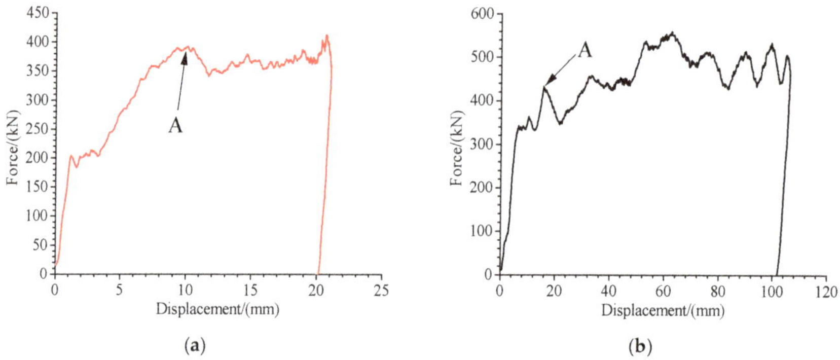

Figure 9. Force–indentation curve of falling weight impact test, (**a**) 0.5 m falling impact condition, (**b**) 3 m falling impact condition.

2.4.3. Comparison of Quasi-Static Test and Dynamic Test

The comparison between Figures 6 and 8 shows that the cruciform structure first undergoes axial compression deformation upon impact. As shown in Figure 10, all three test conditions show a rapid increase in load at this stage. Subsequently, the cruciform structure buckles, producing wrinkling deformation, and the rate of increase in the impact load curve is significantly reduced. The first collision peak value is relatively high in the 3 m falling condition, primarily due to the structural inertia, resulting in a delayed buckling of the cruciform structure. During the initial stage of collision, when the displacement of the indenter is small, the fold deformation range is also small. In dynamic collision scenarios, the length of the cruciform structure fold deformation area along the web girder's direction is smaller than in quasi-static conditions. As the indenter descends, the degree of fold deformation increases, and the length along the direction of the web girder gradually increases as well. The height of the fold in the axial direction at the initial stage of collision is similar to that after the fold is flattened. Once the indenter moves to a certain extent, the front panel of the specimen tears, and the panel's failure to support the cruciform structure results in tear deformation in the direction of the cruciform structure axis. In the falling weight impact test, the load curve exhibits clear dynamic fluctuation.

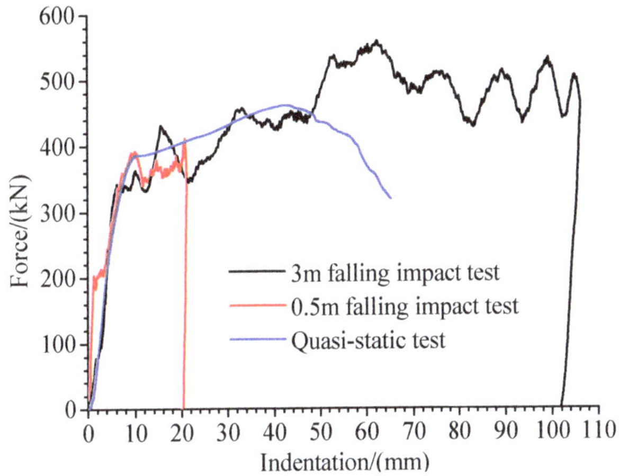

Figure 10. Comparison of load–indentation curves in quasi-static and falling weight impact tests.

3. Failure Criteria

With the development of computer performance, the finite element method has become an important tool for assessing the safety of engineering structures. The failure criterion is a judgment method for assessing whether a unit produces damage in finite element calculations. For the damage to the structure during collision stranding, the selection of an appropriate failure criterion is a key factor to ensure that the numerical simulation results are reasonable and accurate. The following failure criteria are commonly used in this field:

3.1. EPS Criterion

The EPS failure criterion posits that material failure occurs when the critical value of equivalent plastic strain is reached. Due to its simple definition, this failure criterion was commonly adopted in numerical simulations for ship collision scenarios in the early stage [31].

$$\varepsilon_f = \bar{\varepsilon}_{cr} \quad (1)$$

3.2. BWH Criterion

The BWH failure criterion was proposed by Alsos and Amdahl [6]. This failure criterion integrates the theoretical necking instability model of Hill [32] and the shear stress criteria of Bressan Williams [33] and transforms them into stress space. When a plate-like structure undergoes significant plastic deformation, local necking will occur at the location where failure is imminent before the plate is damaged. The BWH criterion can be represented as follows:

$$\sigma_1 = \begin{cases} \frac{2K}{\sqrt{3}}\frac{1+0.5\beta}{\sqrt{\beta^2+\beta+1}}\left(\frac{2}{\sqrt{3}}\frac{\widehat{\varepsilon}_1}{1+\beta}\sqrt{\beta^2+\beta+1}\right)^n & \beta \le 0 \\ \frac{2K}{\sqrt{3}}\frac{\left(\frac{2\widehat{\varepsilon}_1}{\sqrt{3}}\right)^n}{\sqrt{1-\left(\frac{\beta}{2+\beta}\right)^2}} & \beta > 0 \end{cases} \tag{2}$$

According to Hill's criteria, the critical strain $\widehat{\varepsilon}_1$ may be equivalent to the power law coefficient n. The strain ratio β can be expressed as $\dot{\varepsilon}_2/\dot{\varepsilon}_1 = \varepsilon_2/\varepsilon_1$ during the calculation, where the parameters K and n are the hardening coefficient and hardening index, respectively, of the power-law material model $\sigma = K\varepsilon^n$.

3.3. RTCL Criterion

The RTCL failure criterion is a failure criterion for ductile metals based on the equivalent plastic strain damage assessment, taking into account the complete stress triaxiality, proposed by Tornqvist [19], combining the Rice-Tracey criterion [34] and Cockcroft-Latham criterion [35]. It can be expressed as:

$$f(T) = \begin{cases} 0 & T \le -1/3 \\ 2\frac{1+T\sqrt{12-27T^2}}{3T+\sqrt{12-27T^2}} & -1/3 \le T \le 1/3 \\ \frac{1}{1.65}\exp(3/2T) & T \ge 1/3 \end{cases} \tag{3}$$

$$D_{RTCL} = \frac{1}{\bar{\varepsilon}_{cr}}\int f(T)d\varepsilon \tag{4}$$

The parameter D_{RTCL} represents the cumulative damage of the element, and when the cumulative damage reaches 1, the element reaches the failure condition and undergoes damage. The parameter $\bar{\varepsilon}_{cr}$ represents the failure strain, which usually exhibits mesh sensitivity. The parameter T denotes the stress triaxiality, whose value is hydrostatic stress divided by the equivalent stress and is dimensionless.

4. Numerical Simulation Setup

4.1. True Stress–Strain Relationship

Before the experiment, the mechanical properties of the low-carbon steel material were obtained through quasi-static tensile tests. The uniaxial tensile test specimen was designed and manufactured following standards [36], and its dimensions are illustrated in Figure 11. The CMT5305 universal testing machine (MTS, Shanghai, China) was employed for the experiment. One end of the specimen was fixed while the other end was pulled at a constant speed of 1 mm/min until a fracture occurred. The data acquisition system integrated into the testing machine was used to measure the tensile force and displacement. An extensometer was attached to the parallel section to determine the elastic modulus and the engineering stress–strain curve of the material. To reduce experimental errors, the tests were carried out three times. The material's relevant parameters are presented in Table 1.

Table 1. Mechanical properties of the steel.

Parameter	Unit	Value
Density	kg/m^3	7850
Elastic modulus	GPa	213
Poisson's ratio	-	0.3
Yield stress	MPa	221
Ultimate tensile stress	MPa	353
Fracture strain	-	0.35

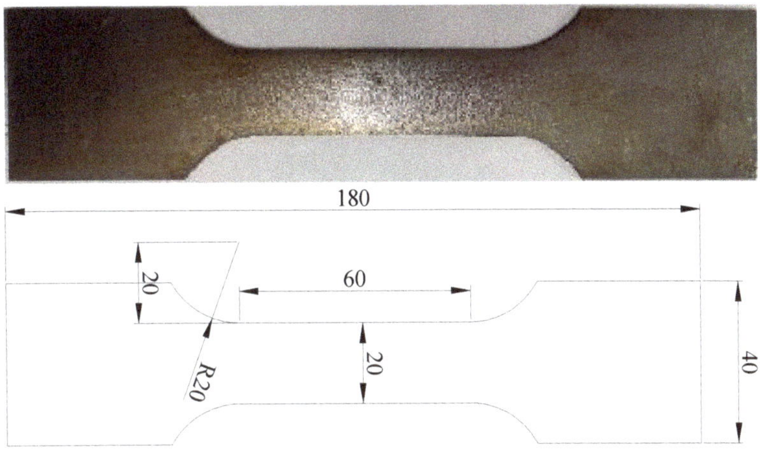

Figure 11. Tensile specimen and dimensional drawing.

The mechanical properties of the steel used in the finite element model were determined through the quasi-static tensile test. Engineering and true stress–strain curves of the front panel and cruciform structure were obtained through uniaxial tensile tests, which are shown in Figures 12 and 13. The true stress–strain curve was divided into two parts. Before the onset of necking, true stress σ_t and true strain ε_t are expressed by engineering stress σ_e and engineering strain ε_e.

$$\sigma_t = \sigma_e(1 + \varepsilon_e) \tag{5}$$

$$\varepsilon_t = \ln(1 + \varepsilon_e) \tag{6}$$

True stress–strain curves beyond necking can be represented by the following formula.

$$\sigma_t = C\varepsilon_t^n \tag{7}$$

where

$$n = \ln(1 + A_g) \tag{8}$$

and

$$C = R_m(e/n)^n \tag{9}$$

R_m is ultimate stress and e is the natural logarithmic constant. The proper A_g can be approximately obtained by the following formula [37].

$$A_g = \frac{1}{0.24 + 0.01395R_m} \tag{10}$$

Part I and Part II in Figure 13 are simply connected by a straight line, i.e., part "C" of the figure. This "combined material" was first proposed by Villavicencio R and Soares C G [37] to predict the true stress–strain relation beyond the necking of the test piece. where I is the true stress−strain curve portion and segment II is the power function portion of Equation (7).

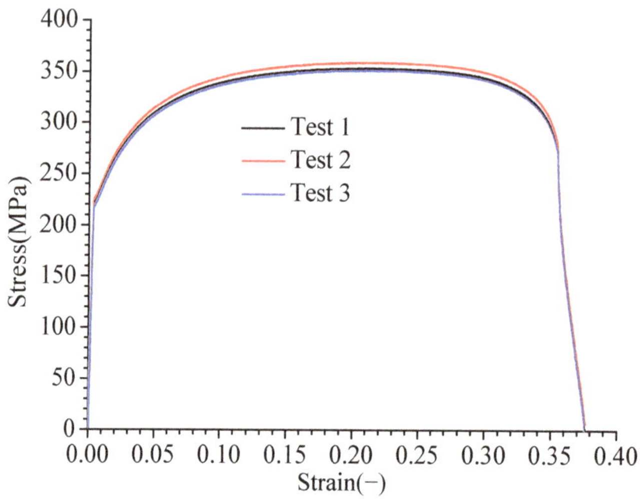

Figure 12. Engineering stress–strain curve.

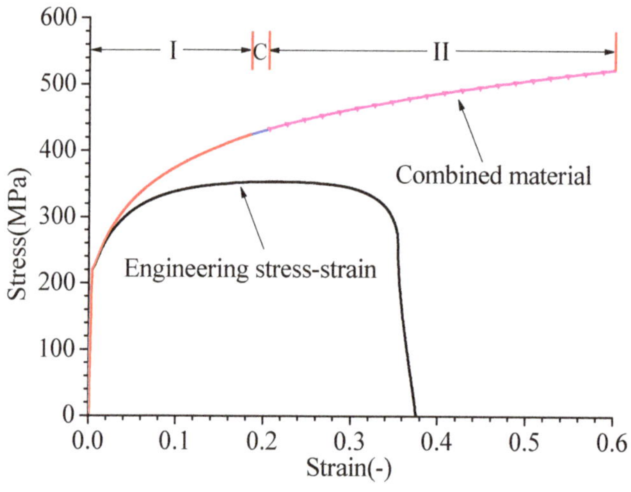

Figure 13. Combined material relationship.

4.2. Element Size Sensitivity

Data from tensile testing and other material properties are found in Table 2. Due to the local necking of the material during uniaxial stretching, the failure strain measured by the extensometer is not accurate and cannot be directly defined in the finite element. Therefore, finite element simulation of the uniaxial tensile test was conducted, with one end of the specimen fixed and the other end stretched at a constant rate of 100 times the loading speed of the testing machine [3,11,38], that is, the loading rate was 100 mm/min. The parallel section of the specimen was divided into meshes of 4 mm, 6 mm, 8 mm, and 10 mm, and the obtained force–displacement curves were compared with the test results. By calculation, the failure strain corresponding to different mesh sizes was obtained when the force–indentation curve of the test and simulation corresponded to the same tensile displacement at fracture, as shown in Table 2.

Table 2. Finite element models of different mesh densities.

	Mesh Size/mm	Finite Element Model	Failure Criteria
Model 1	10 × 10	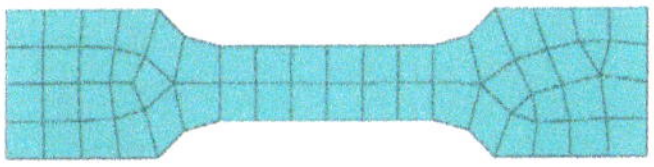	0.293
Model 2	8 × 8	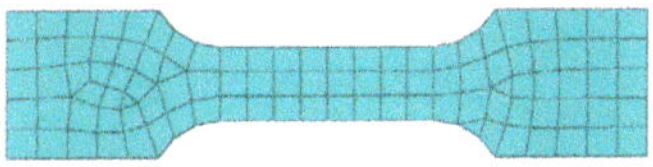	0.306
Model 3	6 × 6	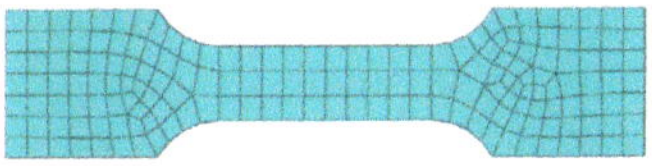	0.33
Model 4	4 × 4	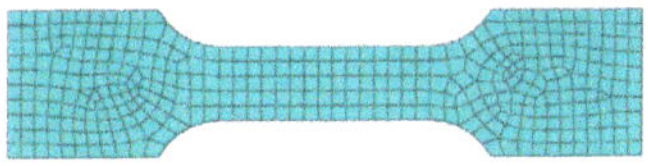	0.382

4.3. *Definition of Failure Criteria*

To compare the applicability of EPS, BWH, and RTCL failure criteria in the corresponding finite element simulations of the structural test, simulations were conducted using three failure criteria. Assuming that the mesh size of each key position element is the same as the overall mesh size, the difference between the set mesh size and the defined failure strain values for each material is ignored. For example, when the mesh size is set to 4 mm, the defined failure strain value for the material is 0.382. The plastic behavior of the material is the combined material relationship curve in Figure 13.

For the EPS and RTCL failure criteria, the defined failure strain values correspond to their mesh sizes, as shown in Table 2. For the BWH failure criterion, the true stress–strain curve is fitted to a power function curve to obtain the BWH failure criterion parameters corresponding to the material. The parameters K and n in Formula (2) are 572×10^6 and 0.1845, respectively.

4.4. *Parameter Settings of Finite Element Model*

The computations were performed using the FEM software ABAQUS, which is suitable for nonlinear explicit dynamic simulations with large deformations. The geometry of the specimen, as shown in Figure 14, was established in ABAQUS, with the wings and waist of the channel steel being equivalent to 8.5 mm and 5.3 mm thick, respectively. The test pieces were modeled using quadrilateral shell elements with reduced integration and large-strain formulation. As the indenter was solid and had high hardness, it was set as a rigid body. When simulating small-scale structural elements, it was necessary to consider the weld joints. This was because weld joints not only increased the thickness of the plate but also smoothed the transverse transition between the web girder and the front panel, thereby increasing the strength of the structure. Studies by Nassiraei, H. [39,40] have shown that reinforcement at structural joints can increase the load-carrying capacity of the structure to a greater extent, and welds are also a form of reinforcement at structural joint locations. The presence of weld joints was accounted for in the simulations by increasing the thicknesses of the front panel and cruciform structure at their intersection, as described in references [11,37]. After measuring the width of the weld joint to be between 5 mm and 7 mm, a width of 6 mm was used in Figure 14, with the thickness of the welding foot being equivalent to the plate on both sides. After trial calculations, a weld joint was set to increase the thickness of the welding element on both sides by 1 mm, resulting in relatively ideal simulation results. As shown in Figure 14, the weld elements on the axis of the cruciform structure were increased by 2 mm, and the weld elements at the intersection of the cruciform structure and the front panel were increased by 1 mm. The mesh size of the indenter in the finite element model was 4 mm, and the mesh sizes of the test pieces were 4 mm, 6 mm, 8 mm, and 10 mm, respectively, for comparing the structural responses using three failure criteria. Figure 15 shows the finite element model with a mesh size of 4 mm. In the finite element model corresponding to the quasi-static test, the moving speed of the indenter was set to 80 mm/s to save simulation time, and the calculation time was set to 1 s to cover the displacement of the indenter in the test. In the simulation of dynamic impact testing, the degrees of freedom other than the Y-direction movement of the indenter were constrained according to the actual test conditions. The corresponding initial velocities of the indenter were 3.13 m/s and 7.67 m/s, and the dynamic response of the indenter from contact with the specimen to leaving it was calculated. The material parameters used in the quasi-static test and the falling weight impact test are shown in Table 1 and Figure 13. Due to the limitation of conditions, the related material strain rate sensitivity test was not carried out. Considering that the Cowper-Symonds parameter, which is commonly used in the material strain rate sensitivity constitutive model, will increase the post-yield stress of the material, the strain rate of 100 s^{-1} in the finite element simulation has a great influence on the Q235 material. In reality, the effect of strain rate on the post-yield stress is small [41], so the strain rate sensitivity of the material is not considered here.

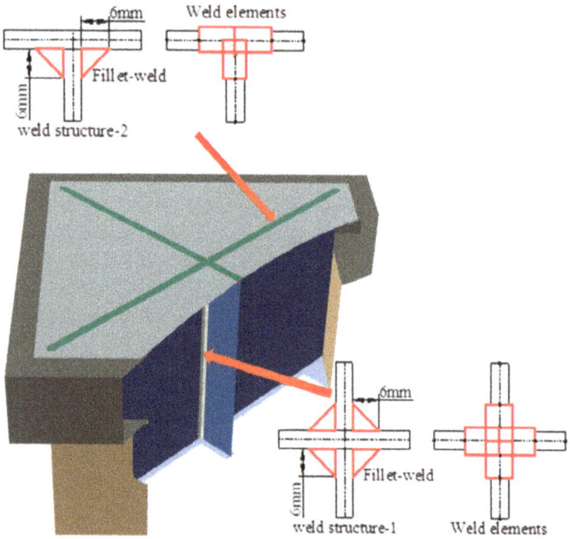

Figure 14. An equivalent method of welding elements.

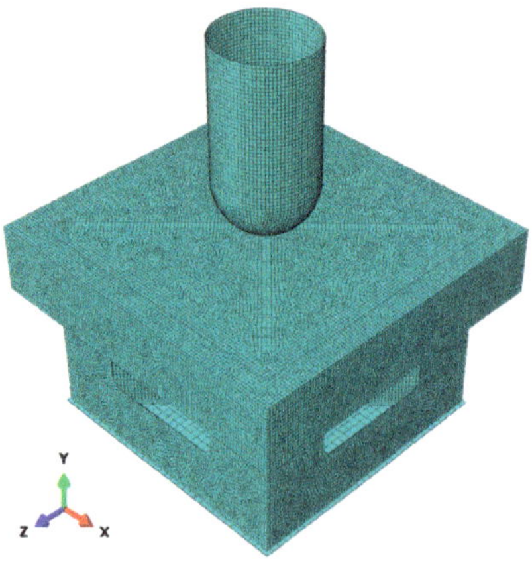

Figure 15. Finite element model (mesh size 4 mm).

5. FE Results and Discussion

The displacement cloud maps of the front panel and the cruciform structure for the quasi-static test were generated using the BWH, EPS, and RTCL failure criteria, as well as four mesh sizes in finite element analysis, shown in Figure 16. The cloud maps used the same scale to compare the degree and range of structural deformation. The corresponding force–indentation curves are shown in Figure 17.

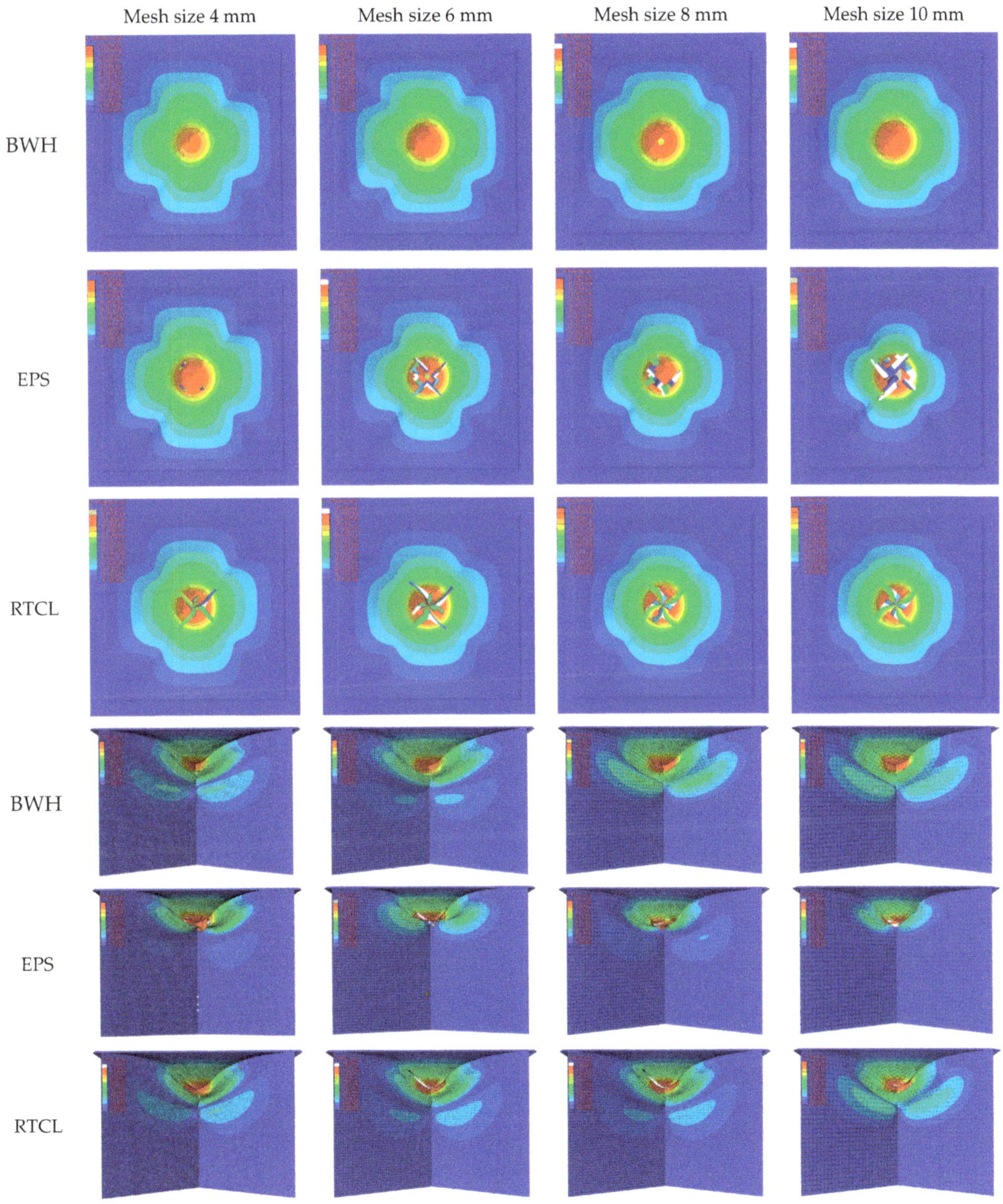

Figure 16. Simulation results for the front panel and cruciform structure in quasi-static test.

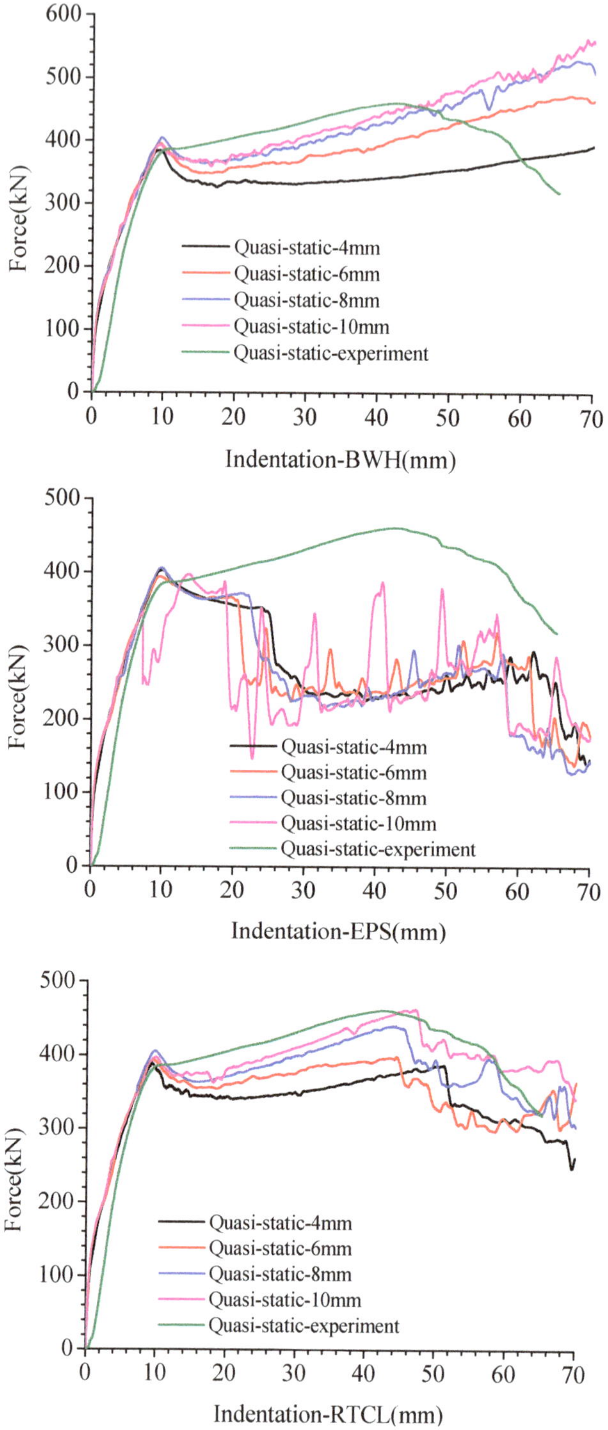

Figure 17. Comparison of force–indentation curves from experiment and simulation in the quasi-static test.

By combining the force–indentation curves in Figure 17 and the deformation cloud maps in Figure 16, it can be seen that when using the BWH failure criterion, the structure did not fail. For a mesh size of 4 mm, the concave range of the front panel was small. From the deformation of the cruciform structure, it can be seen that the height of the wrinkles in the axial direction of the cruciform structure was relatively small when using a small mesh size. In a smaller mesh size, the cruciform structure had one more group of wrinkle deformations compared to the larger mesh size. From the force–indentation curve in Figure 17, it can be seen that during the stable deformation of the wrinkles, the larger the height of the wrinkles, the greater the deformation range of the cruciform structure and the faster the force–indentation curve rises. Based on the initial peak value in the force–indentation curve and the structural deformation in Figure 16, it can be seen that the initial buckling load of the cruciform structure was approximately the same, and the initial height of the wrinkles was also similar.

As shown in Figure 17, when using the EPS and RTCL failure criteria, both the front panel and the cruciform structure failed. When using the EPS failure criterion, the cruciform structure below the hammer head buckled. As the displacement of the hammer head increased, the strain of the lower element gradually increased. When the failure strain was reached, the element was deleted, and the load quickly dropped. When the front panel made contact with the undamaged elements again, the load continued to rise. From the changes in the force–indentation curve in the later stage, it can be seen that using a smaller element size resulted in less curve fluctuation, the more frequent appearance of wave peaks, and relatively smaller wave peaks. Notably, when the element size was 10 mm, the deformation of the cruciform structure along the axial direction was mainly concentrated on the element directly below the hammer head, and the element was deleted rapidly, causing the load to drop quickly before the cruciform structure buckled. When using the RTCL failure criterion, the center of the cruciform structure mainly undergoes compression deformation, and at this position, the element does not accumulate damage and will not fail. However, the connection between the front panel and cruciform structure undergoes bi-directional tensile deformation due to the presence of a weld joint and section change, resulting in stress concentration and damage accumulation. As a result, when using the RTCL failure criterion, the simulation results show four tears that are consistent with the experiment. Compared with the BWH and EPS failure criteria, the RTCL failure criterion is more suitable for simulating the failure of this structure.

The deformation maps of the structure with three failure criteria and four mesh sizes are shown in Figure 18 when the height of the hammer drop was 0.5 m. From the deformation cloud maps of the front panel, it can be seen that when using the EPS failure criterion with a mesh size of 10 mm, the concave deformation of the panel is relatively large, and a layer of elements on the upper end of the corresponding cruciform structure fails and is deleted. From the force–indentation curves in Figure 19, it can be seen that when using the EPS failure criterion with a mesh size of 10 mm, the shell element directly below the welded element at the upper end of the cruciform structure fails before the cruciform structure buckles, resulting in a rapid decrease in the load curve. As the mesh size increases, the height of the wrinkles formed on the cruciform structure increases to a certain extent. The force–indentation curves for the other computational cases are similar and show the opposite pattern to that of the quasi-static tests, i.e., as the mesh size increases, the height of the wrinkles along the axial direction of the cruciform structure increases, the number of structures involved in deformation absorption increases, and the maximum displacement and residual deformation of the indenter are relatively small.

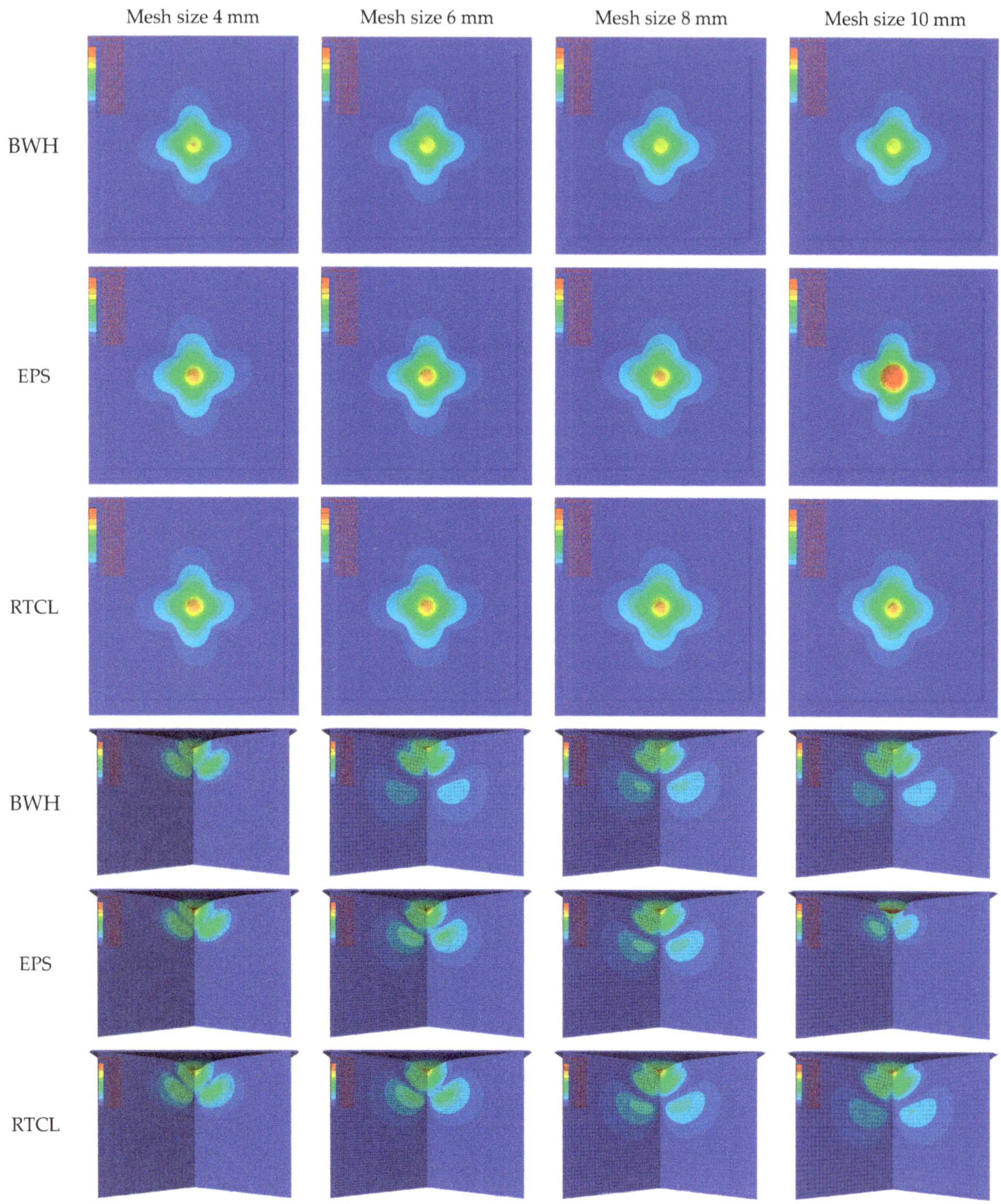

Figure 18. Simulation results for the front panel and cruciform structure in 0.5 m falling impact test.

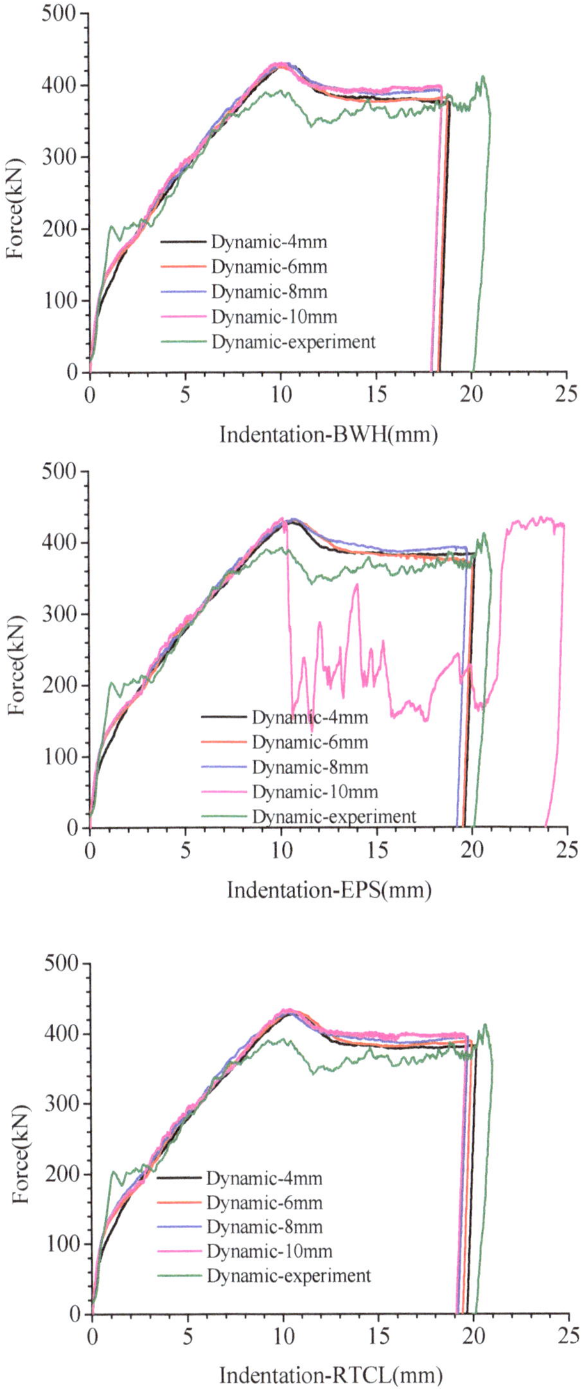

Figure 19. Comparison of force–indentation curves from experiment and simulation in 0.5 m falling impact test.

Figure 20 shows the structural deformation diagrams using three failure criteria and four grid sizes when the indenter falls from a height of 3 m. From the deformation cloud diagram of the front panel, it can be seen that when using the BWH failure criterion, there will be some damage to the panel at the 4 mm and 6 mm grid sizes, but there is no obvious damage to the cruciform structure. When using the EPS failure criterion, both the top panel and the cruciform structure suffered significant damage. The top panel showed tearing at the outer edge of the indenter, and since the indenter penetrated the top panel completely in this condition, the degree of deformation of the top panel was similar. The cruciform structure axis and the connection between the cruciform structure and the front panel both suffered obvious tearing damage. When using the RTCL failure criterion, the tearing form of the front panel was most similar to the test results, with four obvious cracks forming at the connection between the cruciform structure and the panel. The cruciform structure axis suffered obvious damage under all four grid sizes. Comparing the force–indentation curves using the three failure criteria, it can be seen that the BWH failure criterion caused relatively little damage to the structure, resulting in a smaller ultimate penetration depth than the test results. When using the EPS failure criterion, due to the deletion of many elements under the indenter and around the indenter, the structure that absorbed impact energy was reduced, resulting in an ultimate penetration depth far greater than the test results. The RTCL failure criterion accurately simulated the failure of the specimen, reproducing the characteristics of the top panel failure and the cruciform structure not failing under compression. The various curves in Figure 21 indicate that the use of the BWH failure criterion in this structure cannot effectively predict the failure of the structure, resulting in a smaller maximum penetration depth. The use of the EPS failure criterion incorrectly judged the compressive failure of the ductile metal, resulting in poor simulated impact resistance performance of the structure. The use of the RTCL failure criterion, due to the influence of grid size on the deformation characteristics of the structure, resulted in some differences in ultimate penetration depth and residual deformation, but the simulation results were closest to the test results compared to the other two failure criteria.

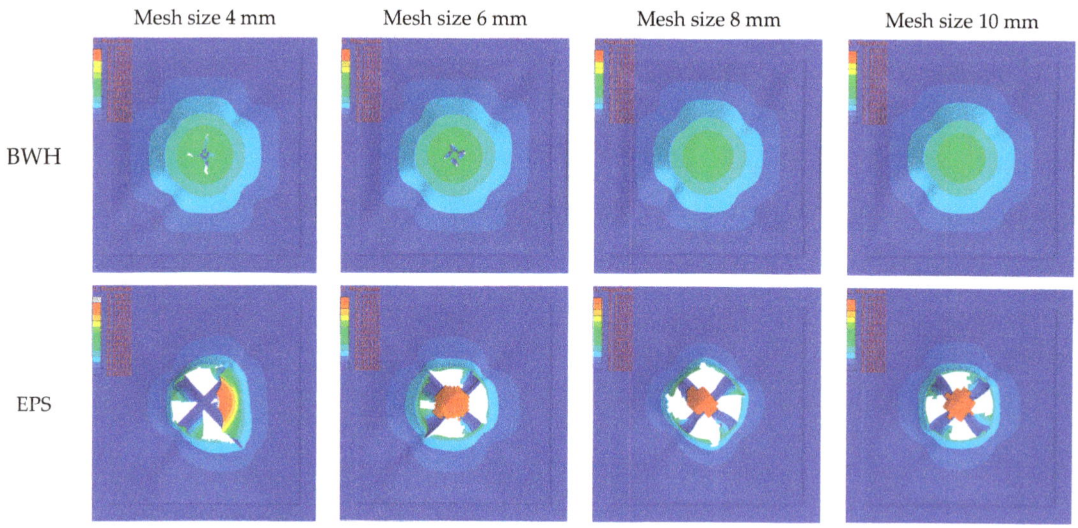

Figure 20. *Cont.*

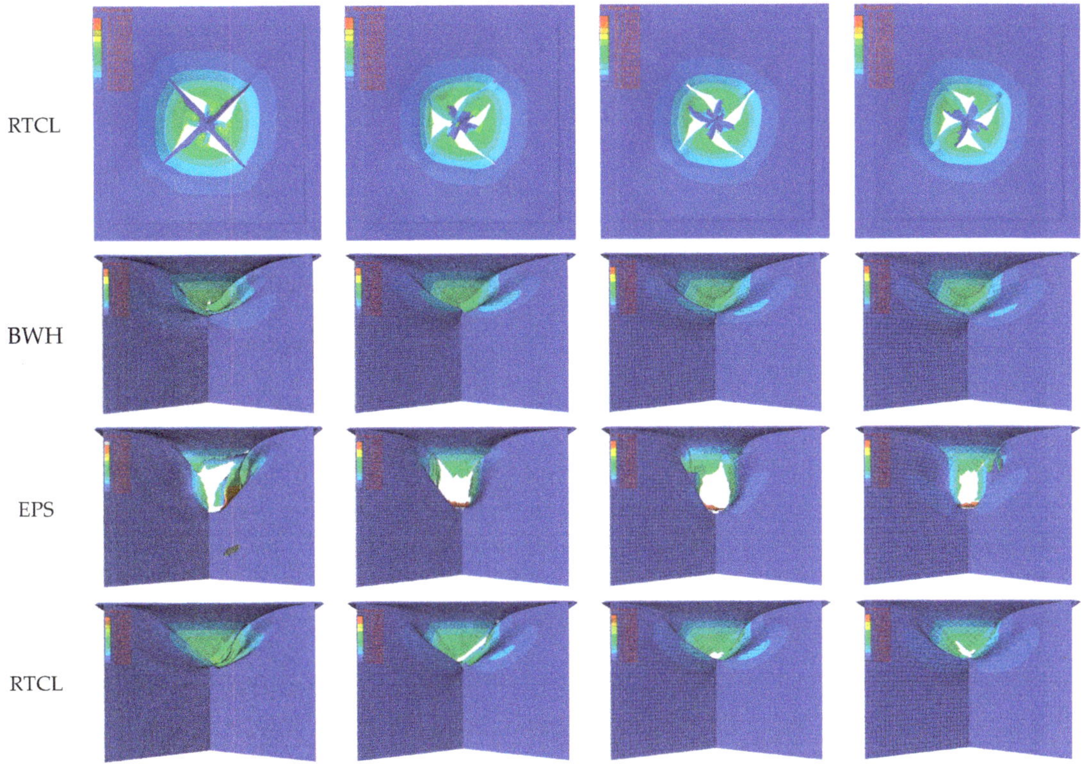

Figure 20. Simulation results for the front panel and cruciform structure in 3 m falling impact test.

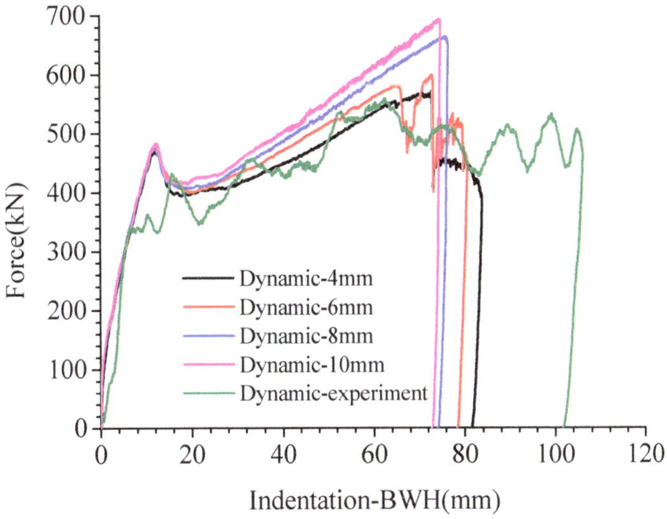

Figure 21. *Cont.*

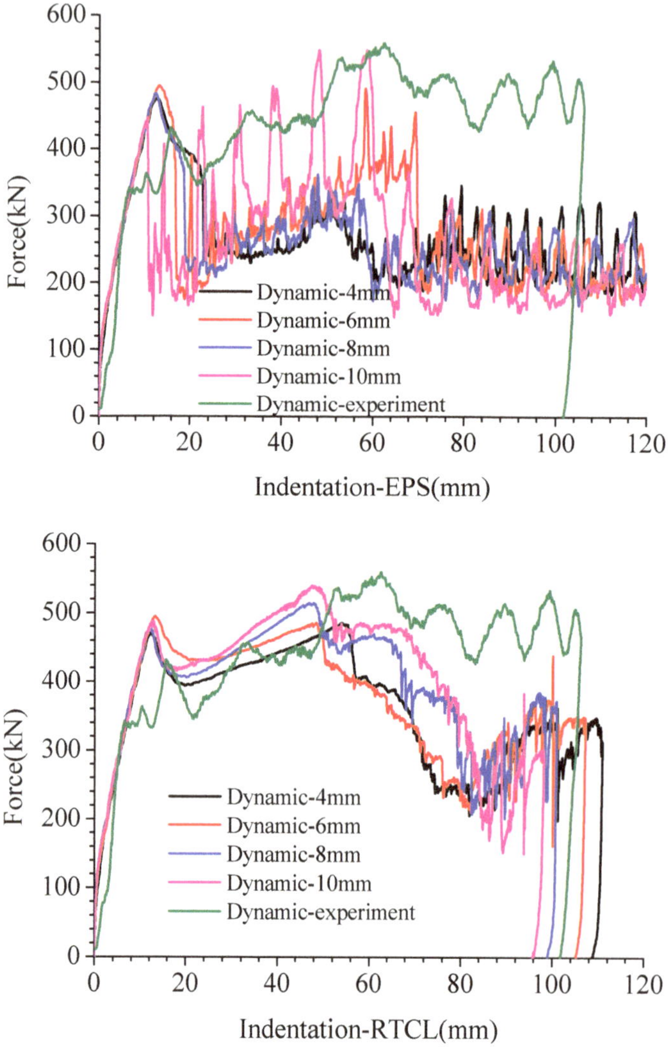

Figure 21. Comparison of force–indentation curves from experiment and simulation in 3.0 m falling impact test.

5.1. *Effect of Failure Criteria on FE Results*

The BWH failure criterion determines the failure of a unit using the necking of the material. It is more difficult to obtain the damage state of a structure accurately in calculation conditions where the structure is complex, the stress state varies greatly, and the stress gradient varies widely. In cases where the stress state is more homogeneous, such as when the structure has a large area in the bi-directional tensile state, the tearing of a ductile metal structure after necking can be predicted more accurately. The EPS failure criterion and the RTCL failure criterion determine the failure of a cell by the cumulative damage to the cell, where the EPS failure criterion defines that the cell fails when the equivalent plastic strain accumulates to the failure strain and the RTCL failure criterion introduces the stress triaxiality to consider the stress state of the cell. The stress state influences the cumulative damage value resulting from the plastic strain.

By comparing the structural deformation and displacement–load curves in Figures 16–21, it is seen that the RTCL failure criterion can better predict the damage of this cross structure under in-plane ramming loads, and the simulation results are closer to the experimental results. Due to the complex and variable stress state of the structure, it is difficult for the BWH failure criterion to determine the failure of the unit by necking down, while the EPS failure criterion produces a large compressive deformation directly below the indenter when the cruciform structure is axially compressed, resulting in a large local strain. The EPS failure criterion determines the failure of the unit early, resulting in the premature removal of the cruciform structure unit directly below the indenter, which in turn affects the deformation of the overall structure and results in a lower impact resistance than is the case.

5.2. Effect of Mesh Size on FE Results

Using different mesh sizes in this structure will lead to different deformation modes of the cruciform structure, and the corresponding collision load will also vary. When the mesh size is larger, the height of the wrinkles generated in the cruciform structure will be relatively large. After the cruciform structure buckles, the range of stable membrane tension deformation of the cruciform structure is relatively large, leading to a faster rate of load increase. Comparing the force–indentation curves in the simulation results under different test conditions, the yield limit of the cruciform structure shows a slight increase with the increase in mesh size.

5.3. Comparison of the Dynamic and Static Response of the Structure

By comparing the load–displacement curves in Figures 17, 19 and 21, it can be observed that in quasi-static conditions, the initial peak load is around 400 kN, while in the 1.5 m drop condition, the initial peak load is around 440 kN, and in the 3 m drop condition, the initial peak load is around 490 kN. When the structure undergoes failure in the 3 m drop condition, the impact load is approximately 13% higher compared to the quasi-static condition. This indicates that in high-speed collision scenarios, both the ultimate buckling load and average impact force of the structure are higher than those in quasi-static collision scenarios.

6. Analysis of the Effect of Friction Coefficient and Indenter Offset

In actual ship accidents, there are many uncertainties in environmental factors such as boundary and load conditions, and it is impossible to include all scenarios in experiments. In this section, the influence of the friction coefficient and collision position between two structures on the structural response will be compared and discussed.

6.1. Effect of Coefficient of Friction

By changing the friction coefficient between the indenter and the specimen from 0.1 to 0.4, the influence of the friction coefficient parameter on the structural response is compared in the same quasi-static accident scenario with the same impact energy collision scenario. The corresponding quasi-static test and 3.0 m falling impact test were performed, the failure criterion used was the RTCL failure criterion, and the mesh size used was 10 mm to minimize the computation time.

As shown in Figure 22, when the friction coefficient is taken as 0.3 as the benchmark, the crack size of the structure increases significantly when the friction coefficient is small in the quasi-static scenario. This is particularly evident when the friction coefficient is 0.1, 0.15, 0.2, and 0.25, which is greater than other friction coefficient calculation conditions. This is because the finite element mesh has a certain size and cannot represent continuous cracks. Moreover, within the range of the friction coefficient variation, the change in structural energy dissipation is within 5% of the benchmark condition. Therefore, it is determined that under the same impact depth, the size of the structure crack decreases slightly with the increase in the friction coefficient. On the contrary, the overall energy dissipation of the structure shows a slight increase, as the presence of friction results in a stronger constraint on the deformation of the structure, resulting in more energy absorption. As shown in

Figure 23. The difference in the maximum energy dissipated by the structure for the same impact energy is small, within 5% of the baseline value. However, the impact depth of the indenter shows a significant decrease with the increase in the friction coefficient, mainly because the decrease in the friction coefficient results in less constraint on the deformation of the structure, and the deformation area is relatively concentrated, resulting in greater damage to the structure.

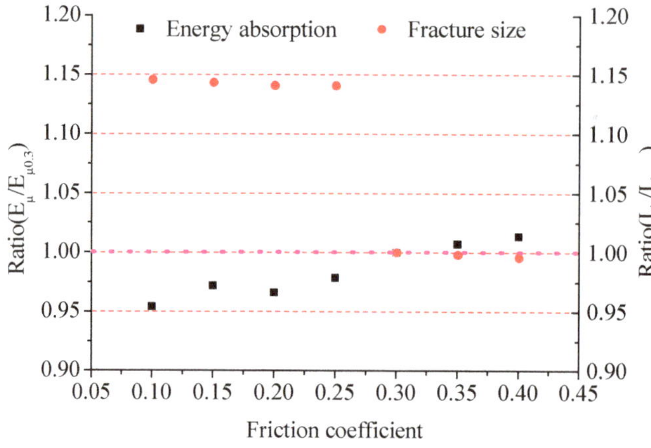

Figure 22. Structural response under the same indentation (quasi-static scenario, change friction coefficient).

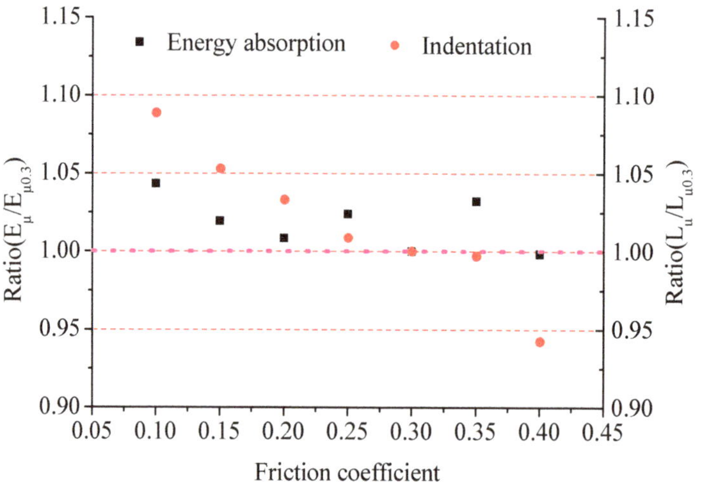

Figure 23. Structural response under the same impact energy (dynamic impact scenario, change friction coefficient).

6.2. *Effect of Indenter Offset*

In addition, as it is well known, changing the collision center position has a significant impact on the structural response and energy absorption. As shown in Figure 24, the indenter is moved in the direction of the rib plate and at a 45-degree angle with the rib plate, with a movement distance from 10 mm to 50 mm. The calculation conditions are the corresponding quasi-static test and the 3.0 m falling impact test, using the RTCL failure criterion, a mesh size of 10 mm, and a friction coefficient of 0.3.

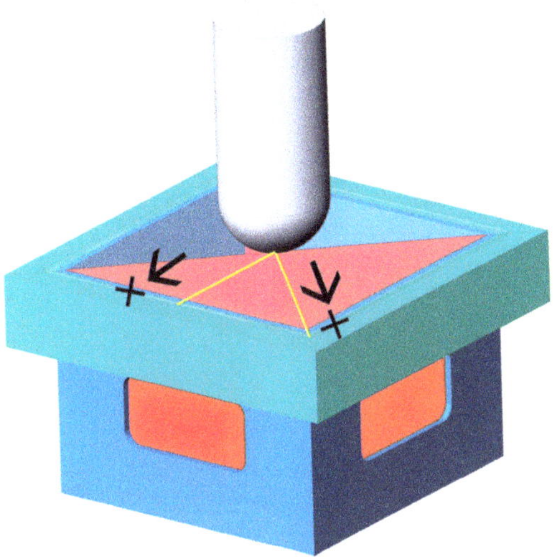

Figure 24. Schematic diagram of indenter deviation direction.

Figure 25 shows the structural response data under quasi-static and impact scenarios when the indenter impact position deviates. From the quasi-static scenario, it can be seen that under the same impact depth conditions, the higher the degree of deviation along the rib direction, the less energy absorbed by the structure. This is due to the reduction in the number of structures directly below the indenter, and the fracture size shows a significant increase when the indenter deviates the most from the axis of the cruciform structure. When the indenter deviates at a 45-degree angle with the rib, the energy absorbed by the structure shows significant random variations, and there is a significant decrease at the moment of the maximum deviation. However, the crack size does not change significantly in this deviation scenario. As shown in Figure 26, under the same conditions of impact energy calculation, the changes in the energy absorption capacity and impact depth of the structure are more obviously affected by the direction of deviation. As the deviation from the axis distance increases, the energy absorption capacity of the structure decreases, and the impact depth increases significantly.

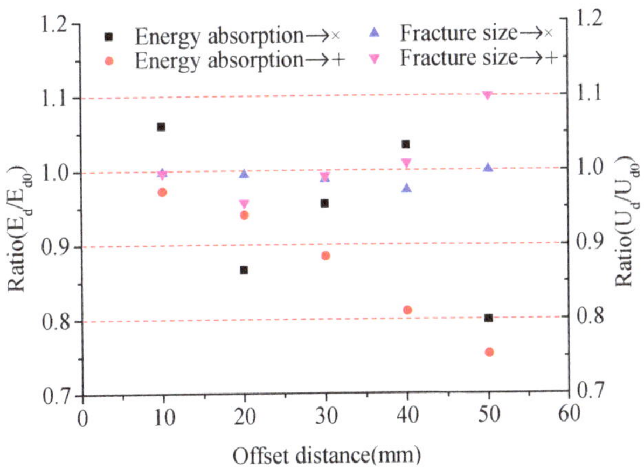

Figure 25. Structural response under the same indentation (quasi-static scenario, change indenter position).

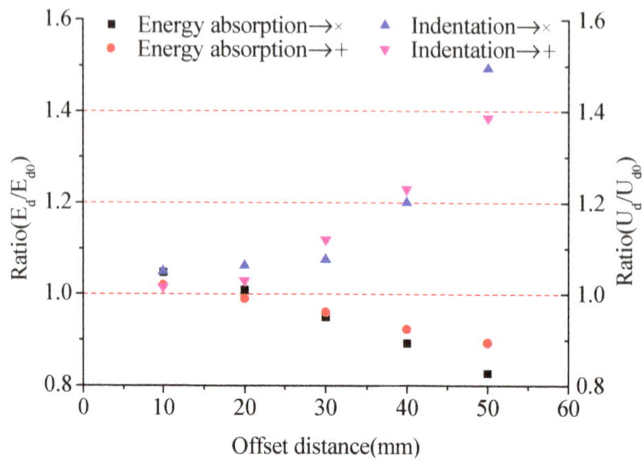

Figure 26. Structural response under the same impact energy (dynamic impact scenario, change indenter position).

From the above comparison, it can be seen that under small-size hammer quasi-static loads, the energy absorption characteristics of the cruciform structure change are slight. In the impact scenario, the reduction of the friction coefficient among the structures will lead to a decrease in the impact resistance performance of the structure and a reduction in the ability to absorb impact energy. With the deviation of the indenter relative to the axis of the cruciform structure, the energy absorption capacity of the structure significantly decreases due to the decrease in the structure directly below the indenter, and the impact depth significantly increases.

7. Conclusions

This study designed and conducted quasi-static and falling weight impact tests on a typical ship's cruciform structure. Material property tests were carried out on the test specimen to obtain the true stress–strain relationship of the material. To determine the parameters for using the EPS and RTCL failure criteria, numerical simulations were conducted on uniaxial tensile tests using Ling's method, and the failure strain corresponding to different mesh sizes was determined. The models of the three test conditions were divided into sizes of 4 mm, 6 mm, 8 mm, and 10 mm, and the EPS, BWH, and RTCL failure criteria were used respectively to compare the applicability of the three failure criteria in the structure's collision condition, as well as the influence of mesh size on the structural response.

The following conclusions were drawn from this study:

- When the cruciform structure in a typical hull structure is subjected to in-plane loading, the structural deformation mainly has the following processes: axial compression deformation of the structure, with a rapid increase in structural resistance; flexural deformation of the structure, with a slight decrease in structural resistance; crease deformation of the cruciform structure, with a smooth increase in structural resistance; and tearing of the front panel, with a decrease in structural load carrying capacity.
- When subjected to the impact load of a rigid indenter, the structure's deformation range is smaller than the quasi-static load, and the deformation is relatively concentrated at the collision location. In the 3 m drop impact condition, the initial peak impact load increases by nearly 20% compared to the quasi-static collision scenario, and the average impact force increases by approximately 13% before structural failure. This indicates that the structure has a higher energy absorption capacity in the 3 m drop impact condition compared to the quasi-static collision scenario.
- The EPS failure criterion in the simulation in this scenario would prematurely determine the failure of the junction due to the localized compressive deformation produced

by the unit. The BWH failure criterion struggles to determine the failure of the structure, and the crashworthiness of the structure evaluated by this criterion is higher than that of the actual scenario. The RTCL failure criterion is more suitable for this complex stress state crash scenario, and the tearing phenomenon of the upper panel is better simulated.

- The height of the wrinkles produced by the cruciform structure is closely related to the mesh size. The height of the wrinkles produced by a large mesh model is relatively high, and the area where the subsequent membrane tensile deformation occurs in the belly beam that makes up the cruciform structure is relatively large. The structure's resistance is also relatively small in the large mesh model. In contrast, a small mesh model produces smaller wrinkles, and subsequent wrinkle formation occurs earlier than in the large mesh model.

Author Contributions: Conceptualization, X.W.; Writing—original draft, X.W.; Writing—review and editing, K.L., M.X. and H.L.; Visualization, K.L. and M.X. All authors have read and agreed to the published version of the manuscript.

Funding: This research was funded by the National Natural Science Foundation of China (Grant No. 52171311; Grant No. 52271279).

Institutional Review Board Statement: Not applicable.

Informed Consent Statement: Not applicable.

Data Availability Statement: Not applicable.

Conflicts of Interest: The authors declare no conflict of interest.

References

1. Oil Tanker Spill Statistics 2022. Available online: https://www.itopf.org/knowledge-resources/data-statistics/statistics/ (accessed on 23 May 2023).
2. Cho, S.-R.; Lee, H.-S. Experimental and analytical investigations on the response of stiffened plates subjected to lateral collisions. *Mar. Struct.* **2009**, *22*, 84–95. [CrossRef]
3. Liu, K.; Wang, Z.; Tang, W.; Zhang, Y.; Wang, G. Experimental and numerical analysis of laterally impacted stiffened plates considering the effect of strain rate. *Ocean Eng.* **2015**, *99*, 44–54. [CrossRef]
4. Liu, B.; Soares, C.G. Effect of strain rate on dynamic responses of laterally impacted steel plates. *Int. J. Mech. Sci.* **2019**, *160*, 307–317. [CrossRef]
5. Yang, L.; Wang, D.-Y. Experimental and numerical investigation on the response of stiffened panels subjected to lateral impact considering material failure criteria. *Ocean Eng.* **2019**, *184*, 193–205. [CrossRef]
6. Alsos, H.S.; Amdahl, J. On the resistance to penetration of stiffened plates, Part I–Experiments. *Int. J. Impact Eng.* **2009**, *36*, 799–807. [CrossRef]
7. Alsos, H.S.; Amdahl, J.; Hopperstad, O.S. On the resistance to penetration of stiffened plates, Part II: Numerical analysis. *Int. J. Impact Eng.* **2009**, *36*, 875–887. [CrossRef]
8. Zhang, M.; Liao, X.; Li, S.; Song, S.; Liu, J.; Hu, Z. Experimental and numerical investigation of the damage characteristics of a ship side plate laterally punched by a scaled raked bow indenter. *Ocean Eng.* **2023**, *280*, 114808. [CrossRef]
9. Wang, G.; Arita, K.; Liu, D. Behavior of a double hull in a variety of stranding or collision scenarios. *Mar. Struct.* **2000**, *13*, 147–187. [CrossRef]
10. Zhang, M.; Liu, J.; Hu, Z.; Zhao, Y. Experimental and numerical investigation of the responses of scaled tanker side double-hull structures laterally punched by conical and knife edge indenters. *Mar. Struct.* **2018**, *61*, 62–84. [CrossRef]
11. Villavicencio, R.; Kim, Y.-H.; Cho, S.-R.; Soares, C.G. Deformation process of web girders in small-scale tanker double hull structures subjected to lateral impact. *Mar. Struct.* **2013**, *32*, 84–112. [CrossRef]
12. Chen, B.-Q.; Liu, B.; Guedes Soares, C. Experimental and numerical investigation on a double hull structure subject to collision. *Ocean Eng.* **2022**, *256*, 111437. [CrossRef]
13. Urban, J. Crushing and Fracture of Lightweight Structures. Ph.D. Thesis, Technical University of Denmark, Kongens Lyngby, Denmark, 2003.
14. Zhou, C.; Li, T.; Ming, S.; Song, Z.; Wang, B. Effects of welding on energy absorption of kirigami cruciform under axial crushing. *Thin Wall Struct.* **2019**, *142*, 297–310. [CrossRef]
15. Haris, S.; Amdahl, J. Crushing resistance of a cruciform and its application to ship collision and grounding. *Ships Offshore Struct.* **2012**, *7*, 185–195. [CrossRef]
16. Haris, S.; Amdahl, J. An analytical model to assess a ship side during a collision. *Ships Offshore Struct.* **2012**, *7*, 431–448. [CrossRef]

17. Hayduk, R.J.; Wierzbicki, T. Extensional collapse modes of structural members. *Comput. Struct.* **1984**, *18*, 447–458. [CrossRef]
18. Yang, P.; Caldwell, J. Collision energy absorption of ships' bow structures. *Int. J. Impact Eng.* **1988**, *7*, 181–196. [CrossRef]
19. Tørnqvist, R. *Design of Crashworthy Ship Strucures*; Technical University of Denmark: Kongens Lyngby, Denmark, 2003.
20. Zhang, L.; Egge, E.-D.; Bruhns, H. *Approval Procedure Concept for Alternative Arrangements*; Germanischer Lloyd: Hamburg, Germany, 2004.
21. Peschmann, J. Energy absorption computations of ship steel structures under collision and grounding (translated from German language). *Schriftenreih e Schiffbau/TU Hambg.-Harbg.* **2001**, *613*, 4269–4355.
22. Lehmann, E.; Peschmann, J. Energy absorption by the steel structure of ships in the event of collisions. *Mar. Struct.* **2002**, *15*, 429–441. [CrossRef]
23. Alsos, H.S. Ship grounding: Analysis of Ductile Fracture, Bottom Damage and Hull Girder Response. Ph.D. Thesis, Norwegian University of Science and Technology, Trondheim, Norway, 2008.
24. Yagi, S.; Kumamoto, H.; Muragishi, O.; Takaoka, Y.; Shimoda, T. A study on collision buffer characteristic of sharp entrance angle bow structure. *Mar. Struct.* **2009**, *22*, 12–23. [CrossRef]
25. Kõrgesaar, M.; Romanoff, J. Influence of softening on fracture propagation in large-scale shell structures. *Int. J. Solids Struct.* **2013**, *50*, 3911–3921. [CrossRef]
26. Kõrgesaar, M.; Remes, H.; Romanoff, J. Size dependent response of large shell elements under in-plane tensile loading. *Int. J. Solids Struct.* **2014**, *51*, 3752–3761. [CrossRef]
27. Marinatos, J.; Samuelides, M. Material characterization and implementation of the RTCL, BWH and SHEAR failure criteria to finite element codes for the simulation of impacts on ship structures. In Proceedings of the International Conference of Collision and Groundings of Ships (ICCGS-2013), Trondheim, Norway, 17–19 June 2013; pp. 57–67.
28. Lu, Y.; Liu, K.; Wang, Z.; Tang, W. Modelling of ductile fracture in ship structures subjected to quasi-static impact loads. *Int. J. Impact Eng.* **2021**, *156*, 103941. [CrossRef]
29. Lu, Y.; Liu, K.; Wang, Z.; Tang, W.; Amdahl, J. Development of ductile fracture modelling approach in ship impact simulations. *Ocean Eng.* **2022**, *252*, 111173. [CrossRef]
30. Costas, M.; Morin, D.; Hopperstad, O.S.; Børvik, T.; Langseth, M. A through-thickness damage regularisation scheme for shell elements subjected to severe bending and membrane deformations. *J. Mech. Phys. Solids* **2019**, *123*, 190–206. [CrossRef]
31. Calle, M.; Alves, M. A review-analysis on material failure modeling in ship collision. *Ocean Eng.* **2015**, *106*, 20–38. [CrossRef]
32. Hill, R. On discontinuous plastic states, with special reference to localized necking in thin sheets. *J. Mech. Phys. Solids* **1952**, *1*, 19–30. [CrossRef]
33. Bressan, J.; Williams, J. The use of a shear instability criterion to predict local necking in sheet metal deformation. *Int. J. Mech. Sci.* **1983**, *25*, 155–168. [CrossRef]
34. Rice, J.R.; Tracey, D.M. On the ductile enlargement of voids in triaxial stress fields*. *J. Mech. Phys. Solids* **1969**, *17*, 201–217. [CrossRef]
35. Cockcroft, M. Ductility and workability of metals. *J. Metals* **1968**, *96*, 2444.
36. *GB/T228. 1-2010*; Metallic Materials-Tensile Testing-Part 1: Method of Test at Room Temperature. China Standard Press: Beijing, China, 2011.
37. Villavicencio, R.; Soares, C.G. Numerical plastic response and failure of a pre-notched transversely impacted beam. *Ships Offshore Struct.* **2012**, *7*, 417–429. [CrossRef]
38. Liu, B.; Villavicencio, R.; Guedes Soares, C. Plastic Response and Failure Prediction of Stiffened Plates Punched by a Wedge. In Proceedings of the ASME 2013 32nd International Conference on Ocean, Offshore and Arctic Engineering. Volume 2B: Structures, Safety and Reliability, Nantes, France, 9–14 June 2013; Volume 2.
39. Nassiraei, H.; Lotfollahi-Yaghin, M.A.; Ahmadi, H. Structural behavior of tubular T/Y-joints with collar plate under static in-plane bending. *J. Constr. Steel Res.* **2016**, *123*, 121–134. [CrossRef]
40. Nassiraei, H.; Lotfollahi-Yaghin, M.A.; Ahmadi, H.; Zhu, L. Static strength of doubler plate reinforced tubular T/Y-joints under in-plane bending load. *J. Constr. Steel Res.* **2017**, *136*, 49–64. [CrossRef]
41. Liu, K.; Liu, B.; Villavicencio, R.; Wang, Z.; Guedes Soares, C. Assessment of material strain rate effects on square steel plates under lateral dynamic impact loads. *Ships Offshore Struct.* **2018**, *13*, 217–225. [CrossRef]

Journal of
Marine Science
and Engineering

MDPI

Article

The Dynamic Response of a Floating Wind Turbine under Collision Load Considering the Coupling of Wind-Wave-Mooring Loads

Shuai Zong [1], Kun Liu [1,*], Yichi Zhang [2], Xingpeng Yan [1] and Yukai Wang [3]

1 School of Naval Architecture and Ocean Engineering, Jiangsu University of Science and Technology, Zhenjiang 212003, China; 220110101103@stu.just.edu.cn (S.Z.); 221110101204@stu.just.edu.cn (X.Y.)
2 State Key Laboratory of Ocean Engineering, Shanghai Jiao Tong University, Shanghai 200240, China; lemon8127269@sjtu.edu.cn
3 China Shipbuilding Industry Corporation No. 704 Research Institution, Shanghai 200240, China; ykai_wang@163.com
* Correspondence: kunliu@just.edu.cn; Tel.: +86-135-1169-2085; Fax: +86-0511-8444-6543

Abstract: As the number of offshore wind turbines continues to rise and their proximity to navigational routes decreases, the risk of collisions between passing vessels and wind turbines increases, thereby presenting serious threats to the safety of personnel and equipment. Given that collisions between floating wind turbines and vessels entail a complex interplay of wind, wave, and mooring loads, this study established a bidirectional fluid-structure coupling simulation methodology based on Star-CCM+ and ABAQUS. Under the combined influences of wind, wave, and mooring loads, the study investigated the dynamic response of floating wind turbines following bow and side impacts from vessels. Analyses were conducted on the structural damage and deformation of floating wind turbines, the transformation of energy during collision processes, and the resultant motion response of the turbines. A sensitivity analysis was performed on parameters such as collision speed, collision angle, wind speed, and wave height. The findings indicate that the amplitude of pitching and heaving motions of the turbine exceed those observed under conditions devoid of collision loads, with the amplitude of motion intensifying with an increase in these parameters. The turbine's floating body absorbed a minimal amount of internal energy, leading to minor damage, with the stress generated predominantly localized in the collision area of the floating body. The impact of a side collision from vessels exerted a larger influence on the structural dynamic response of floating wind turbines. The analysis results indicate that even though the offshore wind turbine structure is not critically damaged by ship impact, the equipment inside may still fail to work due to the high value of acceleration induced by ship impact. The research outcomes can benefit the safety design of offshore wind turbines in engineering practice.

Keywords: Star-CCM+ and ABAQUS Co-simulation; offshore semi-submersible wind turbine; fluid-structure interaction (FSI); multiphysics load

Citation: Zong, S.; Liu, K.; Zhang, Y.; Yan, X.; Wang, Y. The Dynamic Response of a Floating Wind Turbine under Collision Load Considering the Coupling of Wind-Wave-Mooring Loads. *J. Mar. Sci. Eng.* **2023**, *11*, 1741. https://doi.org/10.3390/jmse11091741

Academic Editor: Giuseppe Roberto Tomasicchio

Received: 25 July 2023
Revised: 21 August 2023
Accepted: 22 August 2023
Published: 4 September 2023

1. Introduction

Amid escalating concerns over environmental degradation and energy scarcity, the advancement of wind energy has emerged as a pivotal research area. Predominantly, wind energy extraction is realized through the generation of wind power, classifiable into onshore and offshore categories. A significant number of offshore wind installations are strategically located in proximity to navigational routes, thereby subjecting them to the potential risk of maritime collisions. According to data sourced from the UK's Caithness Windfarm Information Forum [1], there have been multiple instances of structural damage to wind turbines and associated human casualties as a result of such accidents. For example, a collision event at England's Sheringham Shoal offshore wind farm in 2012 resulted in injuries to five sailors. Similarly, in 2014, a collision involving a high-velocity wind farm service vessel near Kinmel Bay in Wales inflicted considerable damage to a wind turbine tower and caused injuries to three staff members. Moreover, in 2020, an accident at the Borkum Riffgrund 1 offshore wind farm in the North Sea, Germany, led to injuries among three crew members. The accelerating global deployment of floating wind turbines is anticipated to contribute to an upward trend in ship-floating wind turbine collisions. The theoretical construct of the Floating Offshore Wind Turbine (FOWT) was initially conceived in 1972 by Professor Heronemus of the Massachusetts Institute of Technology [2]. Nonetheless, due to limitations in technological capabilities and financial feasibility, the concept remained dormant until the 1990s. With the maturation of onshore and nearshore fixed wind turbine technologies, the FOWT concept was rejuvenated and has swiftly evolved into a contemporary focal point in the domain of offshore wind energy research.

At present, the body of research addressing collision dynamics concerning floating wind turbines remains relatively constrained. Biehl et al. [2] amalgamated numerical simulations of vessel-wind turbine collisions with pertinent statistical data and scenario-specific occurrence probabilities to facilitate a comprehensive assessment of turbine safety. Complementary investigations by Dai [3] and Presencia [4] extended into risk evaluations for offshore wind turbines, encompassing an array of conditions involving diverse vessel types. Biehl [5] elucidated the application of the nonlinear finite element program LS-DYNA for simulating vessel-offshore wind turbine collisions, with an emphasis on the dynamic response characterization during impact events. Moulas et al. [6] leveraged the Sourne [7] analysis methodology to delineate damage deformation modalities of monopile and jacket foundations under varying collision circumstances. Collaborative work by Biehl and Lehman [8] focused on the kinematic response analysis of offshore wind turbines equipped with monopile, tripod, and jacket foundations during vessel impact scenarios. Pire et al. [9] undertook examinations of jacket foundations subsequent to vessel-wind turbine collisions, proffering correspondingly simplified methodologies for anti-collision analysis. Comprehensive studies by Song and Jiang [10,11] scrutinized the collision dynamics between a 4600-ton vessel and a 5 MW monopile offshore wind turbine, encompassing evaluations of parameters such as aerodynamic damping, impact velocity, mean wind speed, wind orientation, and bow rigidity, and their implications on collision response. Additionally, Pire et al. [8] further extended their study of jacket foundations following vessel-wind turbine impacts, with an offering of corresponding simplified models for anti-collision scrutiny. Echevery et al. [12] delved into the collision robustness of spar-type floating offshore wind turbines under vessel impact conditions via numerical simulation approaches, placing particular concentration on the internal dynamic behavior of the structure, including facets such as energy dissipation mechanisms, collision-induced force profiles, and structural deformation patterns.

The mooring platform of offshore floating wind turbines is characterized by flexible movements, and the entire motion system is demonstrably sensitive to influencing factors such as wave load, wind load, and mooring load [13]. Petersen [14] pioneered a coupled simulation methodology, deploying strip theory alongside approximation techniques to compute the sectional added mass and damping. This innovative approach aimed to encapsulate hydrodynamic loads during maritime collisions as well as predict vessel motion within the horizontal plane. Building on this foundation, Tabri et al. [15] broadened the scope of analysis to encompass all six degrees of freedom (DOF), applicable to both the colliding and impacted vessels during collision events. Le [7], utilizing the MCOL code [16] in synergy with the super element technique, embarked on a concurrent exploration of structural responses and extensive overall movements. Subsequent advancements were made by Yu et al. [17], who innovatively employed user-defined subroutines within the LS-DYNA framework, forging an effective coupling between a vessel's horizontal three-dimensional motion and structural deformation. This was achieved through a 3-DOF coupled approach, tailored to collision computations between offshore maintenance vessels and floating wind turbine platforms, with comparative analyses against predictions yielded by the decoupled method. These investigations later evolved to incorporate 6-DOF overall motion [18], culminating in the successful implementation of linear potential flow theory through this refined technique [19]. Echevery et al. [11] leveraged numerical simulation protocols to scrutinize the collision resilience of spar-type floating offshore wind turbines under vessel impact conditions. Localized structural deformation was processed via the LS-DYNA solver, with overall motion resolved through the MCOL program, facilitating a real-time conversion of collision forces between the LS-DYNA and MCOL frameworks at successive time steps. Further integrative work by Zhang and Hu et al. [20,21] synthesized analytical models with floating wind turbine simulation tools, generating authentic reproductions of vessel-wind turbine collision scenarios. To cultivate a wind-wave synergized collision simulation approach, Zhang et al. [22] pioneered an air-water coupled method, accurately replicating collisions between vessels and floating wind turbines, and concomitantly calculating both external and internal dynamic phenomena. Notably, their findings illuminated that high-velocity impacts could precipitate immediate tower collapse; even in instances of withstanding initial impact, residual structural integrity may prove inadequate to bear subsequent wind loads, and the inherent flexibility of the floating wind turbine tower was determined to have negligible influence on energy dissipation mechanisms.

Research efforts addressing collision phenomena pertaining to offshore floating wind turbines remain conspicuously limited on both national and international fronts. Distinctive characteristics of these turbines manifest in a marked overall motion response coupled with localized structural deformations subsequent to collision events. The intricate interplay between external forces, including wind, waves, and mooring, and the global motion response gives rise to multifaceted synergistic effects. Moreover, the analysis of localized structural responses necessitates an understanding of coupling dynamics with global motions, introducing substantial complexity into the entire response assessment process. In light of these challenges, the present study deploys a synergistic simulation methodology, utilizing Star-CCM+ in conjunction with ABAQUS, to undertake an in-depth examination of the dynamic behavior of floating wind turbines under the influence of impact loads within a wind, wave, and mooring system coupled framework. The overarching objective of this research initiative is to furnish a comprehensive insight into the post-collision dynamic behavior of floating wind turbines. Furthermore, this investigative pursuit aspires to contribute substantively to the theoretical underpinnings and technological scaffolding requisite for the independent design and proliferation of deep-sea floating wind turbines within the domestic context.

2. Method for Collision Analysis of Ship-Floating Wind Turbine under the Coupled Effect of Wind, Wave, and Mooring

2.1. Fluid-Structure Interaction (FSI) Method between STAR-CCM+ and Abaqus

Within the domain of marine engineering, naval and offshore structures including vessels and ocean platforms are subject to continuous and multifaceted interactions with the diverse loading conditions prevailing in their aquatic environment. Such loads encompass wind-induced forces above the waterline and wave-induced forces at the water surface interface. The cumulative effect of these disparate loading conditions gives rise to a complex fluid-structure interaction (FSI) problem. The governing principles underpinning this interaction can be delineated as follows:

1. Taking into account the cumulative effect of a multitude of load factors, including wind, current, waves, and collision inertia forces, on the foundation platform of the floating wind turbine, a precise computation must be conducted to ascertain the magnitude of wind, wave, current, and inertia loads exerted on various integral structural components of the wind turbine system.

2. The pressure is transmitted from the fluid field in the Star-CCM+ domain to the structural domain within ABAQUS for the computation of the corresponding structural response.
3. Within the ABAQUS structural domain, considering the effects of wind, wave, current loads, and collision inertia forces on the floating wind turbine, ABAQUS is employed to compute the nodal pressures and displacements of the floating wind turbine system.
4. Utilizing ABAQUS, the nodal pressures and displacements within the structural domain are transferred to Star-CCM+ for the updating of fluid field boundaries. Subsequently, a recalculation is performed to determine the magnitude of the loads exerted on various structural components of the floating wind turbine.

Utilizing the synergistic capabilities of Star-CCM+ and ABAQUS, a cohesive integration of transient dynamic structural analysis and Computational Fluid Dynamics (CFD) is realized. This unification has enabled an in-depth examination of the motion response and structural deformation of a semi-submersible floating wind turbine subject to collision loads within the context of the compounded effects of wind, wave, and mooring forces. This innovative methodology provides a robust simulation platform for capturing the dynamic behavior of a semi-submersible floating wind turbine during an impact event. Figure 1 presents the schematic of the bi-directional coupling methodology between the semi-submersible floating wind turbine models in Star-CCM+ and ABAQUS. Figure 2 illustrates the kinematic sequence of the offshore wind turbine during a ship collision under the influences of wind-wave-mooring load.

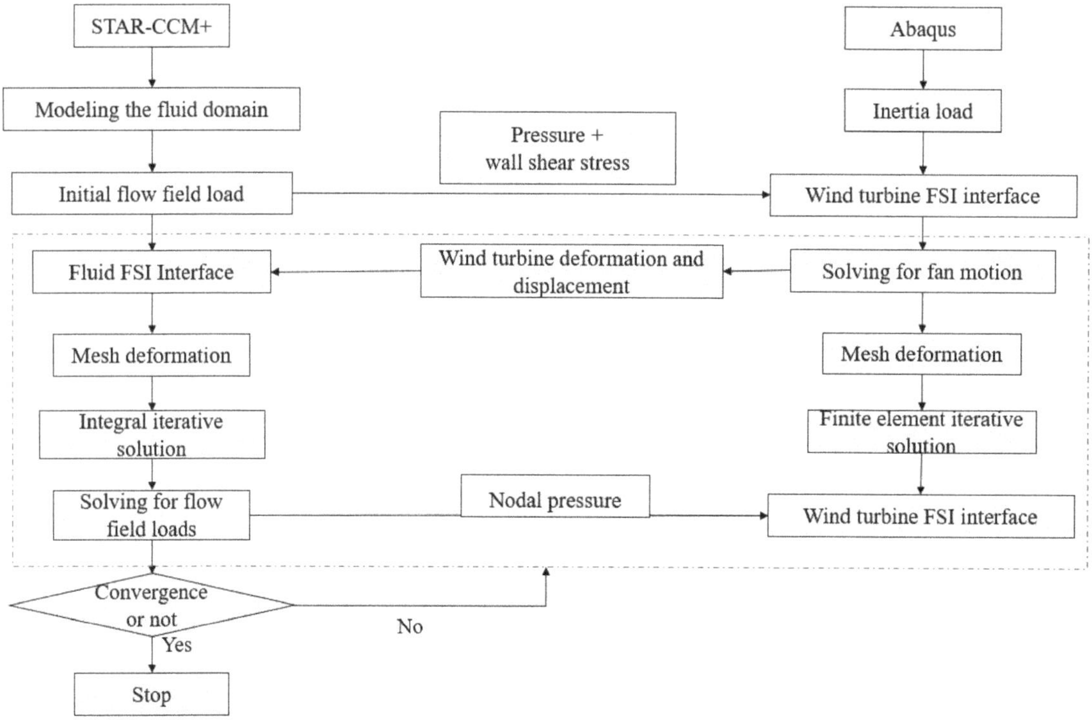

Figure 1. Fluid-Structure Interaction (FSI) Flowchart between Star-CCM+ and ABAQUS.

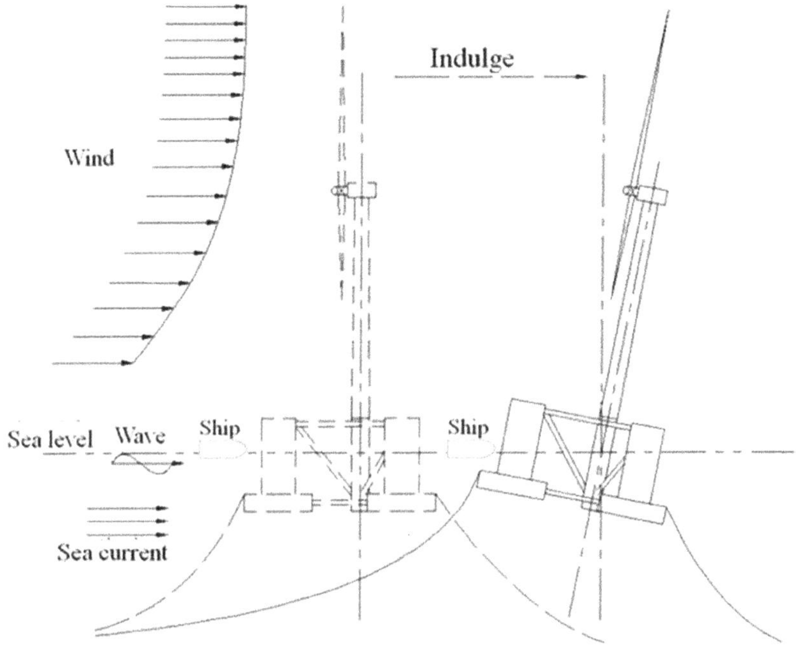

Figure 2. Influence of platform motion.

2.2. Boundary Condition

Given the specific characteristics of semi-submersible wind turbines, the mooring system is treated as a boundary condition within the computational framework. To balance computational efficiency with requisite accuracy, this study has judiciously undertaken a simplification of the mooring system inherent to the semi-submersible wind turbine. In scenarios where the semi-submersible wind turbine undergoes only minimal angular motion and associated displacements remain comparatively restricted, the mooring cable may be appropriately modeled as a linear elastic system. Within the STAR-CCM+ environment, partial derivatives are derived by employing the central difference method, resulting in the formulation of the elasticity stiffness matrix for the mooring cable as delineated in Equation (1):

$$C_{ij}^{Lines} = \begin{bmatrix} 7.08\times10^4\ \mathrm{N/m} & 0 & 0 & 0 & -1.08\times10^5\ \mathrm{N/rad} & 0 \\ 0 & 7.08\times10^4\ \mathrm{N/m} & 0 & 1.08\times10^5\ \mathrm{N/rad} & 0 & 0 \\ 0 & 0 & 1.91\times10^4\ \mathrm{N/m} & 0 & 0 & 0 \\ 0 & 1.07\times10^5\ \mathrm{N/m} & 0 & 8.73\times10^7\ \mathrm{Nm/rad} & 0 & 0 \\ -1.07\times10^5\ \mathrm{Nm/m} & 0 & 0 & 0 & 8.73\times10^7\ \mathrm{Nm/rad} & 0 \\ 0 & 0 & 0 & 0 & 0 & 1.17\times10^8\ \mathrm{Nm/rad} \end{bmatrix} \quad (1)$$

In the context of the ABAQUS simulation environment, the mooring system of the semi-submersible wind turbine is represented through the application of three rotational and three translational spring elements at each of the trio of mooring connection points, a configuration schematically illustrated in Figure 3.

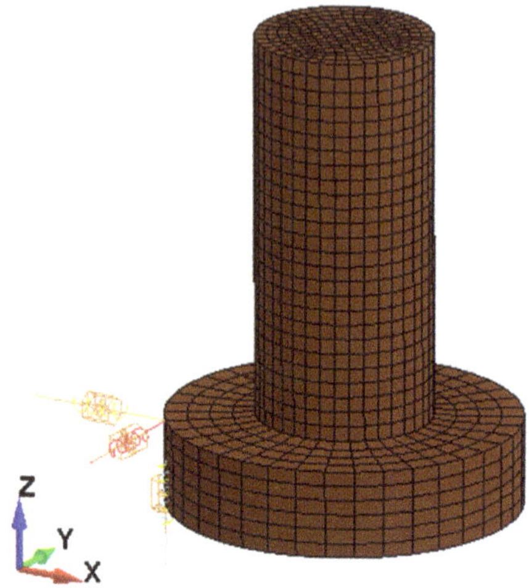

Figure 3. FEM of spring elements.

3. Wind Turbine Models and Environmental Conditions

3.1. Parameters of OC4-DeepCwind Semi-Submersible Wind Turbine and Impacting Vessel Model

In the present investigation, the object of study is the OC4-DeepCwind semi-submersible wind turbine system, a product of development by the National Renewable Energy Laboratory (NREL) in the United States [23]. The upper structure features the NREL-designed 5 MW wind turbine unit, with the complete model visualized in Figure 4a, and the fundamental parameters enumerated in Table 1. Constituting the base of the semi-submersible wind turbine is a supporting platform, fundamentally comprising a central hull (incorporating the wind turbine), a trio of peripheral buoyancy compartments, and an array of transverse and diagonal support rods. The architectural modeling is methodically executed using information extracted from pertinent scholarly literature, with the primary and plan views of the semi-submersible platform delineated in Figure 4b.

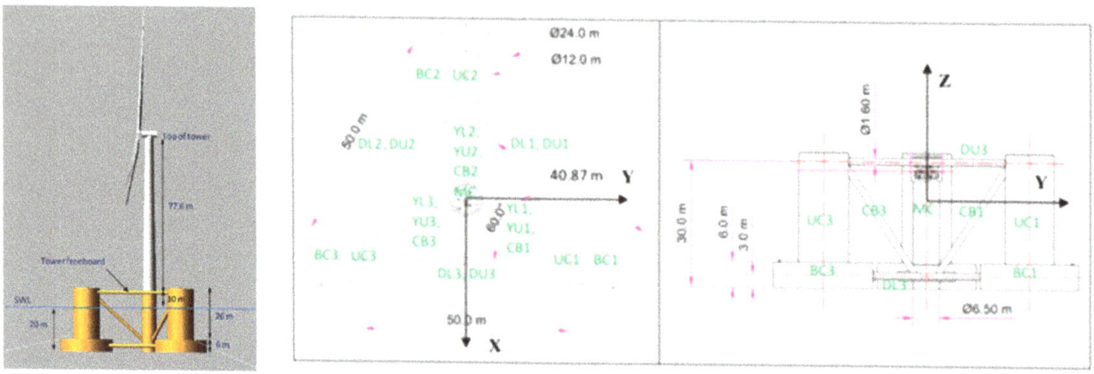

(a) (b)

Figure 4. (**a**) DeepCwind floating wind system (**b**) DeepCwind Floating platform top view (**left**) and main view (**right**).

Table 1. Main parameters of DeepCwind [23].

Principal Dimensions (m)		Weight Parameter	
Bottom buoy diameter/height	24/6	Engine room quality (kg)	240,000
Upper buoy diameter/height	12/26	Platform quality (ballast water) (kg)	1.3473 × 107
Main buoy diameter/height	6.5/30	Platform roll inertia moment (kg·m^2)	6.827 × 109
Diameter of horizontal and inclined support rod	1.6	Platform pitch moment of inertia (kg·m^2)	6.827 × 109
Tower height	77.6	Rocking platform moment of inertia (kg·m^2)	1.226 × 1010
Number of leaves (pieces)	3	Displacement volume (m^3)	13,874.255
		Overall draught (m)	20

The blades of the floating wind turbine system are conceptualized on the 5 MW wind turbine model promulgated by the National Renewable Energy Laboratory (NREL) in the United States. The fundamental attributes of this model encompass a tri-blade configuration, horizontal axis, and upwind orientation. Specific parameters pertinent to the blades are delineated and tabulated in Table 2.

Table 2. Gross properties of DeepCwind.

Parameter Name	Numerical Value
leaf quality (kg)	16,450
The distance between blade centroid and root (m)	23.4
Blade moment of inertia (kg·m^2)	1.39 × 10^7

In the context of this investigation, a wind farm service vessel has been designated as the impacting vessel. The principal parameters characterizing the impacting vessel are itemized in Table 3. To economize on computational resources and truncate the computational duration, the superstructure was consciously excluded during the vessel model formulation, and the inherent structure of the hull was judiciously simplified. The finite element representation and mesh topology of the vessel are depicted in Figure 5, wherein the mesh cell dimensions are precisely configured to 0.5 m. Quadrilateral cells are employed in the mesh, culminating in an aggregate of 6202 cells and 6804 nodal points.

Table 3. Main scale of the colliding ship.

Vessel	Length (m)	Width (m)	Height (m)	Load Draught (m)	Total Weight (t)
support vessel	28	10.8	3.7	2.2	238

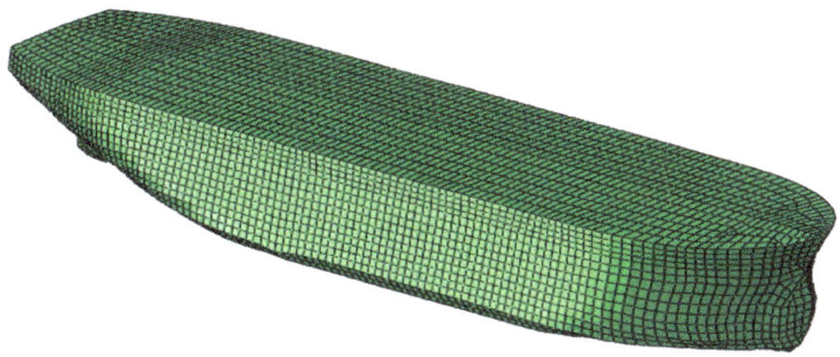

Figure 5. Finite element model and mesh generation of colliding ship.

3.2. Selection of Material Model Parameters

Due to the primary focus on the structural dynamic response of floating wind turbines during collisions, in order to conserve computational resources and reduce calculation duration, the upper structure of the colliding vessel was omitted and the internal structure of the vessel hull was simplified during the ship modeling for collision. The materials used for the semi-submersible wind turbine and the colliding vessel are marine-grade high-strength steel (Q345) and low-carbon steel (Q235), respectively. The material parameters for these components are provided in the Table 4.

Table 4. The material parameters.

Material	Density (kg/m^3)	Elasticity Modulus (GPa)	Poisson's Ratio	Yield Strength (MPa)
Q345	7850	206	0.28	345
Q235	7850	206	0.3	235

3.3. Anchor Chain Parameters

The mooring configuration of the semi-submersible wind turbine comprises three sets of catenary mooring chains, equidistantly aligned along the central pillar's longitudinal axis. Each chain in this configuration is affixed at its lower extremity to an anchorage point embedded within the seabed and at its upper end to the fairlead aperture of the semi-submersible platform. A comprehensive listing of the relevant parameters governing this mooring system is methodically delineated in Table 5.

Table 5. Mooring parameters.

Mooring Parameters		Mooring Parameters	
Parameter	Mooring value	Parameter	Mooring value
Number of anchor chains	3	Anchor chain diameter (m)	0.092
Angle between each anchor chain	120	Anchor chain equivalent density (kg/m)	113.35
Staggered point depth (m)	200	Equivalent density of anchor chain in water (kg·m)	108.63
Guide cable hole height (m)	14	Equivalent tensile stiffness (MN)	753.6
Mooring point radius (m)	837.6	Breaking Strength (kN)	6090
Radius of guide cable hole (m)	40.868	Pretension (kN)	1100
Anchor chain length (m)	835.5		

3.4. Fluid Domain Model

In the collaborative simulation utilizing Star-CCM+ and ABAQUS, the interplay between fluid and structure manifests as an implicitly bi-directional coupled process. The turbulence is modeled via the K-ω SST formulation, while wave simulation is achieved through the application of the Volume of Fluid (VOF) method. The overlapping grid technique is employed to facilitate the movement analysis of the wind turbine. This approach is depicted in Figure 6, where Figure 6a delineates the boundary conditions governing the fluid domain, comprising velocity inlets at the top and left, a pressure outlet to the right, and a wall at the bottom. The water depth, representing the vertical separation between the still water surface and the seabed, is specified as 200 m. For the overlapping area (170 m × 170 m × 230 m), the model surface is stipulated as a physical surface condition, while the interface between the overlapping and background areas is designated as an overset boundary condition. Figure 6b illustrates the cross-sectional view of the grid configuration. In the context of both the vessel and the floating platform base, a refined mesh is generated across the entire surface region. Guided by the model convergence analysis results of Zhang et al. [24], the grid dimensions are determined with maximum and minimum sizes of 2 m and 0.25 m, respectively. Adjacent to the platform's surface, three layers of boundary grids are constructed, amounting to a total thickness of

0.2 m, with a stretching ratio for the boundary layer of 1.2. The grid generation process described herein results in a total count of approximately 6.15 million cells, of which the interior overlapping grids constitute 80% of the aggregate.

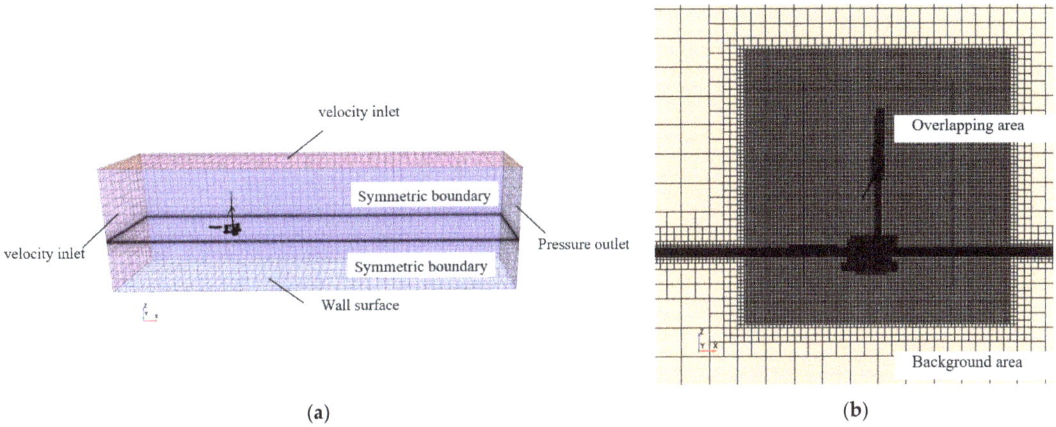

Figure 6. Overlaid mesh area for offshore floating wind turbines. (**a**) Computational domain boundary conditions. (**b**) Grid section diagram.

3.5. Model Experimental Verification

The attenuation observed in the amplitude of motion during free decay motion principally originates from the viscous damping effect imparted by the fluid and the dynamics of the mooring system. The numerical simulations of this free decay motion not only provide insights into the intrinsic frequency and additional characteristics of the wind turbine platform but also serve as an essential validation tool for the reliability of the coupled numerical simulation technique. This validation is particularly critical in the context of addressing hydrodynamic challenges associated with floating structures [25].

In the present investigation, numerical simulations were executed to analyze the free decay motion of the platform in three distinct degrees of freedom: surge, heave, and roll. During the simulation of free decay in each singular degree of freedom, substantial initial displacements were allocated (22 m for surge, 6 m for heave, and 8° for roll), whilst the remaining degrees of freedom were constrained. Subjected to the forces exerted by the wind turbine's inherent gravitational pull and the intricate mooring system, the platform endeavors to regain equilibrium following any deviation from its stabilized position. Modulated by fluid damping, the amplitude of the platform's motion progressively attenuates, asymptotically approaching equilibrium. The congruence between the simulation outcomes and those presented in reference [26] is further corroborated by the analogous free decay curves illustrated in Figure 7.

The outcomes of the bi-directional coupled method simulation presented in this manuscript were juxtaposed with the MARIN tank experimental findings as reported by Coulling et al. [25], along with the direct computational results obtained using FAST V8.10. The semi-submersible platform was adapted for the purpose of modeling, employing the NREL-5MW wind turbine design. Within the context of the MARIN tank experiment, an approximation congruent with the NREL-5MW served as the wind turbine representation. This test model of the wind turbine exhibited slight variations in mass and blade configuration when compared to the archetype, leading to minor disparities in the computation of the platform motion's natural period. According to the delineated natural periods of the platform in Table 6, the coupled analysis model established through the methodology presented herein reveals a more pronounced discrepancy in the surge direction, quantified by an error rate of 5.79%. Nevertheless, the cumulative error rate of the model remains confined within a 6% margin, a deviation considered to be within acceptable boundaries.

The hydro-aerodynamic coupled analysis model, forged by the means delineated in this study, aligns favorably with extant experimental data, thereby corroborating the fidelity of the software-based modeling and the validity of the numerical simulation approach for the semi-submersible wind turbine.

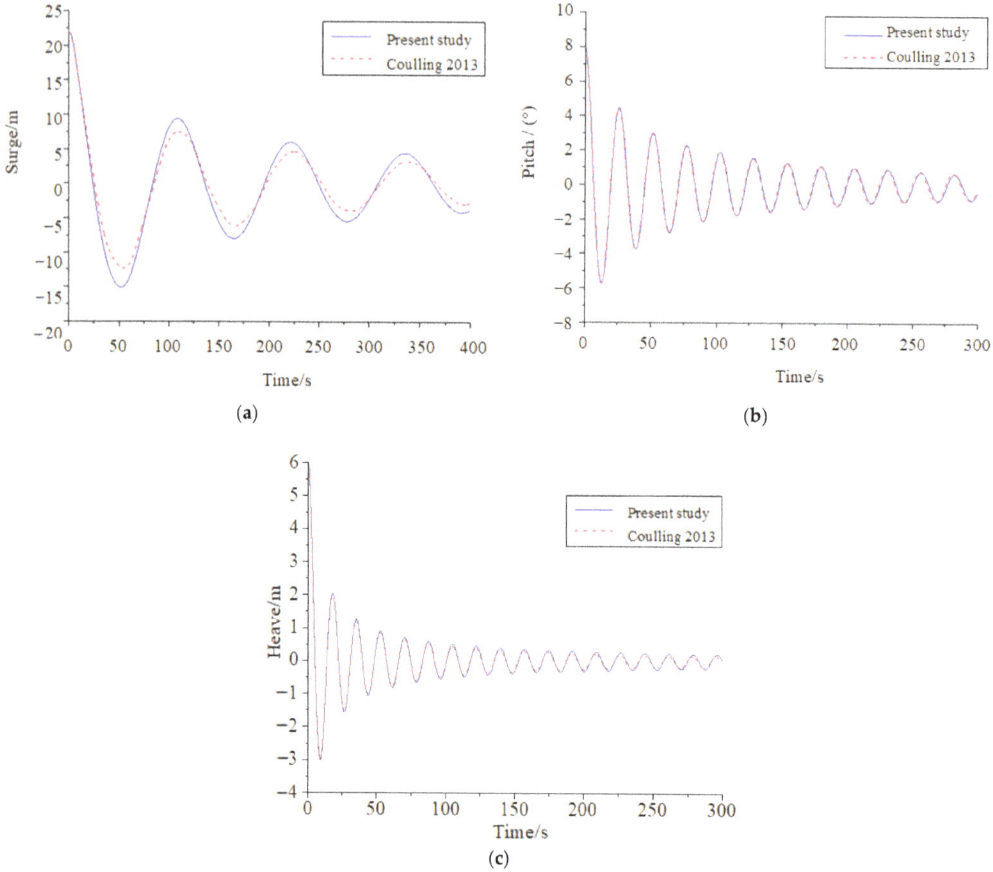

Figure 7. Platform free decay motion contrast: (**a**) Surge; (**b**) Pitch; (**c**) Heave.

Table 6. Comparison of calculation results of platform natural period test.

Parameter	Platform Inherent Period/(s)			Error (Compared to Test)
	Experiment (Coulling)	FAST V8.10	Calculates	
Surge	107.0	120.0	113.2	5.79%
Heave	17.5	17.7	17.7	1.14%
Pitch	26.8	25.0	25.7	4.1%

4. Scenarios for Collision

In compliance with the standards stipulated in DNVGL-ST-0126 and DNV-ST-0119 by the ship classification society, the requisite velocity of the ship during a collision event is mandated to be no less than 0.5 m/s [27,28]. Drawing from empirical collision incidents, the impact velocities of the colliding ship were systematically evaluated at 2.0 m/s, 1.5 m/s, and 1.0 m/s. This analysis was conducted to scrutinize the ramifications of varied collision velocities on the dynamic response characteristics of the floating wind turbine structure.

Both frontal and lateral collisions were established at an impact speed of 2.0 m/s. To faithfully emulate the circumstances preceding a collision with a floating wind turbine in motion, an initial separation distance of 20 m between the colliding vessel and the wind turbine was designated. The locus of collision was meticulously chosen near the draft line of the floating wind turbine. Throughout the collision dynamics involving the maintenance ship and the offshore wind turbine, a definable angular relationship typically exists at the moment of impact. Collisions were simulated at a speed of 1.5 m/s, encompassing the collision angles of 0°, 45°, and 90°—angles representative of conventional collision scenarios. Subsequent analyses were conducted to decipher the influence of these distinct collision angles on the wind turbine's structural response. Environmental parameters were calibrated to a wind speed of 11.4 m/s, corresponding to the rated wind speed for the floating wind turbine, accompanied by a wave height of 2 m. Both the wind and wave directions were aligned with the negative x-axis. Figure 8 elucidates these conditions, wherein subfigure (a) illustrates the schematic of the bow collision, and subfigure (b) portrays the side collision schematic. Table 7 enumerates the specific collision conditions pertinent to offshore wind turbines. The temporal domain of the simulation extended over 100 s, discretized using a time increment of 0.02 s.

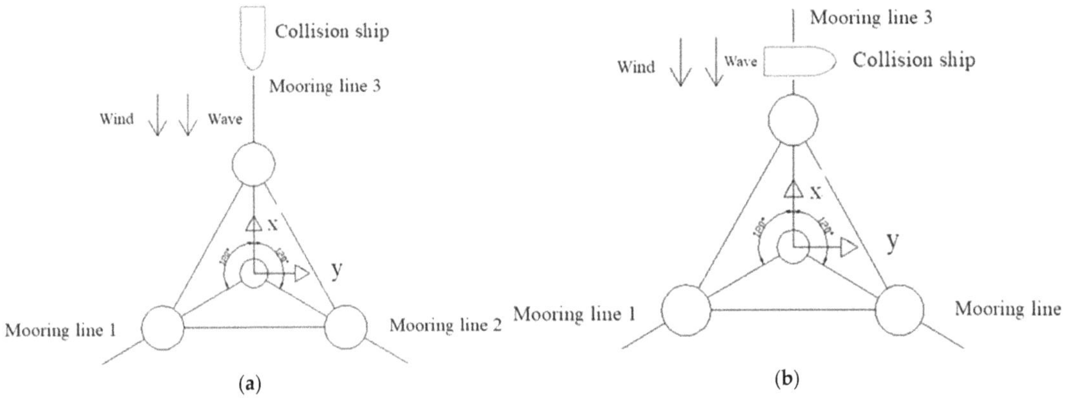

Figure 8. Simplification of collision scenarios: (**a**) bow collision scene and (**b**) ship-side collision scene.

Table 7. Case definitions.

Case	Impact Velocity (m/s)	Impact Angle (°)	Wind (m/s)	Wave (m)
LC1	1	0	11.4	2
LC2	1.5	0	11.4	2
LC3	2	0	11.4	2
LC4	1.5	45	11.4	2
LC5	1.5	90	11.4	2

To meticulously investigate the influence of environmental parameters on the collision response dynamics of floating wind turbines, a comprehensive examination was conducted utilizing selected wind speeds of 11.4 m/s, 10.0 m/s, and 8.0 m/s. This examination aimed to elucidate the consequences of disparate wind speeds on the dynamic response characteristics of the turbines. Concurrently, a spectrum of wave heights—namely 2.5 m, 2.0 m, and 1.5 m—was chosen for the analysis, intending to shed light on the impact of variations in wave heights on the turbines' dynamic behavior. Standardized conditions for the collision scenario were established, incorporating a typical collision angle of 0° and a vessel speed of 2 m/s. The orientation of both wind and wave directions was presumed to coincide with the negative x-axis direction. The numerical simulation extended over a temporal span of 100 s, adopting a timestep resolution of 0.02 s. A detailed depiction of the specifications pertinent to these carefully defined scenarios can be found in Table 8.

Table 8. Case definitions.

Case	Impact Velocity (m/s)	Impact Angle (°)	Wind (m/s)	Wave (m)
LC6	2	0	11.4	2
LC7	2	0	10	2
LC8	2	0	8	2
LC9	2	0	11.4	2.5
LC10	2	0	11.4	1.5

5. Dynamic Simulation Analysis of Semi-Submersible Wind Turbine under Multi-Field Coupling Collision Load

To systematically investigate the influence of collision loads on the dynamic response of floating wind turbines, comparative analyses were conducted, juxtaposing the responses of turbines subjected to both head-on and lateral collisions with those devoid of collision-induced loads. Within the scope of this rigorous examination, during the discrete scenarios encompassing head-on and lateral collisions, the vessel's collision velocity was meticulously fixed at 2.0 m/s. As for the environmental parameters under consideration, the wind velocity was calibrated to correspond to the designated rated wind speed of the floating turbines, precisely 11.4 m/s, while the wave height was uniformly stipulated at 2 m.

5.1. Results and Analysis of Bow Collision Simulation

5.1.1. Collision Force

Figure 9 elucidates the temporal progression of the collision force during the bow impact, when the impacting vessel engages the floating wind turbine. As depicted in the figure, the entire collision sequence transpires within an approximate time frame of 0.15 s, manifesting a marked non-linearity in the force profile. Upon the initiation of the contact between the vessel's bow and the floating wind turbine's pontoon, a pronounced surge in collision force occurs, escalating rapidly within a mere 0.04 s. Concurrently, as the interface between the ship and the wind turbine pontoon progressively expands, the bow of the impacting vessel encounters a commensurate increase in resistance, culminating in a peak force magnitude of 4.913 × 1064.913 × 106 N at 0.04 s. In the ensuing interval spanning from 0.04 s to 0.15 s, the collision force commences a decremental phase. This reduction is primarily attributed to the attenuating velocity of the impacting vessel as it engages with the wind turbine pontoon. This engagement precipitates a corresponding depletion in kinetic energy, culminating in the termination of the collision process, whereupon the collision force abates to nullity.

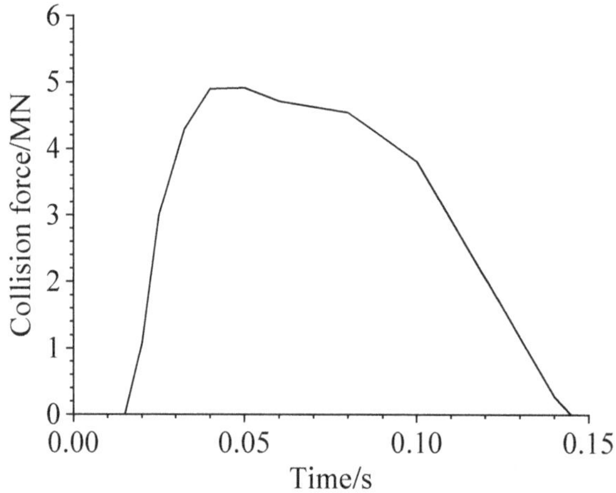

Figure 9. Curves of collision force-time.

5.1.2. Energy Analysis

Figure 10 delineates the comprehensive energy conversion process of the wind turbine. Considering that the spatial orientation of the floating wind turbine system is constrained by the mooring system, the structure is primarily subject to aerial and wave-induced loads prior to collision. Consequently, only a nominal quantity of kinetic and internal energy is manifest. During the collision phase, a predominant fraction of the impacting vessel's kinetic energy is transmuted into the aggregate kinetic energy of the floating wind turbine. This is complemented by minor contributions from internal energy, hourglass energy, damping energy, and frictional energy components. Owing to their diminutive magnitudes, the contributions from damping and friction energies are excluded from the representation in the figure. The kinetic energy attains an apex value of $1.37 \times 1081.37 \times 108$ J. This observation underscores the substantial influence of the impact on the motion response characteristics of the floating wind turbine during the collision dynamics with the ship. Nevertheless, the collision culminates in negligible structural deformation and damage to the system.

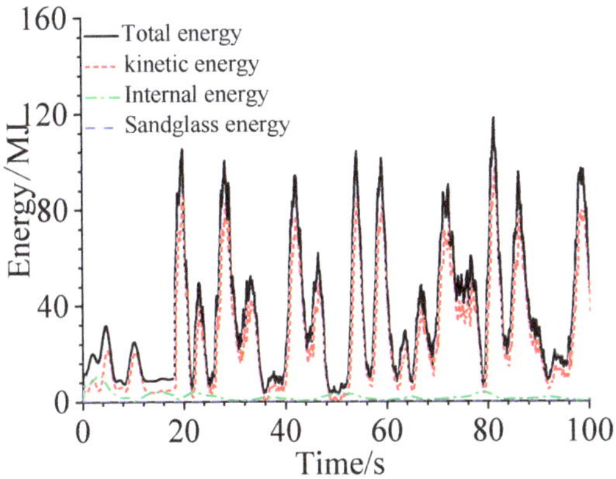

Figure 10. Curve of energy conversion of system.

5.1.3. Structure Stress Analysis

Figure 11 portrays the cumulative structural stress contour diagram in the course of a head-on collision between the vessel and the DeepCwind floating wind turbine at 18.14 s. A scrutiny of the diagram reveals that the stress engendered in the collision interaction between the ship and the floating wind turbine is principally localized within the pontoon's collision zone, exhibiting a marked spatial specificity. Subsequent sections are dedicated to a comprehensive examination of the stress-strain response of the pontoon subjected to impact dynamics.

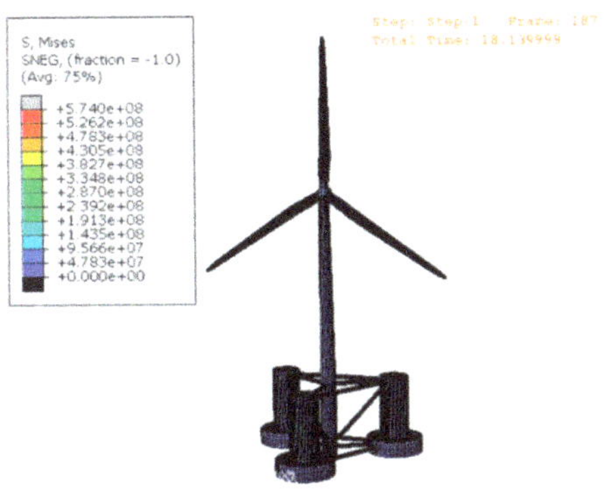

Figure 11. Effective stress contours of front collision between the DeepCwind and vessel.

Figure 12 presents the contour plot of the pontoon's structural stress during the frontal collision between the vessel and the floating wind turbine. A comparative analysis of the figures within the temporal span from 18.14 s to 18.20 s reveals an escalating trend in the peak value of the pontoon's structural stress, culminating in a zenith of 3.546×10^8 Pa at 18.16 s. Subsequent to this peak, the stress commences a reduction phase, corresponding to the dissipation of the collision force, and the spatial distribution of the structural stress displays pronounced localization.

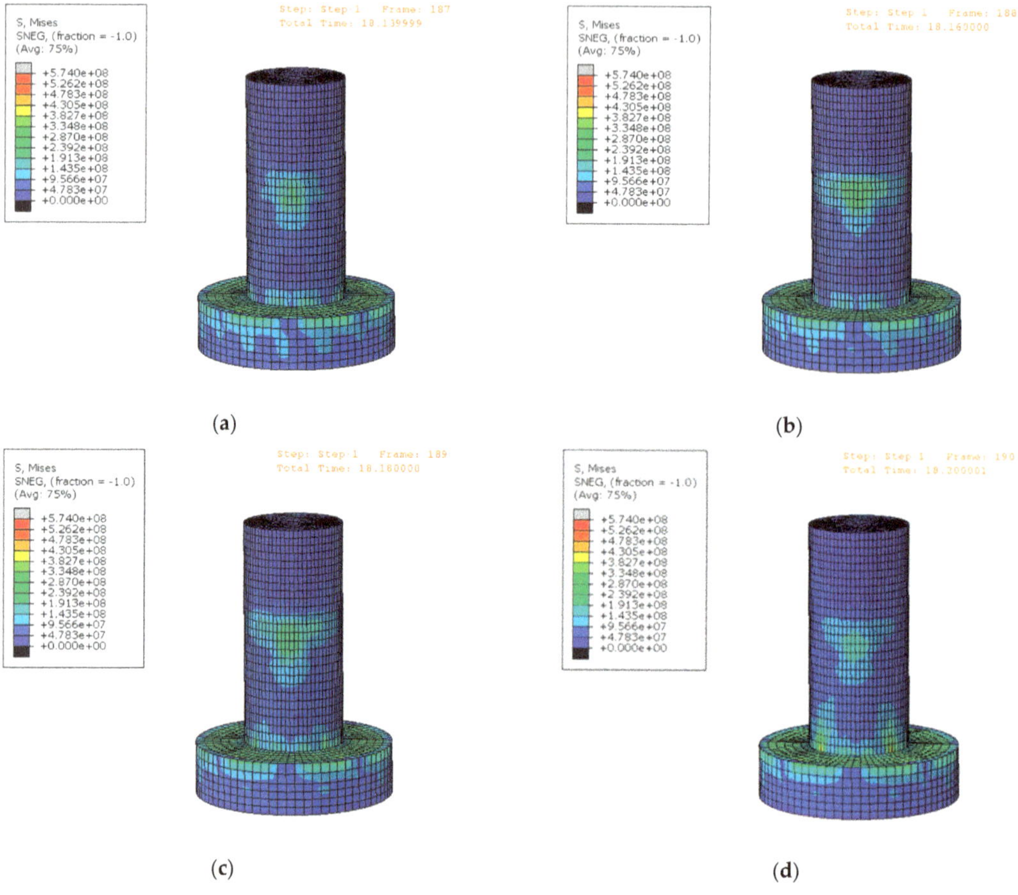

Figure 12. Effective stress contours of upper column: (**a**) 18.14 s; (**b**) 18.16 s; (**c**) 18.18 s; (**d**) 18.20 s.

5.1.4. Motion Response

Utilizing the motion response of the floating wind turbine under the combined influences of wind, wave, and mooring loads (without the effect of ship collision) as a benchmark, a more insightful juxtaposition of the floating wind turbine's motion responses can be conducted, both with and without the imposition of impact loads subsequent to the collision event. Figure 13 furnishes a comparative portrayal of the pitch and heave motion responses of the floating wind turbine under these differing scenarios. Owing to the orientation of the wind, waves, and ship impact, all of which are aligned towards the negative direction of the x-axis, the roll and sway motion components manifest with relatively diminutive magnitudes and are therefore omitted from the analysis.

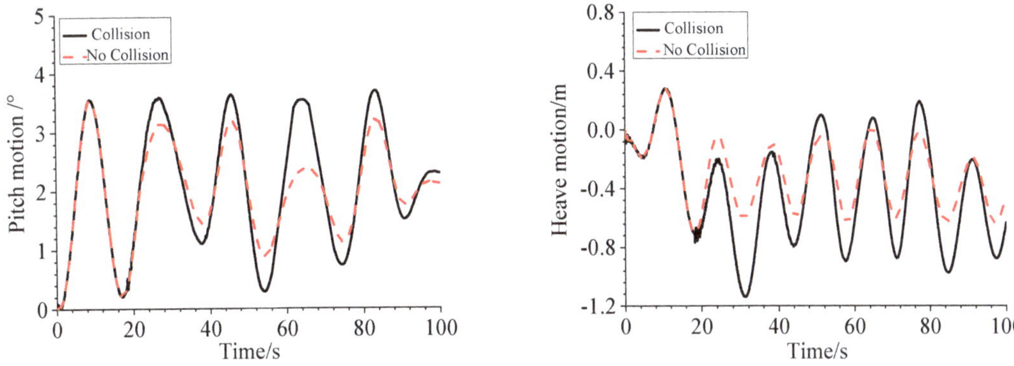

(a) (b)

Figure 13. Comparison of motion response: (**a**) comparison diagram of Pitch and (**b**) comparison diagram of Heave.

As depicted in the figure, the collision transpired at 18.14 s. Subsequent to the collision, the motion response curves corresponding to pitch and heave degrees of freedom both manifest conspicuous nonlinearities, with the amplitudes of pitch and heave motions in the impacted state exceeding those in the non-impacted state. The maximum pitch angle attained a value of 3.71°, while the maximum heave amplitude reached 1.12 m. These observations elucidate that the impact load exerts a pronounced influence on the motion response of the floating wind turbine. This phenomenon can be attributed to the transformation of the majority of the kinetic energy generated by the colliding vessel into the kinetic energy of the wind turbine, culminating in an augmented amplitude of the pitch and heave motions of the floating wind turbine.

5.2. Results and Analysis of Ship Side Collision Simulation

In contrast to a bow collision, a side collision with a floating wind turbine typically involves a larger contact area, which may consequently affect both the structural integrity and motion response of the floating wind turbine under the interconnected influence of wind, wave, and mooring system. Hence, this section is devoted to a detailed numerical simulation analysis, scrutinizing the ramifications of a side impact exerted by a ship on the floating wind turbine structure.

5.2.1. Collision Force

Figure 14 delineates the temporal progression of the collision force during a side impact from a service vessel on the floating wind turbine. In this scenario of side impact, parameters such as the kinetic energy of the impact vessel and wind speed are maintained constant, yet the magnitude of the impact wave fluctuates as a result of variations in the timing of the collision. The duration of the collision under the side impact conditions extends to 0.17 s, constituting an increase of 0.05 s in comparison to the bow impact. The force exerted during the collision reaches its maximum value at 0.09 s, culminating in a peak force of 5.195 × 106 N, a value that surpasses the corresponding force observed in the bow impact. This phenomenon indicates that the side collision mode prolongs the interaction time between the impacting vessel and the floating wind turbine, thereby amplifying the collision force.

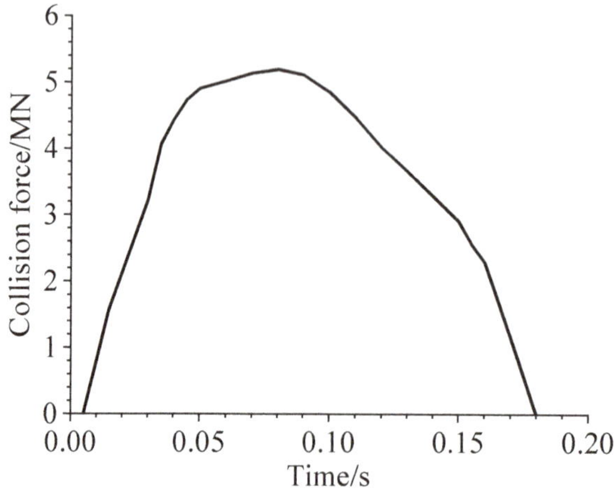

Figure 14. Curves of collision force-time.

5.2.2. Energy Analysis

Figure 15 delineates the comprehensive energy transformation within the system during a side collision event. As manifestly demonstrated by the figure, the collision commences at 19.30 s, and, akin to the scenario of the bow collision, the energy conversion curve in the case of a side collision exhibits a pronounced nonlinear behavior. The preponderance of the kinetic energy originating from the colliding vessel is transmuted into the floating wind turbine's kinetic energy, with a subordinate fraction being diverted into the internal energy and hourglass energy of the turbine. It is noteworthy that the aggregate energy of the system during a side collision slightly surpasses that during a bow collision, and the occurrence of peak values is more frequent. This leads to an elevation in both the kinetic energy transmitted to the wind turbine and the absorbed internal energy in comparison to the bow collision, thus underscoring the distinct dynamical consequences of the collision orientation.

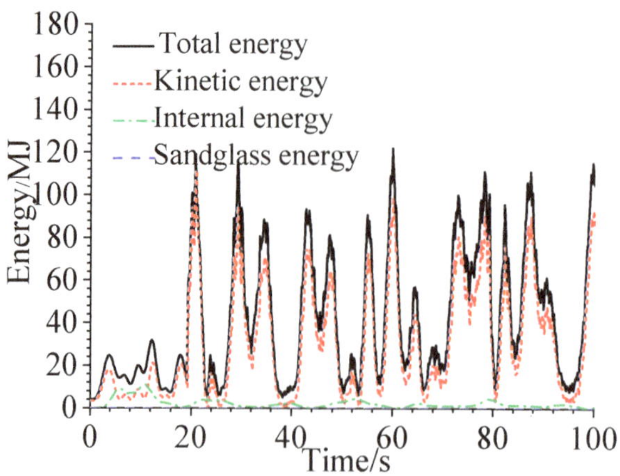

Figure 15. Curve of energy conversion of system.

5.2.3. Structure Stress Analysis

Figure 16 exhibits the global equivalent stress cloud diagram at the temporal instance of 19.29 s during a side collision between the ship and the DeepCwind floating wind turbine. The distribution of equivalent stress ensuing from the ship's side collision predominantly localizes at the impacted pontoon region of the wind turbine. A comparative analysis elucidates that the magnitude of the equivalent stress attributed to the side collision is notably inferior to that induced by the frontal collision. This observation underscores the intrinsic disparity in the stress response behavior contingent on the direction of collision, with the side collision yielding a less intense stress concentration within the specified structural components.

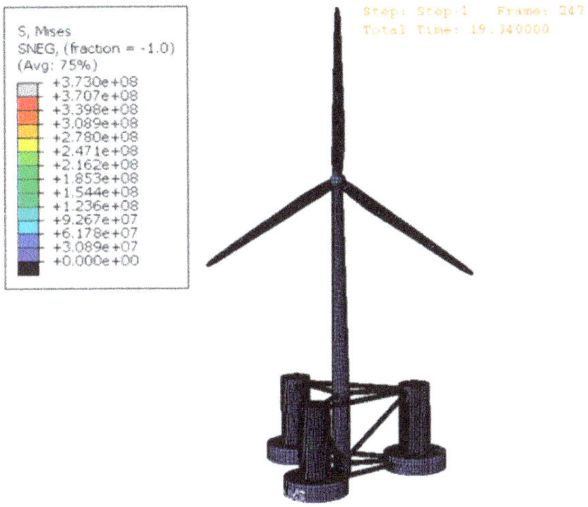

Figure 16. Effective stress contours of side collision between the DeepCwind and vessel.

Figure 17 delineates the stress contour map of the pontoon during the side collision process between the vessel and the floating wind turbine. The visualization manifestly reveals that the equivalent stress engendered by the vessel's lateral impact is principally localized at the point of collision on the pontoon. Commencing at 19.30 s, the collision stress initiates at a value of 2.475 × 108 Pa, subsequently augmenting to its zenith of 3.73 × 108 Pa by 19.32 s. Contrasted with the evidence presented in Figure 16, it can be deduced that the equivalent stress resultant from a side collision is somewhat attenuated relative to that emanating from a bow collision. Nevertheless, the extended temporal duration of the collision in the side impact case potentially amplifies the deleterious influence on the structural integrity and functional performance of the floating wind turbine. This insight accentuates the complex interplay of collision geometry and mechanical response, necessitating a nuanced understanding for comprehensive structural assessment and design.

5.2.4. Motion Response

Figure 18 provides an illustrative comparison of the roll and heave motion responses of the floating wind turbine under conditions with and without side impact. From the graphical representation, it is evident that both the roll and heave motion responses subsequent to the lateral engagement of the vessel with the floating wind turbine are accentuated relative to their non-impact counterparts. When juxtaposed with Figure 13, which elucidates the motion responses following a bow collision, an increase in both roll and heave motion amplitudes is discernible in the context of a side collision. This empirical observation underscores that a side collision imparts a more pronounced influence on the floating wind turbine's motion response. This comparative analysis not only elucidates the differential response mechanisms of the structure under various collision orientations but also reinforces the necessity for an in-depth examination of collision dynamics, especially in the scenario of lateral impacts. Such insights bear significant implications for the design, evaluation, and risk mitigation strategies of floating wind turbine structures, catering to the multifold complexity of maritime collision phenomena.

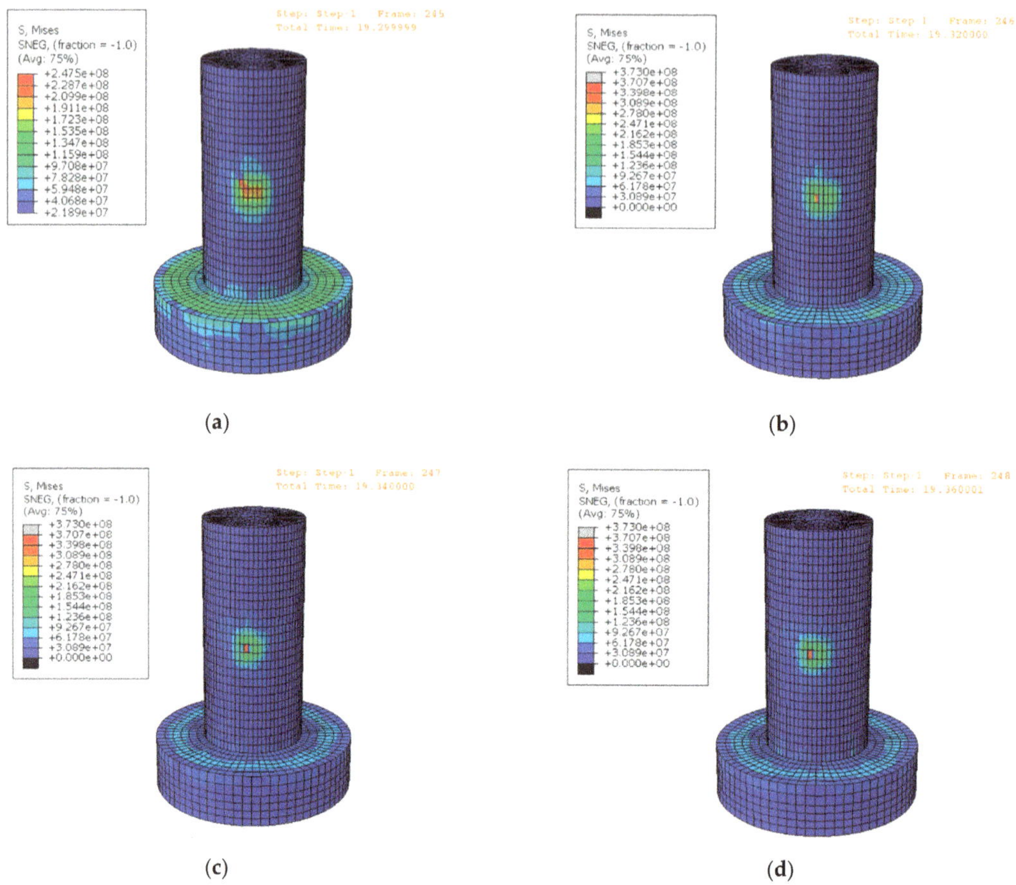

Figure 17. Plastic strain contours of upper column: (**a**) 19.30 s; (**b**) 19.32 s; (**c**) 19.34 s; (**d**) 19.36 s.

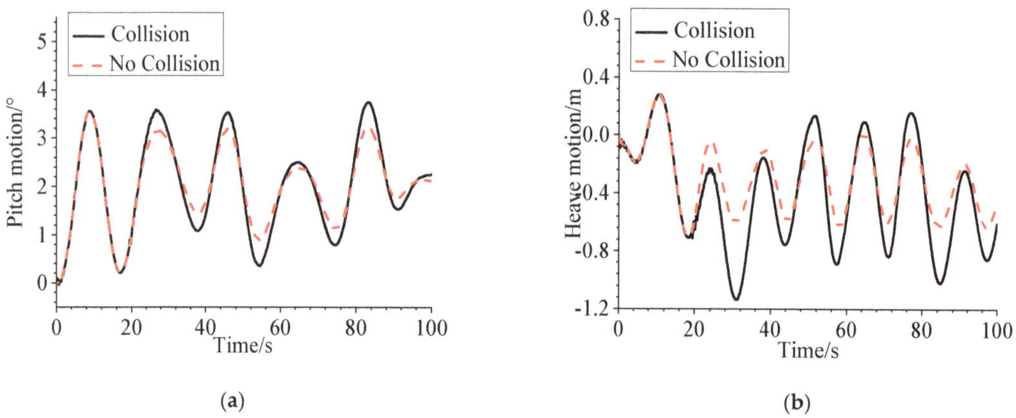

Figure 18. Comparison of motion response: (**a**) comparison diagram of Pitch and (**b**) comparison diagram of Heave.

6. Results and Discussions

6.1. Sensitivity Analysis of Different Collision Parameters on Floating Wind Turbine Collision

6.1.1. Collision Velocity

(1) Collision force

Figure 19 delineates the time-history curve of the collision force exerted on the floating wind turbine, specifically capturing the dynamics at varying impact speeds. Within the visual representation, the onset of the collision phase is marked by a significant generation of collision force in the localized contact area between the vessel and the floater component of the wind turbine. Through a comparative analysis of the collision force curves encapsulated in Figure 19, distinct response patterns can be discerned across different impact velocities. Specifically, at impact speeds of 2.0 m/s, 1.5 m/s, and 1.0 m/s, the corresponding collision durations are measured at 0.15 s, 0.12 s, and 0.11 s, with the resultant peak collision forces registered at 4.913×10^6 N, 3.487×10^6 N, and 2.001×10^6 N, respectively. This analytical observation clearly underscores the direct correlation between vessel impact speeds and both the magnitude and duration of the collision process. In scenarios where the vessel impact speed is elevated (v = 2 m/s), the collision event between the vessel and the wind turbine floater manifests heightened severity. This amplification can be attributed to the corresponding increase in kinetic energy associated with the elevated vessel velocity, culminating in both an augmented peak collision force and an extended collision process. These findings impart critical insights into collision dynamics, offering quantitative evidence of the sensitivity of collision parameters to the varying speed of impact. Such understanding is imperative for the optimization of design, safety assessment, and risk management strategies in the realm of floating wind turbine installations, further emphasizing the necessity for thorough investigation and characterization of collision phenomena within marine environments.

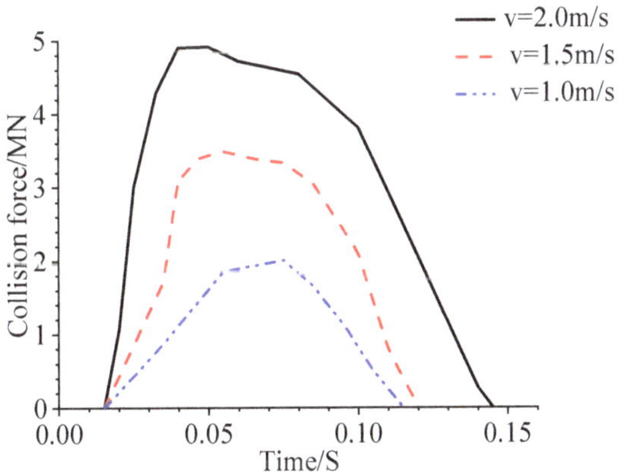

Figure 19. Curves of collision force-time under different impact velocities.

(2) Energy analysis

In the event of a collision between a vessel and a floating wind turbine, the kinetic energy of the impacting vessel is predominantly transmuted into the kinetic energy of the floating wind turbine, while only a fractional conversion into the system's internal energy and other energy modalities is observed. Figure 20 offers a comprehensive visual representation of the transformation ratio between kinetic and internal energy within the floating wind turbine system, conditioned on various collision speeds. The data presented in Figure 20 unequivocally illustrates the consequential influence that different impact speeds exert on both the kinetic and internal energy curves. Specifically, when the impacting vessel's speed is registered at 1.0 m/s, the conversion of kinetic and internal energy transferred to the floating wind turbine is at its nadir. Conversely, as the speed of the impacting vessel incrementally ascends, a corresponding amplification is observed in the kinetic and internal energy imparted to the floating wind turbine. This relationship elucidates a critical insight into the dynamics of collision phenomena: the motion response and structural integrity of the floating wind turbine are intrinsically linked to the velocity of the impact. The escalating nature of the transferred energies as a function of increasing impact speed implies a progressive exacerbation in both the motion response and the structural duress experienced by the floating wind turbine. The understanding gleaned from this analysis provides an imperative foundation for the engineering, design, and risk mitigation strategies pertinent to floating wind turbine systems. Recognizing the kinetic dependencies and the transformation characteristics of collision energies contributes to a more robust predictive modeling capability, further enabling the development of resilient structures and adaptive response mechanisms to mitigate the multifaceted effects of vessel collisions within the maritime renewable energy sector.

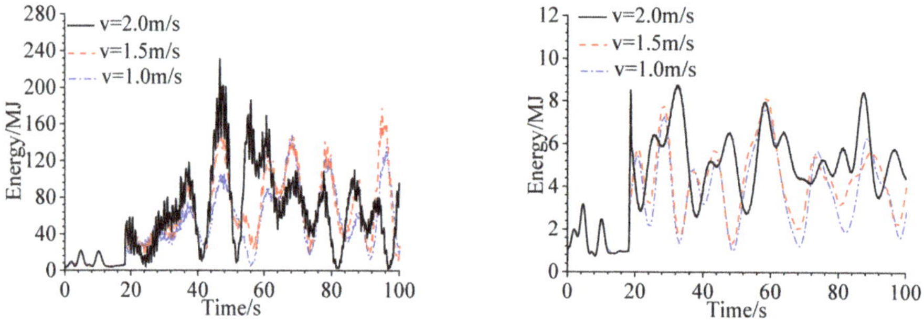

(a) (b)

Figure 20. Curves of energy of system under different impact velocities: (**a**) Kinetic Energy and (**b**) Internal Energy.

(3) Structure stress analysis

Figure 21 shows the distribution of equivalent stress in the floating wind turbine system under different collision speeds. From the information depicted in the figure, it is evident that variations in the collision speed result in different deformation characteristics of the spar. When the collision speeds are 1.0 m/s and 1.5 m/s, the peak equivalent stress in the spar occurs at 18.24 s, with magnitudes of 2.795×10^8 N and 3.347×10^8 N, respectively. For a collision speed of 2.0 m/s, the peak equivalent stress in the spar is reached at 18.14 s with a magnitude of 3.846×10^8 N. These observations indicate that an increase in the collision speed of the vessel leads to an augmentation in the structural stress in the spar, thereby causing a larger deformation.

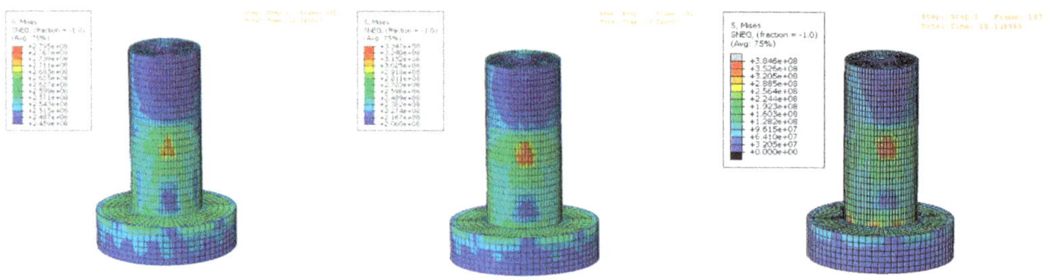

(a) (b) (c)

Figure 21. Effective stress contours of upper column under different impact velocities: (**a**) v = 1 m/s; (**b**) v = 1.5 m/s; (**c**) v = 2 m/s.

(4) Motion response

Figure 22 portrays the dynamic response of the floating wind turbine system in terms of pitch and heave movements under the impact of varying collision velocities. It can be unequivocally discerned from the figure that as the collision speed of the vessel escalates, the amplitude of the pitch and heave motions of the floating wind turbine also proportionally amplifies. This phenomenon can be attributed to the increased kinetic energy of the impacting vessel due to its augmented speed. This extra kinetic energy, transferred to the floating wind turbine during the collision, subsequently leads to an increased amplitude in the pitch and heave motions of the turbine.

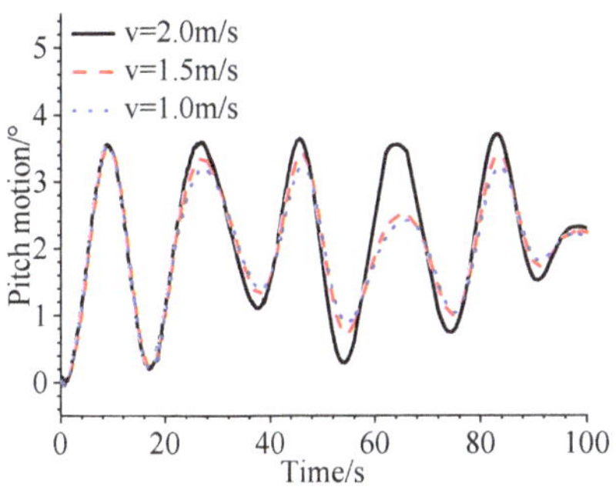

(**a**)

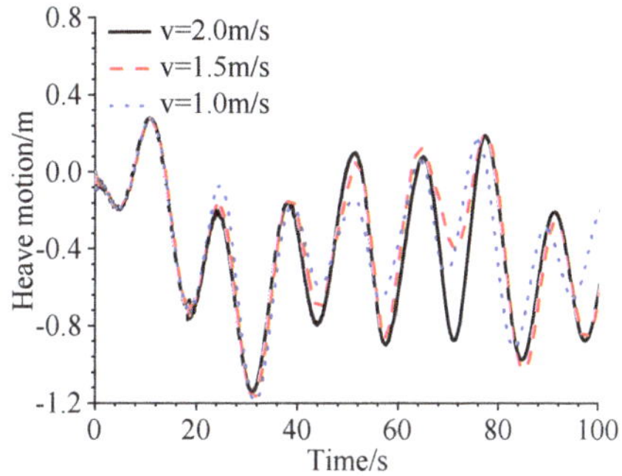

(**b**)

Figure 22. Comparison of motion response under different impact velocities: (**a**) comparison diagram of Pitch and (**b**) comparison diagram of Heave.

6.1.2. Collision Angle

(1) Collision force

Figure 23 elucidates the chronological evolution of collision forces under disparate impact angles in the context of a collision between a maritime vessel and a floating wind turbine. As delineated by the illustration, the temporal trajectories of the collision forces, despite variations in impact angles, exhibit nominal disparity, consistently demonstrating an immediate acceleration followed by a subsequent decrement. The temporal duration of the collision procedure maintains an approximate constancy across the spectrum of varying impact angles, thereby substantiating the observation that the magnitude of the collision force is relatively invariant with alterations in the impact angle.

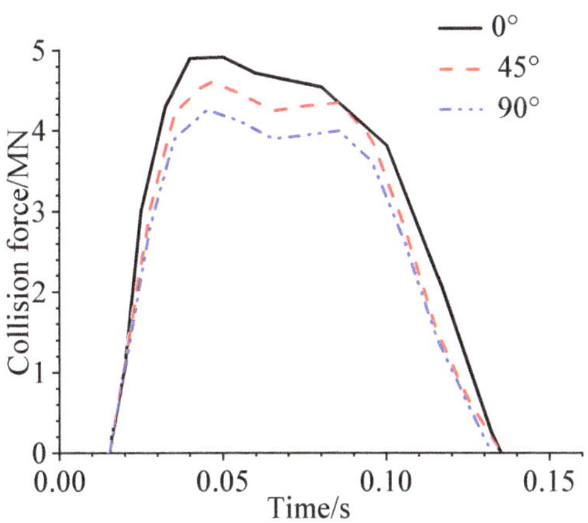

Figure 23. Curves of collision force-time under different impact angles.

(2) Energy analysis

Figure 24 delineates the juxtaposed distribution of kinetic and internal energy within the floating wind turbine system when exposed to collisions under multifarious angular trajectories. As can be inferred from the illustration, the variations in the impact angle impose only a negligible influence upon the kinetic and internal energy contours within the system. This marginal effect can be principally attributed to the invariability of the kinetic energy of the colliding vessel, assuming a constant velocity, in conjunction with unaltered meteorological conditions such as wind velocity and wave amplitude. Collectively, these factors converge to substantiate the minimal perturbation of different impact angles on the integral energy components within the floating wind turbine system.

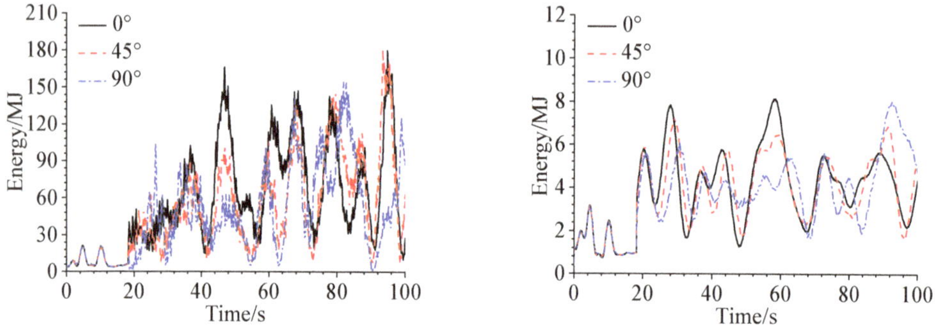

(**a**) (**b**)

Figure 24. Curves of energy of system under different impact angles: (**a**) Kinetic Energy and (**b**) Internal Energy.

(3) Structure stress analysis

Figure 25 graphically illustrates the salient disparities in the deformational damage experienced by the buoyant wind turbine system when subjected to impacts at distinct angular orientations. As discerned from the figure, the exertion of maximum stress on the cylindrical structure is moderately reduced at an impact angle of 90°, compared to alternative angular configurations, achieving its zenith at 3.530 × 108 N at precisely 18.02 s, subsequently culminating in a relatively attenuated structural impairment. Conversely, an impact angle of 45° induces the superlative stress level among the investigated angles, reaching an apex of 3.996 × 108 N at 18.34 s. When the system is exposed to an impact angle of 0°, the zenithal stress on the cylinder peaks at 3.846 × 108 N at 18.14 s. Cumulatively, these empirical observations underscore that the structural damage inflicted upon the floating wind turbine system is most pronounced and deleterious at an impact angle of 45°.

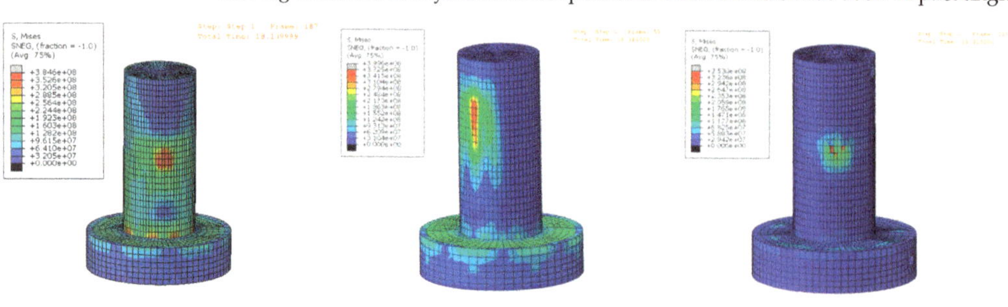

(**a**) (**b**) (**c**)

Figure 25. Effective stress contours of upper column under different impact angles: (**a**) 0°; (**b**) 45°; (**c**) 90°.

(4) Motion response

Figure 26 methodically delineates the dynamic response of the floating wind turbine system in both pitch and heave motion subjected to disparate impact angles. The graphical representation elucidates that the amplitude of pitch motion consequent to an impact at 0° perceptibly surpasses that resultant from impacts at 45° and 90°. In parallel, variations in the angle of impact manifest a marginal influence on the heave motion response within the floating wind turbine system. This seemingly subdued effect can be ascribed to the relatively diminutive disparities in the collision forces engendered under various angular impacts, compounded by invariant meteorological conditions such as wind velocity and wave state. Collectively, these factors coalesce to exert only minimal perturbations on both the pitch and heave motion response characteristics within the floating wind turbine system.

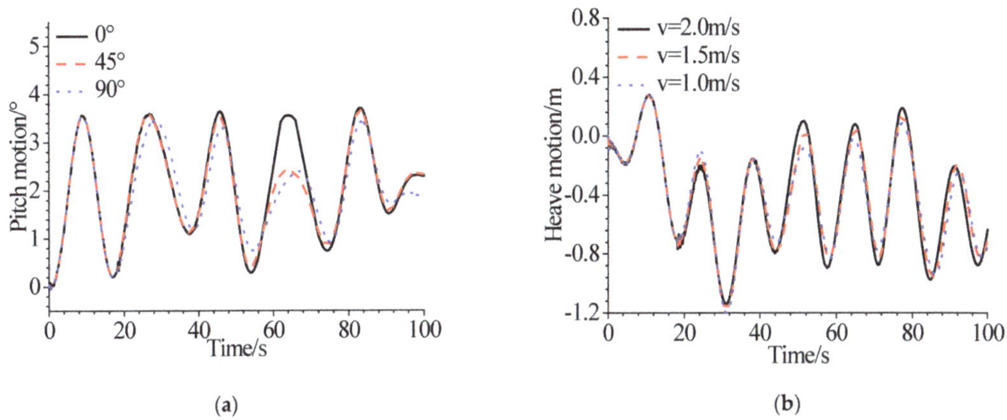

Figure 26. Comparison of motion response under different impact angles: (**a**) comparison diagram of Pitch and (**b**) comparison diagram of Heave.

6.2. Sensitivity Analysis of the Impact of Different Environmental Parameters on the Collision of Floating Wind Turbines

6.2.1. Wind Velocity

(1) Collision force

Figure 27 provides a detailed illustration of the collision force-time curves corresponding to different wind speeds within a comprehensive framework. From a critical analysis of the depiction, it becomes manifest that the trajectories of the collision force-time curves under varying wind speeds exhibit a remarkable consistency, characterized by an expeditious augmentation followed by a subsequent diminution. Nevertheless, a concomitant elevation in wind speed is correlated with a proportional increase in both the collision force and the duration of the collision event. This phenomenon can be systematically attributed to the fact that the responsive dynamics in pitch, roll, and sway degrees of freedom escalate in direct correspondence with the wind speed. This escalation engenders an augmented resistance encountered by the bow of the colliding vessel, thereby culminating in a consequential amplification in the peak collision force. This analytical observation serves to deepen the understanding of the complex interplay between wind speed and the corresponding collision dynamics within the studied system.

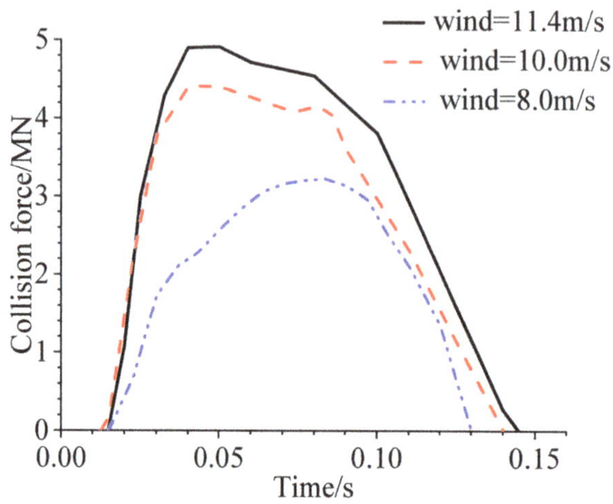

Figure 27. Curves of collision force-time under different wind speeds.

(2) Energy analysis

Figure 28 meticulously illustrates a comparative analysis of the kinetic and internal energy distribution within a floating wind turbine system, consequent to collisions under varied speed conditions. A scrupulous examination of the figure reveals that divergent wind speeds exert a palpable and significant influence on both the kinetic and internal energy profiles of the system. Particularly, when the wind speed reaches a magnitude of 11.4 m/s, the transferred kinetic and internal energy to the floating wind turbine system attain their respective zeniths. Conversely, a decrement in wind speed is concomitantly associated with a proportional diminution in the transferred kinetic and internal energy. This systematic observation accentuates the integral relationship between wind speed and the ensuing motion response and structural integrity of the floating wind turbine system. It establishes a clear corollary that an augmented wind speed invariably leads to an enhancement in motion responsiveness and a corresponding exacerbation in structural damage within the floating wind turbine assembly.

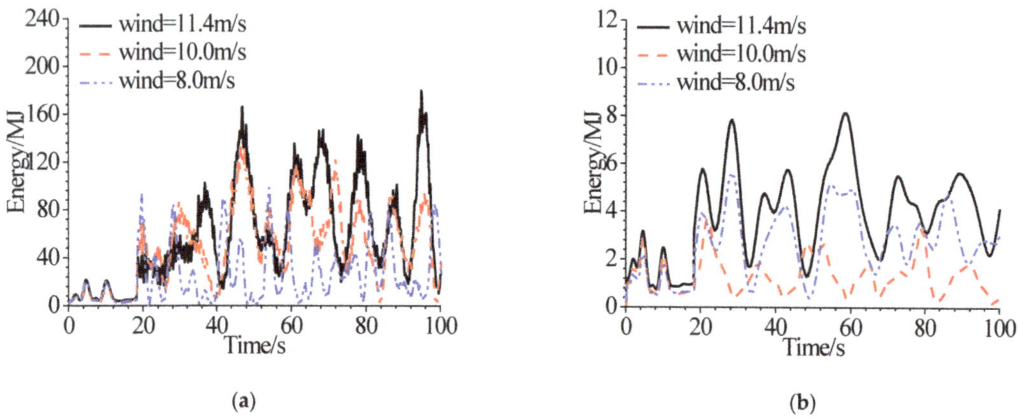

Figure 28. Curves of energy of system under different wind speeds: (**a**) Kinetic Energy and (**b**) Internal Energy.

(3) Structure stress analysis

Figure 29 illustrates the equivalent stress cloud diagrams for the floating wind turbine system under different wind speeds. As seen from the figure, different wind speeds result in varying degrees of deformation in the buoyancy can. When the wind speed is 11.4 m/s, the equivalent stress on the buoyancy can peaks at 3.846×10^8 N at 18.14 s. At a wind speed of 10.0 m/s, the buoyancy can reaches its maximum equivalent stress of 3.676×10^8 N at 18.12 s. When the wind speed is 8.0 m/s, the maximum equivalent stress of the buoyancy can is 3.542×10^8 N at 17.98 s. These results indicate that the greater the wind speed acting on the floating wind turbine, the higher the structural stress on the buoyancy can, suggesting an increase in structural stress with the increase in wind speed.

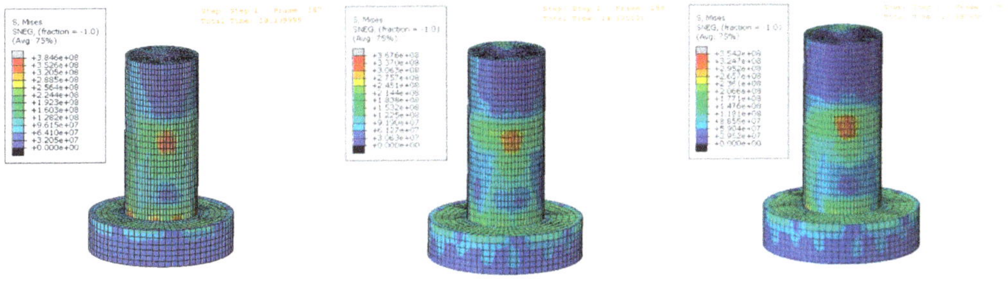

(a) (b) (c)

Figure 29. Effective stress contours of upper column under different wind speeds: (**a**) wind = 11.4 m/s; (**b**) wind = 10.0 m/s; (**c**) wind = 8.0 m/s.

(4) Motion response

Figure 30 elucidates the roll and heave motion response characteristics of the floating wind turbine system subjected to variations in wind speed within a controlled experimental framework. An analytical inspection of the figure reveals a discernible correlation, wherein both the apex roll angle and the maximal heave displacement escalate with the increment of wind speed, attaining their respective maxima at an average wind speed proximate to the rated operational wind speed (11.4 m/s). This observed trend is intrinsically attributable to the concomitant amplification of the system's kinetic energy as a function of wind speed. Consequently, this systematic analysis permits the inference that the roll and heave motions of the floating wind turbine, in the aftermath of a collision event, manifest augmented magnitudes in direct correspondence with an increase in wind speed.

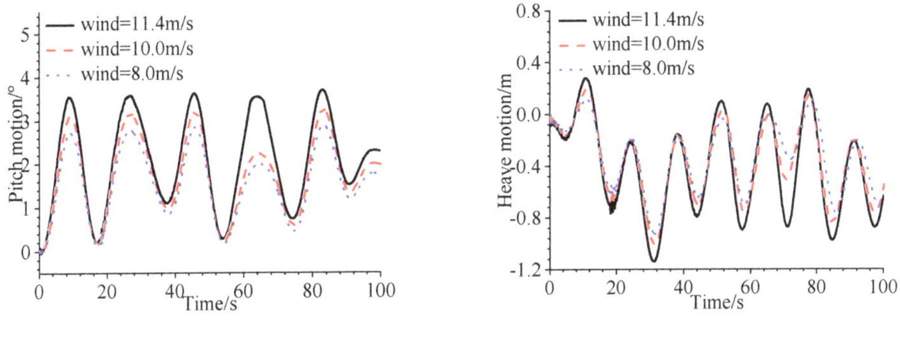

(a) (b)

Figure 30. Comparison of motion response under different wind speeds: (**a**) comparison diagram of Pitch and (**b**) comparison diagram of Heave.

6.2.2. Wave Height

(1) Collision force

Figure 31 illustrates the collision force-time curves for the interaction between a ship and a floating wind turbine under different wave heights. As depicted in the figure, the collision force-time curves for the floating wind turbine differ under distinct wave heights, with both peak collision force and collision duration increasing with wave height. When wave heights are 2.0 m, 1.5 m, and 1.0 m, the collision durations are 0.15 s, 0.14 s, and 0.13 s, respectively, and peak collision forces reach 5.792×10^6 N, 4.912×10^6 N, and 4.412×10^6 N, respectively. Therefore, as wave height increases, the motion response of the floating wind turbine also increases. Consequently, the resistance force on the ship's bow heightens, leading to a rise in the peak collision force.

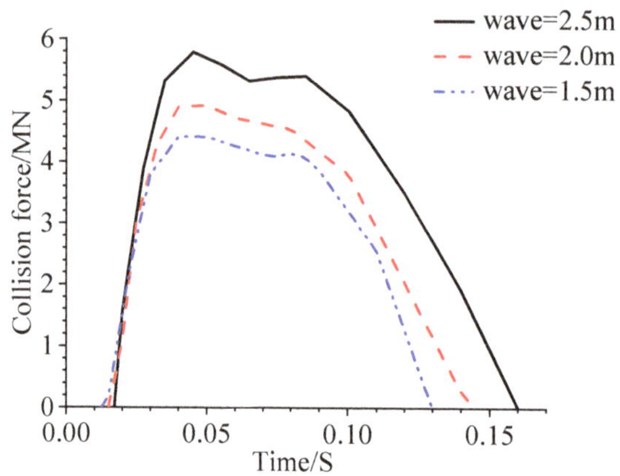

Figure 31. Curves of collision force-time under different wind speeds.

(2) Energy analysis

Figure 32 presents a comparison of kinetic and potential energy responses of the floating wind turbine system subject to varying wave heights. The graph illustrates a significant influence of differing wave heights on both kinetic and potential energy response curves. At a wave height of 2 m, the kinetic and potential energy transferred to the floating wind turbine is at its maximum. As wave height diminishes, the energy transferred to the floating wind turbine—both kinetic and potential—decreases correspondingly. This implies that the larger the wave height the greater the motion response and structural damage the floating wind turbine experiences.

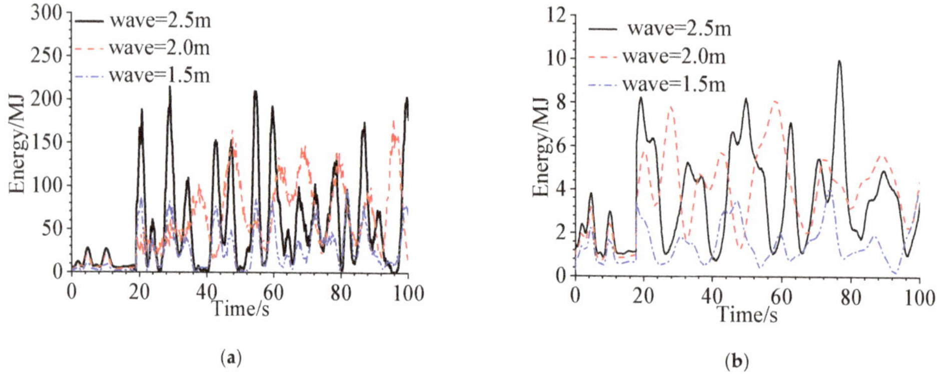

Figure 32. Curves of energy of system under different wave heights: (**a**) Kinetic Energy and (**b**) Internal Energy.

(3) Structure stress analysis

Figure 33 exhibits the cloud diagrams of equivalent stress experienced by the floating wind turbine system under varying impact velocities. The patterns of deformation damage to the buoy differ under varying wave heights. When the wave height is 2.5 m, the equivalent stress value peaks at 4.084×10^8 N at 17.82 s. When the wave height is 2.0 m, the equivalent stress value peaks at 3.846×10^8 N at 18.14 s. Lastly, when the wave height is 1.5 m, the equivalent stress value peaks at 3.696×10^8 N at 18.44 s. This suggests that higher wave heights impose greater structural stress on the floating wind turbine system, thus intensifying deformation damage to the structure with an increase in wave height.

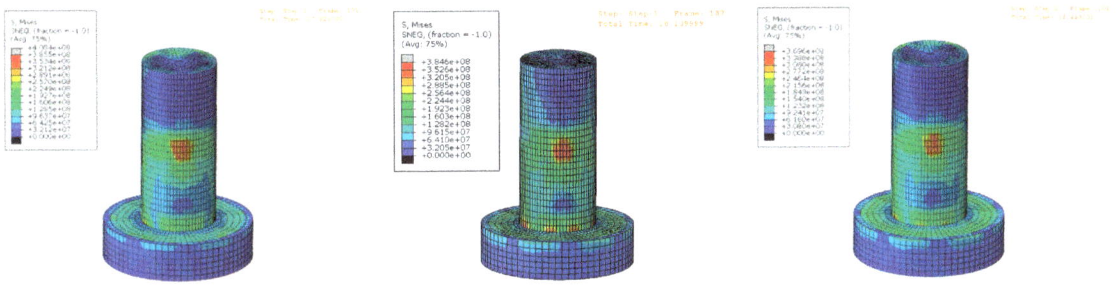

(**a**) (**b**) (**c**)

Figure 33. Effective stress contours of upper column under different wave heights: (**a**) wave = 2.5 m; (**b**) wave = 2 m; (**c**) wave = 1.5 m.

(4) Motion response

Figure 34 illustrates the pitch and heave responses of the floating wind turbine system under different collision speeds. As clearly observed from the figure, with the increment in wave heights experienced by the floating wind turbine system, the amplitudes of both pitch and heave motions also increase correspondingly. This could be attributed to the fact that the kinetic energy of the floating wind turbine system escalates with increasing wave height, which in turn augments the motion amplitudes in both the pitching and heaving directions.

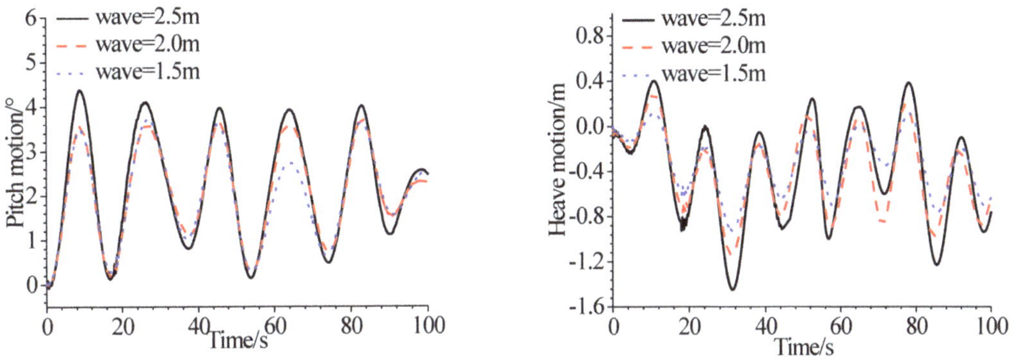

(**a**) (**b**)

Figure 34. Comparison of motion response under different wave heights: (**a**) comparison diagram of Pitch and (**b**) comparison diagram of Heave.

7. Conclusions

A sophisticated bidirectional fluid-structure coupling joint simulation methodology has been formulated, utilizing both Star-CCM+ and ABAQUS. This methodology enables the precise determination of the collision dynamics response analysis procedure for floating wind turbines, considering the comprehensive and coupled effects of wind, wave, and mooring forces. An exhaustive study on the dynamic response of floating wind turbines under collision loads, taking into account the aforementioned coupled effects, was conducted, employing the joint simulation capabilities of Star-CCM+ and ABAQUS. The outcomes of this research provide essential insights into the specific conditions leading to turbine structural damage and the associated motion response patterns. In addition, a detailed analysis was undertaken to elucidate the governing rules for the impact of varying collision parameters and environmental conditions on the dynamic response of floating wind turbines. A focused investigation was performed to assess the influences of disparate collision velocities, collision angles, wind velocities, and wave amplitudes on the structural response dynamics of floating wind turbine systems. The principal conclusions derived from this comprehensive research are as follows:

1. Due to the implementation of a mooring system for floating wind turbines, the floating platform possesses certain flexibility. Hence, when a ship collides with a floating wind turbine, a substantial part of the kinetic energy of the colliding ship is transformed into the kinetic energy of the floating wind turbine. The pitch and heave motion amplitudes of the turbine are all greater than the motion amplitudes in the absence of collision load.

2. The turbine's floating column absorbed minimal internal energy; thus, it did not sustain extensive damage. The stress generated was primarily distributed in the collision region of the floating column. Furthermore, a comparison of bow collision and side collision reveals that side collision has a more significant impact on the dynamic response of the floating wind turbine structure.

3. For different collision angles, given the identical kinetic energy of the colliding ship, there is minimal difference in the trend and peak of the collision force time history curve. However, the structural stress is slightly higher at a collision angle of 45° compared to the other two angles. The pitch and heave motion responses of the floating wind turbine under a 0° collision are larger than those under 45° and 90° collisions.

Author Contributions: Conceptualization, S.Z. and Y.Z.; Writing—original draft, S.Z.; Writing—review and editing, X.Y. and K.L.; Visualization, K.L. and Y.W. All authors have read and agreed to the published version of the manuscript.

Funding: This research was funded by the National Natural Science Foundation of China (Grant No. 52171311; Grant No. 52271279).

Data Availability Statement: MDPI Research Data Policies.

Conflicts of Interest: The authors declare no conflict of interest.

References

1. Caithnness Wind Farm Information Forum. Wind Turbine Accident Compilation 2014. Available online: http://www.caithnesswindfarms.co.uk (accessed on 30 September 2014).
2. Ladeira, I.; Márquez, L.; Echeverry, S.; Le Sourne, H.; Rigo, P. Review of methods to assess the structural response of offshore wind turbines subjected to ship impacts. *Ships Offshore Struc.* **2023**, *18*, 755–774. [CrossRef]
3. Dai, L.; Ehlers, S.; Rausand, M.; Utne, I.B. Risk of collision between service vessels and offshore wind turbines. *Reliab. Eng. Syst. Safe* **2013**, *109*, 18–31. [CrossRef]
4. Presencia, C.E.; Shafiee, M. Risk analysis of maintenance ship collisions with offshore wind turbines. *Int. J. Sustain. Energy* **2018**, *37*, 576–596. [CrossRef]
5. Biehl, F. Collision safety analysis of offshore wind turbines. In Proceedings of the 4th LSDYNA European Conference, Ulm, Germany, 12 September 2005; pp. 28–30.
6. Moulas, D.; Shafiee, M.; Mehmanparast, A. Damage analysis of ship collisions with offshore wind turbine foundations. *Ocean Eng.* **2017**, *143*, 149–162. [CrossRef]

7. Le Sourne, H.; Besnard, N.; Cheylan, C.; Buannic, N. A ship collision analysis program based on super-element method coupled with large rotational ship movement analysis tool. *J. Appl. Math.* **2012**, *2012*, 375686. [CrossRef]
8. Biehl, F.; Lehmann, E. Collisions of Ships and Offshore Wind Turbines: Calculation and Risk Evaluation. In Proceedings of the 25th International Conference on Offshore Mechanics and Arctic Engineering, Hamburg, Germany, 4–9 June 2006; pp. 281–304.
9. Pire, T.; Le Sourne, H.; Echeverry, S.; Rigo, P. Analytical formulations to assess the energy dissipated at the base of an offshore wind turbine jacket impacted by a ship. *Mar. Struct.* **2018**, *59*, 192–218. [CrossRef]
10. Song, M.; Jiang, Z.; Yuan, W. Numerical and analytical analysis of a monopile-supported offshore wind turbine under ship impacts. *Renew. Energy* **2021**, *167*, 457–472. [CrossRef]
11. Jiang, Z.; Gao, Z.; Ren, Z.; Li, Y.; Duan, L. A parametric study on the final blade installation process for monopile wind turbines under rough environmental conditions. *Eng. Struct.* **2018**, *172*, 1042–1056. [CrossRef]
12. Echeverry, S.; Márquez, L.; Rigo, P.; Le Sourne, H. Numerical crashworthiness analysis of a spar floating offshore wind turbine impacted by a ship. In *Developments in the Collision and Grounding of Ships and Offshore Structures*; CRC Press: Boca Raton, FL, USA, 2019; pp. 85–95.
13. Chen, P.; Chen, J.; Hu, Z. Software-in-the-Loop Combined Reinforcement Learning Method for Dynamic Response Analysis of FOWTs. *Front. Mar. Sci.* **2021**, *7*, 628225. [CrossRef]
14. Petersen, M.J. Dynamics of ship collisions. *Ocean Eng.* **1982**, *9*, 295–329. [CrossRef]
15. Tabri, K.; Varsta, P.; Matusiak, J. Numerical and experimental motion simulations of nonsymmetric ship collisions. *J. Mar. Sci. Technol.* **2010**, *15*, 87–101. [CrossRef]
16. Ferry, M. MCOL User's Manual. *Principia Mar.* **2002**.
17. Yu, Z.; Amdahl, J.; Storheim, M. A new approach for coupling external dynamics and internal mechanics in ship collisions. *Mar. Struct.* **2016**, *45*, 110–132. [CrossRef]
18. Yu, Z.; Amdahl, J. Full six degrees of freedom coupled dynamic simulation of ship collision and grounding accidents. *Mar. Struct.* **2016**, *47*, 1–22. [CrossRef]
19. Yu, Z.; Shen, Y.; Amdahl, J.; Greco, M. Implementation of linear potential-flow theory in the 6DOF coupled simulation of ship collision and grounding accidents. *J. Ship Res.* **2016**, *60*, 114–119. [CrossRef]
20. Zhang, Y.; Hu, Z.; Ng, C.; Jia, C.; Jiang, Z. Dynamic responses analysis of a 5 MW spar-type floating wind turbine under accidental ship-impact scenario. *Mar. Struct.* **2021**, *75*, 102885. [CrossRef]
21. Chen, J.; Hu, Z.; Liu, G.; Wan, D. Coupled aero-hydro-servo-elastic methods for floating wind turbines. *Renew. Energy* **2019**, *130*, 139–153. [CrossRef]
22. Zhang, Y.; Hu, Z. An aero-hydro coupled method for investigating ship collision against a floating offshore wind turbine. *Mar. Struct.* **2022**, *83*, 103177. [CrossRef]
23. Jonkman, J.; Butterfield, W. *Definition of a 5-MW Reference Wind Turbine for Offshore System Development*; (No. NREL/TP-500-38060); National Renewable Energy Lab. (NREL): Golden, CO, USA, 2009.
24. Zhang, H.; Ding, X.; Zhou, B.; Zhang, L.; Yuan, Z. Hydrodynamic Performance Study of Wave Energy—Type Floating Breakwaters. *J. Mar. Sci. Appl.* **2019**, *18*, 64–71. [CrossRef]
25. Coulling, A.J.; Goupee, A.J.; Robertson, A.N.; Jonkman, J.M.; Dagher, H.J. Validation of a FAST semi-submersible floating wind turbine numerical model with DeepCwind test data. *J. Renew. Sustain. Energy* **2013**, *5*, 557–569. [CrossRef]
26. Tran, T.T.; Kim, D.-H. The Coupled Dynamic Response Computation for a Semi-Submersible Platform of Floating Offshore Wind Turbine. *J. Wind. Eng. Ind. Aerodyn.* **2015**, *147*, 104–119. [CrossRef]
27. *D.N.V., G.L. A.S. Standard DNVGL-ST-0126*; Support Structures for Wind Turbines. DNV: Oslo, Norway, 2016.
28. *D.N.V., AS. Offshore Standard DNV-ST-0119*; Floating Wind Turbines Structures. DNV: Oslo, Norway, 2018.

Journal of
Marine Science and Engineering

MDPI

Article

Dynamic Mechanical Properties and Damage Parameters of Marine Pipelines Based on Johnson–Cook Model

Xiao Tian [1,*], Jingjing Pei [1,*] and Jingjing Rong [2]

[1] School of Engineering and Technology, China University of Geosciences (Beijing), Beijing 100083, China
[2] Institute of Higher Education, North China Institute of Science and Technology, Langfang 065201, China; zhang@ncist.edu.cn
* Correspondence: tianx@cugb.edu.cn (X.T.); peijj@cugb.edu.cn (J.P.)

Abstract: A comprehensive understanding of the dynamic behavior of materials and structures under impact loads is paramount for the design and maintenance of reliable marine pipelines and associated structures. However, there is a lack of comprehensive research on the full characterization of constitutive and failure models of carbon steels, which are commonly used in marine pipelines. In this paper, Q235 steel was subjected to quasi-static tensile tests at room temperature on smooth specimens to obtain the constitutive parameters using the Johnson–Cook (J-C) model. Subsequently, quasi-static tensile tests were conducted on notched specimens, and dynamic tensile tests were performed on smooth round bars to obtain stress triaxiality and failure strain. The acquired data were then utilized to fit the failure parameters using the Johnson–Cook (J-C) damage model, a widely accepted constitutive model employed in high-strain rate applications through the least squares method. Finally, the tensile test is numerically simulated based on the acquired experimental parameters. The obtained results reveal a remarkable agreement between the curve fitted by the J-C constitutive model and the experimental tensile curve. Additionally, a high degree of correlation between the load-displacement curves of the tests and simulations provides robust validation of the accuracy of the dynamic mechanical parameters for Q235 steel. These findings contribute valuable insights into the behavior of carbon steels commonly used in marine pipelines, enhancing the overall understanding of their response to impact loads and informing more reliable design and maintenance practices.

Keywords: marine pipelines; dynamic behavior; Johnson–Cook (J-C) model; tensile test; numerical simulation

Citation: Tian, X.; Pei, J.; Rong, J. Dynamic Mechanical Properties and Damage Parameters of Marine Pipelines Based on Johnson–Cook Model. *J. Mar. Sci. Eng.* **2023**, *11*, 1666. https://doi.org/10.3390/jmse11091666

Academic Editor: Bruno Brunone

Received: 5 August 2023
Revised: 22 August 2023
Accepted: 23 August 2023
Published: 24 August 2023

1. Introduction

The exploitation and utilization of marine oil and gas resources are of utmost importance for a country's energy supply. Marine pipelines play a critical role as essential conduits for transporting these valuable marine resources. The marine environment presents numerous challenges, as pipelines are exposed to potential impacts from a range of sources, including natural disasters such as tsunamis, storm surges, and earthquakes, as well as human activities like falling objects and trawling [1,2]. Consequently, subsea pipelines and thin-walled metal structures undergo high-speed, dynamic processes that result in significant deformation [3,4]. When subjected to impact loads, these structures are susceptible to local dents and cracking damages, posing direct threats to the safety and reliability of the pipelines. In severe cases, such damage can lead to pipeline leakage and explosion accidents. Therefore, comprehensive research on the dynamic mechanical properties of these structures is essential to accurately characterize material failure behavior, thereby enhancing their overall safety and performance.

Ellinas [5] proposed a semiempirical formula for plastic damage at the impact point of a pipeline due to impact loads, drawing from classical ultimate plasticity theory. However, they omitted consideration of the pipeline's contact with the seabed. Wierzbick et al. [6] explored the influence of initial axial force, providing empirical formulas for dent depth and absorbed energy, alongside impact force formulas that overlooked shear force conditions. Bai [7] examined the mutual contact between falling objects and the pipeline, presenting a semiempirical formula for impact loads and damage depth at the impact point. Wang et al. [8] delved into energy conversion during the impact of falling objects on marine pipelines but overlooked the seabed's influence on energy. DNV synthesized prior research, creating regulations for marine pipeline damage from impact loads and offering a formula to calculate dent depth caused by falling object impacts [9,10]. Nonetheless, this formula disregarded the impact process's nonlinear aspects and failed to consider energy absorption by the seabed, falling objects, and backfill soil above the pipeline. Theoretical analysis can yield straightforward calculation formulas, but their limitations prevent them from comprehensively addressing real-world situations, resulting in notable deviations in calculations. Chen et al. [11] utilized hammer impact experiments to explore how internal pressure affects marine pipeline impact damage. They found that internal pressure can partly suppress pipeline deformation but may also lead to overall pipeline failure. Notably, these experiments did not account for the effects of seawater and the seabed on the pipeline. Building upon this work, Andrew [12,13] enhanced the experiments by incorporating the influence of concrete protective layers and soil. They established a correlation between dent depth, impact energy from falling objects, and energy absorbed by the pipeline. Summarizing common pipeline impact tests across diverse fields, Zhang [14] designed experimental setups for lateral and vertical hammer impacts on pipelines. Given the costliness of impact experiments, contemporary research predominantly relies on numerical simulation methods to examine the process of falling object impacts on marine pipelines. This approach can address nonlinear concerns between various objects and visually depict the entire impact process, producing similarly accurate outcomes. Taking into consideration the interaction between the seabed and the pipeline, Zeinoddini [15] conducted a comparative analysis of experiments and numerical simulations. This analysis explored variations in impact force, displacement, dent depth, and energy for an X70 pipeline under distinct conditions, including different impact models and internal pressures. Yan [16] delved into the damage caused to marine pipelines by objects dropped from platforms or supply ships. They introduced a simple and feasible method to study the effects of pipeline damage under impact loads. In a similar vein, Huang [17] conducted marine pipeline damage experiments and numerical simulations. Their study examined the influence of falling object mass, height, and shape on mechanical damage to marine pipelines, and they adjusted the Ellinas–Walker formula based on experimental findings.

The impact on marine pipelines involves a rapid deformation and damage process. The material's constitutive relationship differs markedly from static scenarios. The majority of the aforementioned studies gauge damage based on the extent of macroscopic pipeline deformation, disregarding material damage attributed to strain rate effects. Further research is warranted to investigate the dynamic mechanical properties of pipeline materials.

In the research of material dynamic mechanical properties, the application of the Hopkinson bar has become relatively mature. The Hopkinson Bar Tensile Test involves applying a sudden impact or stress wave to a specimen to induce tensile stress. The stress wave is generated by impacting the input bar (also called the striker bar) with a projectile, which then travels through the bar and the specimen. The stress wave's characteristics, such as its amplitude, duration, and velocity, are monitored and used to calculate the mechanical properties of the material. It allows testing at strain rates in the range of 10^2 to 10^4 s^{-1}. Within this range, establishing an appropriate constitutive model can predict the true stress–strain relationship of materials under high strain rates. Commonly used constitutive models include the Johnson–Cook (J-C) model, the Zerilli–Armstrong model, the Steinberg model, etc. Among these, the J-C model is widely adopted by many researchers due to its ability to consider the combined effects of stress state, strain rate, and temperature on material failure. The model is known for its simple form, making its parameters easy to test and calibrate [18–21]. Additionally, the Johnson–Cook model is often integrated into finite element simulation software to predict the behavior of metals during forming processes, such as forging, stamping, rolling, and extrusion.

Previous research on the dynamic impact response of metals has predominantly centered on the use of alloy materials. Through a combination of experiments and numerical simulations, researchers have determined parameters for the J-C model, while also evaluating the validity of the J-C constitutive relationship and failure criteria [22–28]. However, marine pipelines are primarily made of carbon steel. Wei et al. studied the mechanical behavior of quenched and tempered 45# steel at temperatures ranging from room temperature to 1000 °C and strain rates from 10^{-4} to 10^3 s^{-1}. They calibrated the model parameters by combining the back-calculation of critical cracking from Taylor impact tests [29]. Zhu et al. conducted experiments and numerical simulations on Q355B steel, determining 10 parameters for the J-C model [30]. However, Q355B steel is commonly used in construction engineering applications. Some scholars have also studied the impact resistance of Q235 steel [31–33], but they did not provide all the parameters for the constitutive and failure models. Moreover, they did not validate the accuracy of the experimental results. The comprehensive evaluation of the precision exhibited by the J-C constitutive model encountered inherent complexities stemming from two distinct rationales. First, the deficiency in an exhaustive set of test data pertaining to a designated material across varied loading circumstances, encompassing authentic stress-authentic strain interdependencies, ramifications of strain-rate variations, temperature fluctuations, and structural failure mechanisms, engendered impediments to a holistic assessment. Second, the cost would be prohibitively high if a complete set of the test data were obtained.

Therefore, this paper employs a universal material testing machine and the Hopkinson bar (SHTB) experimental system to comprehensively investigate the quasi-static and dynamic tensile properties of Q235 steel commonly used for marine pipelines at room temperature. Based on the J-C model, the parameters are determined. Furthermore, numerical simulations are conducted to validate the accuracy of the J-C model, providing valuable reference for the impact dynamic design and damage assessment of marine pipelines. The main variable parameters in this study are summarized in Table 1.

Table 1. Nomenclature.

Parameters	Description
σ	The equivalent stress
σ'	The engineering stress
σ_m	The average stress
$\sigma_1, \sigma_2, \sigma_3$	The principal stresses
σ^*	the stress triaxiality
ε	The equivalent strain
ε'	The engineering strain
$\dot{\varepsilon}$	The strain rate
$\dot{\varepsilon}_0$	The reference strain rate
ε^*	Dimensionless strain rate
ε_f	The effective fracture strain
A	The yield stress of the material
B, n	The strain hardening constants
C	The strain-rate strengthening coefficient
D_1, D_2, D_3, D_4	Material damage parameters

2. Constitutive Model and Damage Model

2.1. Johnson–Cook Constitutive Model

The J-C constitutive relation is shown in Equation (1) [34,35],

$$\sigma = (A + B\varepsilon^n)\left(1 + C\ln\frac{\dot{\varepsilon}}{\dot{\varepsilon}_0}\right)\left[1 - \left(\frac{T - T_{room}}{T_{melt} - T_{room}}\right)^m\right] \quad (1)$$

Figure 5 presents the stress–strain curves of specimens with different notch radii. With increasing notch radius, the material exhibits improved plastic behavior.

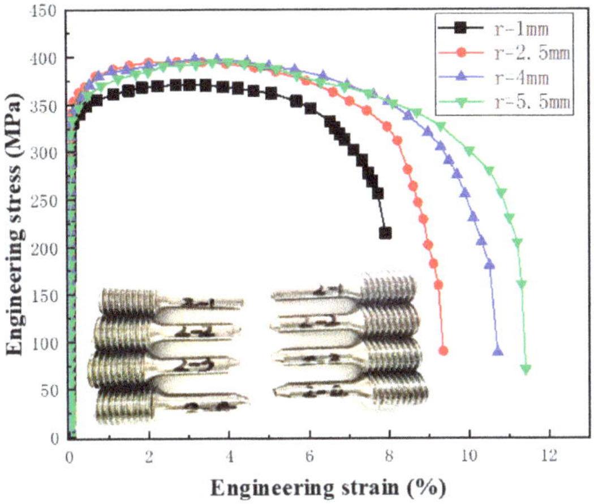

Figure 5. Stress–strain curve of specimens with different notches.

In the tensile tests of smooth specimens, the stress triaxiality is $-1/3$. However, for the tensile tests of notched specimens, the stress triaxiality can be calculated by Equation (15):

$$\sigma^* = -\frac{1}{3} - \ln\left(1 + \frac{a}{2R}\right) \tag{15}$$

where, a is the specimen radius, and R is the notch radius.

Figure 6 presents the data of failure strain for different stress triaxiality conditions. The tensile tests results are shown in Table 3. In quasi-static tensile tests, failure strain decreases as the stress triaxiality increases. Based on the data presented in Figure 6, a fitting analysis was performed using Equation (11) to determine the parameters D_1 to D_3, with values of $D_1 = 0.2918$, $D_2 = 4.6156$, and $D_3 = 6.1566$.

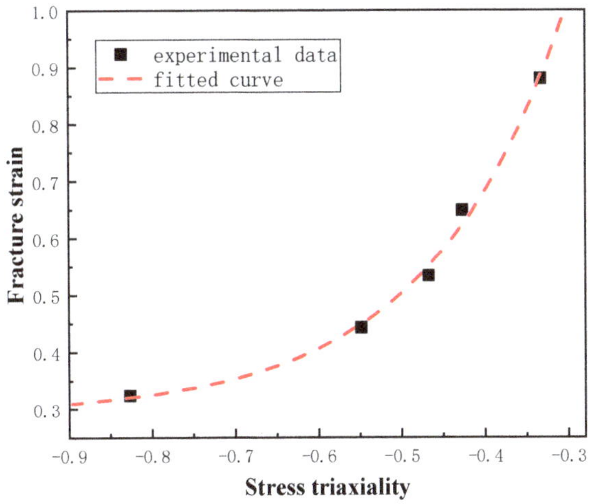

Figure 6. Failure strain versus stress-state curve.

Table 3. Tensile test results of notched specimens.

The Notch Radius /mm	Initial Specimen Diameter /mm	Specimen Diameter after Fracture /mm	The Initial Stress Triaxiality	The Stress Triaxiality after Fracture	Reduction of Area /%
1	3	2.55125	−0.8929	−0.8267	13.83
2.5	3	2.40375	−0.5957	−0.5487	17.89
4	3	2.2975	−0.5052	−0.4675	20.66
5.5	3	2.16875	−0.4612	−0.4274	23.85

3.4. Hopkinson Tensile Test at Different Strain Rates

The split Hopkinson bar test is a widely employed technique for investigating the dynamic mechanical properties of metal materials, particularly their dynamic tensile behavior, at high strain rates. This versatile test method encompasses a broad strain-rate range, typically spanning from 10^2 to 10^4 s^{-1}, enabling comprehensive analyses of material response under dynamic loading conditions. Through the measurement of stress–strain curves at various strain rates, this experiment aims to obtain the fundamental parameters required for the J-C material model.

To comprehensively characterize the dynamic tensile properties of Q235 steel, Hopkinson dynamic tensile tests were performed at three distinct strain rates: 500 s^{-1}, 1500 s^{-1}, and 2500 s^{-1}. The experimental principle is shown in Figure 7.

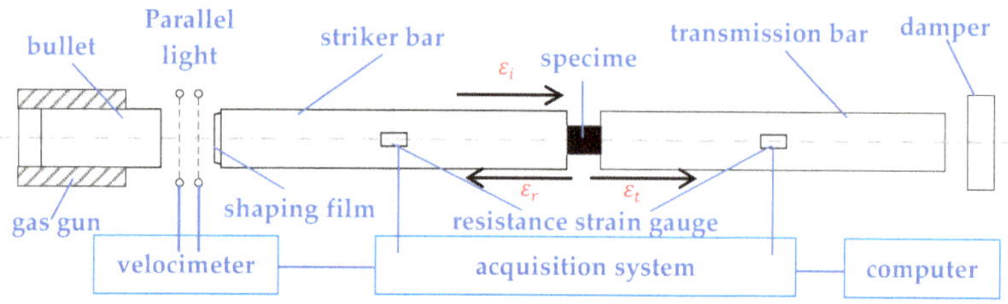

Figure 7. Test principle of split Hopkinson tensile bar device.

The test setup consists of three main components: the striker bar (or input bar), the specimen, and the transmission bar (or output bar). The striker bar is a high-strength bar that is struck by a projectile (usually a gas gun) to generate the stress wave. It is in contact with the specimen. The test material is in the form of a thin cylinder or tube placed between the striker bar and the transmission bar. When the striker bar is impacted, it transmits a stress wave to the specimen, inducing tension. A fraction of the pulse propagates through the specimen, inducing rapid plastic deformation. Simultaneously, a portion of the pulse traverses the specimen, permeating the transmission bar, where it is subsequently dissipated by the buffering mechanism. Additionally, a residual fraction is retro-reflected back through the striker bar. The transmission bar transmits the stress wave to a set of strain gauges or other measuring devices. These strain gauges measure the deformation of the specimen under tension, such as the incident strain ε_i, the reflection strain ε_r, and the transmission strain ε_t. By analyzing the stress and strain data, the material's dynamic tensile properties can be determined, such as stress–strain curves, strain-rate sensitivity, and fracture behavior at high strain rates.

The specimen dimensions are shown in Figure 8. Three distinct experimental series were conducted, encompassing a dual configuration wherein two series employed projectiles of 600 mm in length, subject to strain rates of 500 s^{-1} and 1500 s^{-1}, respectively. Meanwhile, the third series featured a projectile of 400 mm length, subjected to a strain rate of 2500 s^{-1}. Notably, each experimental was repeated to ensure robustness of the results. Subsequent to data acquisition, the empirical findings were analyzed employing the venerable classical two-wave method as delineated in reference [36], culminating in the derivation of precise engineering stress–strain relationships. To counteract the potential influence stemming from transverse deformation of the specimens, Equations (12) and (13) were employed to effectuate a transformation of the recorded data, resulting in the extraction of true stress and strain manifestations.

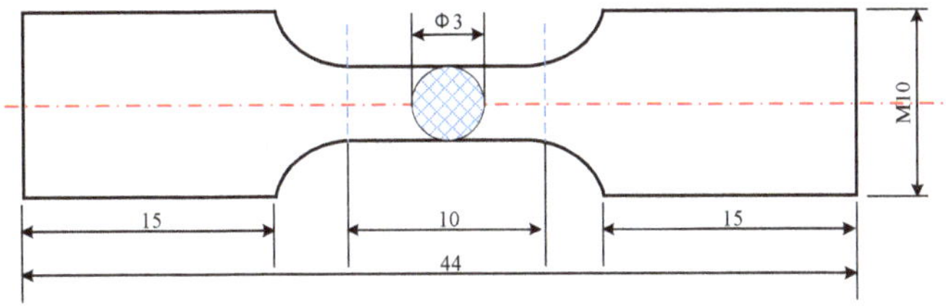

Figure 8. The dimensions of the Hopkinson tensile specimens (Unit: mm).

At the reference strain rate $\dot{\varepsilon}_0 = 0.001\ s^{-1}$, the variation of Q235 steel's equivalent stress and failure strain with the natural logarithm of the dimensionless strain rate $\ln \varepsilon^*$ was plotted at a consistent plastic strain level. These plots are illustrated in Figures 9 and 10, respectively. Our findings highlight the substantial influence of strain rate on Q235 steel's equivalent stress and failure strain. Notably, the material displays distinct linear relationships at low and high strain rates. Particularly, at higher strain rates, the equivalent stress and failure strain exhibit heightened sensitivity to variations in strain rate.

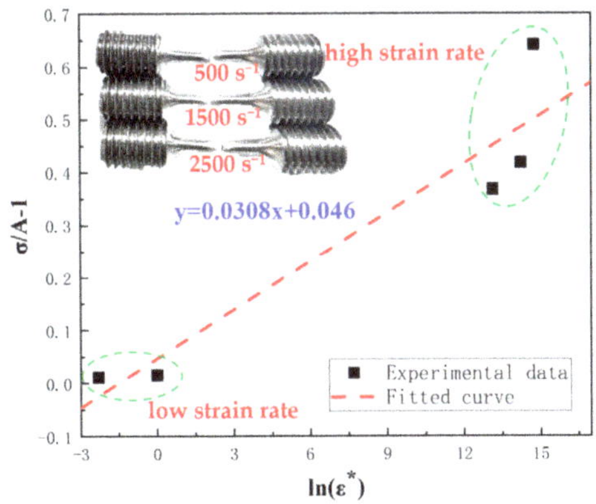

Figure 9. The fitted curve of equivalent stress vs. dimensionless strain rate.

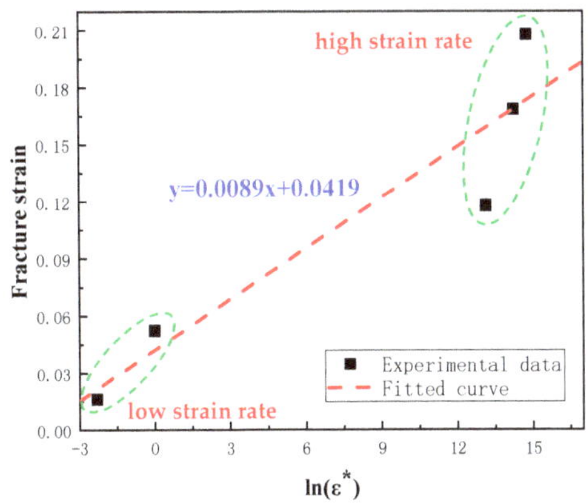

Figure 10. The fitted curve of failure strain vs. dimensionless strain rate.

Through fitting the equivalent stress and failure strain data at different strain rates, we successfully determined the values of the parameters C and D_4, illustrated by the dashed lines in Figures 9 and 10. The linear fitting outcomes revealed parameter values of $C = 0.0308$ and $D_4 = 0.0089$.

4. Validation of the Johnson–Cook Constitutive Model and Failure Parameters through Numerical Simulations

4.1. The Finite Element Models

Finite element analysis (FEA) using the Johnson–Cook model enables the visualization of stress and strain distributions, as well as the prediction of potential defects like cracks or wrinkles. To validate the precision of the eight parameters proposed for the J-C constitutive model and failure model in the preceding section, numerical simulations of Q235 steel tensile tests were conducted.

A three-dimensional solid finite element model of the tensile specimen was developed, adhering to the dimensions as depicted in Figure 1. In the material property module, the Johnson–Cook strengthening model was chosen for plasticity, and the Johnson–Cook damage model was selected for ductile metal damage. The parameters are listed in Table 4, which were obtained from the tensile tests.

Table 4. J-C parameters of Q235 steel.

A	B	C	n	D_1	D_2	D_3	D_4
361.2 MPa	526 MPa	0.0308	0.58	0.2918	4.6156	6.1566	0.0089

A failure displacement of 0.001 mm was set for damage evolution. The material density is 7850 kg/m^3. Computational accuracy was ensured by setting the mesh size in the gauge section to 0.5 mm, with appropriate size increment in the grip section. The model comprised a total of 32,240 elements, as depicted in Figure 11.

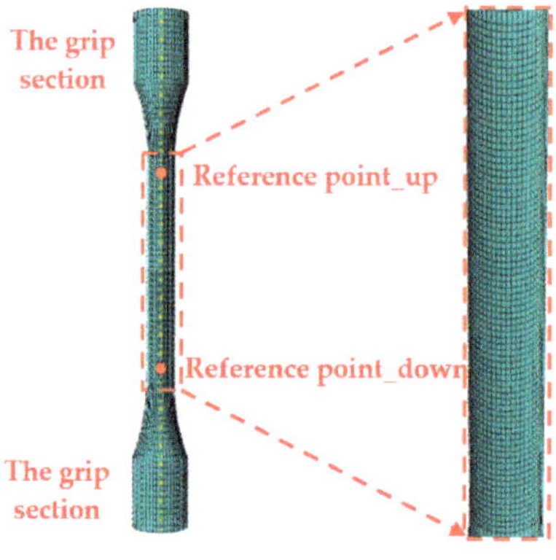

Figure 11. The meshed model.

Two reference points were set at the top and bottom ends of the specimen, respectively. Coupling constraints were applied between the reference points and the grip sections. While the bottom reference point was kept fixed, a controlled vertical upward velocity of 0.04 mm/s was applied to the top reference point.

To precisely determine the strain values, two reference points were selected on the gauge section of the model, as shown in Figure 11. The distance between the two points is 34 mm. Their displacements along the tensile direction were set in the History Output section. In addition, the reaction force of the top reference point was also output in the vertical direction.

4.2. Finite Element Analysis

The explicit dynamic analysis was executed with a loading time of 684 s. Mass scaling technique was used in the analysis to improve the stability and efficiency of the simulation. This allows for larger time steps and can significantly reduce the computational cost. The scaling factor was set to 2×10^4.

Figure 12 illustrates the variation curves of the internal energy and the kinetic energy throughout the tensile process. The maximum kinetic energy is 343 J, much less than 5% of the internal energy. Therefore, it is reasonable to set the scaling factor to 2×10^4.

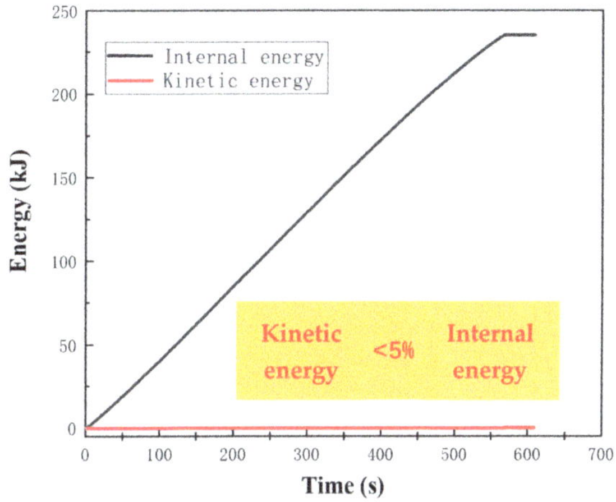

Figure 12. Curves of internal energy and kinetic energy.

The equivalent stress contours at different stages are shown in Figure 13. At the beginning of tensile process, the equivalent stress at gauge section distributes uniformly and is relatively low. As time passes, the specimen gradually elongates while contracting in circumferential direction. The equivalent stress in the gauge range is no longer uniformly distributed and decreases from the center to both ends. When the time reaches 558 s, the center of the specimen is significantly necked with stress concentration. Subsequently, fracture occurs at 570 s.

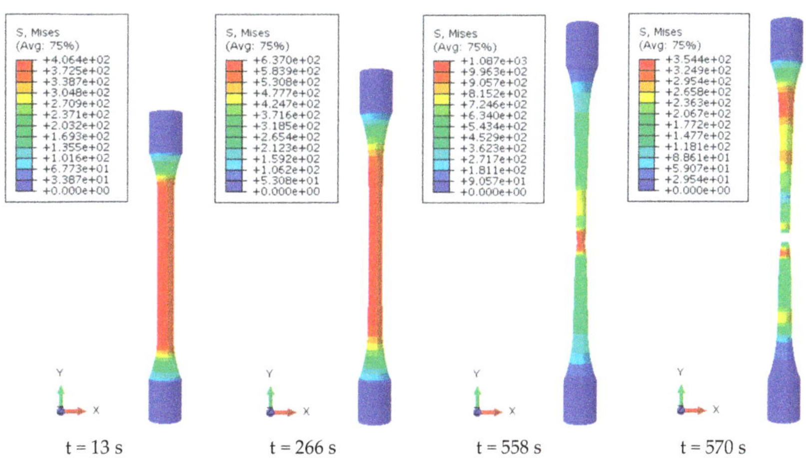

Figure 13. Contours of equivalent stress at different stages.

The strain was calculated based on the relative change in displacement between the two points in Figure 11, before and after tensile loading, with respect to the original gauge length. Subsequently, the load-displacement curve obtained from the simulations was plotted and compared with the corresponding experimental results, demonstrating excellent agreement (Figure 14). Significant fluctuations were observed in the simulation results, shown at the end of the curve. This was due to the occurrence of specimen fracture at this point, and the fracture pattern showed distinct necking, consistent with the experimental results, thus confirming the accuracy of the J-C model parameters.

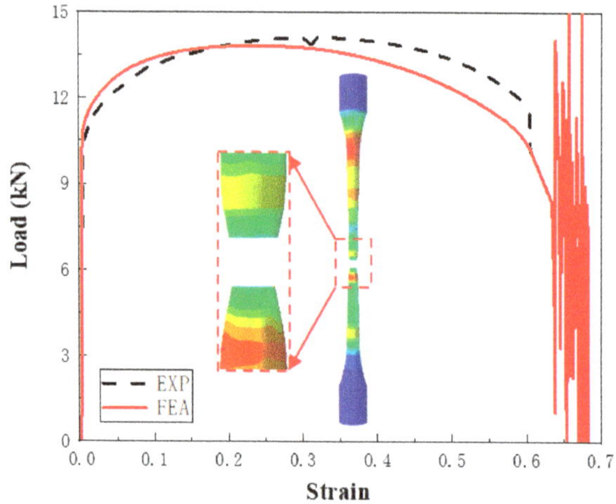

Figure 14. Comparison of load–strain curves between experimental and numerical results.

4.3. Discussions

The validated Johnson–Cook constitutive and damage models' parameters can be directly applied to the simulation of marine pipeline damage under impact loads. By collecting existing experimental data or conducting tensile tests on different pipe materials, a Johnson–Cook model parameter database can be established. When defining material properties, opting for the Johnson–Cook plasticity and damage model facilitates automatic determination of deformation and fracture occurrences within the model. This choice contributes to a more authentic representation of the pipeline damage process. The identification of the critical damage threshold serves as the basis for establishing the ultimate load capacity of the pipeline. Expanding on this foundational knowledge, an exploration of the interaction between the pipeline, seabed, and water flow forms the groundwork for conducting parameter analyses. These analyses are instrumental in the development of a holistic dynamic damage assessment methodology for marine pipelines.

5. Conclusions

In this paper, quasi-static tensile tests on both smooth round bar and notched round bar and dynamic Hopkinson tensile rod tests were conducted for Q235 steel, which is a commonly used material in marine pipelines. The study successfully determined the eight parameters associated with the J-C constitutive model and failure model, characterizing the mechanical behavior of Q235 steel. The numerical simulations of Q235 steel tensile tests yielded results that corroborated the accuracy of the proposed eight parameters for the J-C constitutive model and failure model. The close agreement between the load–strain curves of the simulations and experimental results, coupled with the observation of distinct necking patterns during fracture, serves as strong evidence supporting the accuracy of the J-C model parameters. The study findings significantly enhance our comprehension of Q235 steel's mechanical response and failure behavior, especially in diverse loading scenarios, including dynamic events. The insights provide essential inputs for the design and assessment of marine pipelines and other engineering structures, ensuring their reliability and safety in dynamic environments.

Author Contributions: Conceptualization, J.P.; methodology, X.T.; software, X.T.; validation, X.T.; formal analysis, X.T. and J.R.; investigation, X.T.; resources, X.T.; writing—original draft preparation, X.T.; writing—review and editing, J.P. and J.R.; supervision, J.R.; funding acquisition, X.T. and J.P. All authors have read and agreed to the published version of the manuscript.

Funding: This research was funded by Fundamental Research Funds for the Central Universities grant number 2-9-2021-015 and National Natural Science Foundation of China grant number 52004258. And the APC was funded by Fundamental Research Funds for the Central Universities.

Institutional Review Board Statement: Not applicable.

Informed Consent Statement: Not applicable.

Data Availability Statement: Data sharing not applicable.

Acknowledgments: This work was supported by Fundamental Research Funds for the Central Universities (2-9-2021-015) and National Natural Science Foundation of China (52004258). The reviewers whose comments and suggestions helped improve our research are gratefully appreciated.

Conflicts of Interest: The authors declare that they have no known competing financial interest or personal relationships that could have appeared to influence the work reported in this paper.

References

1. Gao, Q.; Duan, M.; Liu, X.; Wang, Y.; Jia, X.; An, C.; Zhang, T. Damage assessment for submarine photoelectric composite cable under anchor impact. *Appl. Ocean. Res.* **2018**, *73*, 42–58. [CrossRef]
2. Longva, V.; Sævik, S.; Levold, E.; Ilstad, H. Dynamic simulation of subsea pipeline and trawl board pull-over interaction. *Mar. Struct.* **2013**, *34*, 156–184. [CrossRef]
3. Ren, J.; Wang, C.; Zahng, X.; Suo, T.; Li, Y.; Tang, Z. Johnson-Cook constitutive model and failure parameters determination of B280VK high-strength steel. *J. Plast. Eng.* **2023**, *30*, 116–125.
4. Vannucchi de Camargo, F. Survey on Experimental and Numerical Approaches to Model Underwater Explosions. *J. Mar. Sci. Eng.* **2019**, *7*, 15. [CrossRef]
5. Ellinas, C.P. Ultimate strength of damaged tubular bracing members. *J. Struct. Eng.* **1984**, *110*, 245–259. [CrossRef]
6. Wierzbicki, T.; Suh, M.S. Indentation of tubes under combined loading. *Int. J. Mech. Sci.* **1988**, *30*, 229–248. [CrossRef]
7. Bai, Y.; Pedersen, P.T. Elastic-plastic behaviour of offshore steel structures under impact loads. *Int. J. Impact Eng.* **1993**, *13*, 99–115. [CrossRef]
8. Wang, H.M. Action of Ship Anchoring on Submarine Pipeline in Bohai Bay. *Navig. China* **2015**, *38*, 56–59+68.
9. *DNVGL-ST-F101*; Submarine Pipeline Systems. DNV: Bærum, Norway, 2021.
10. *DNVGL-RP-F107*; Risk Assessment of Pipeline Protection. DNV: Bærum, Norway, 2017.
11. Chen, K.; Shen, W.Q. Further experimental study on the failure of fully clamped steel pipes. *Int. J. Impact Eng.* **1998**, *21*, 177–202. [CrossRef]
12. Palmer, A.; Neilson, A.; Sivadasan, S. Impact resistance of pipelines and the loss-of-containment limit state. In Proceedings of the International Pipeline Pigging Integrity Assessment and Repair Conference, Calgary, AB, Canada, 4–8 October 2004.
13. Palmer, A.; Touhey, M.; Holder, S.; Anderson, M.; Booth, S. Full-scale impact tests on pipelines. *Int. J. Impact Eng.* **2006**, *32*, 1267–1283. [CrossRef]
14. Zhang, R. Dynamic Response and Failure Mechanism of Structural Tubular Steel Members Subjected to Transverse Impact. Ph.D. Thesis, Harbin Institute of Technology, Shenzhen, China, 2018.
15. Zeinoddini, M.; Harding, J.E.; Parke, G.A.R. Effect of impact damage on the capacity of tubular steel members of offshore structures. *Mar. Struct.* **1998**, *11*, 141–157. [CrossRef]
16. Yan, S.W.; Tian, Y.H. Analysis of pipeline damage to impact load by dropped objects. *Trans. Tianjin Univ.* **2006**, *12*, 138–141.
17. Huang, Q.F.; Guo, H.Y.; Li, W. Numerical simulation and experimental study on the damage of submarine pipeline impacted by dropped objects. *Ocean. Eng.* **2018**, *36*, 84–88.
18. Zhang, D.N. A modified Johnson-Cook model of dynamic tensile behaviors for 7075-T6 aluminum alloy. *J. Alloys Compd.* **2015**, *619*, 186–194. [CrossRef]
19. Banerjee, A.; Dhar, S.; Acharyya, S.; Datta, D.; Nayak, N. Determination of Johnson cook material and failure model constants and numerical modelling of Charpy impact test of armour steel. *Mater. Sci. Eng. A* **2015**, *640*, 200–209. [CrossRef]
20. Cao, Y.; Zhen, Y.; Song, M.; Yi, H.; Li, F.; Li, X. Determination of Johnson–Cook parameters and evaluation of Charpy impact test performance for X80 pipeline steel. *Int. J. Mech. Sci.* **2020**, *179*, 105627. [CrossRef]

21. Wang, Z.; Hu, Z.; Liu, K.; Chen, G. Application of a material model based on the Johnson-Cook and Gurson-Tvergaard-Needleman model in ship collision and grounding simulations. *Ocean. Eng.* **2020**, *205*, 106768. [CrossRef]
22. Wang, X.M.; Shi, J. Validation of Johnson-Cook plasticity and damage model using impact experiment. *Int. J. Impact Eng.* **2023**, *60*, 67–75. [CrossRef]
23. Aktürk, M.; Boy, M.; Gupta, M.K.; Waqar, S.; Krolczyk, G.M.; Korkmaz, M.E. Numerical and experimental investigations of built orientation dependent Johnson-Cook model for selective laser melting manufactured $AlSi_{10}Mg$. *J. Mater. Res. Technol.* **2021**, *15*, 6244–6259. [CrossRef]
24. Deshpande, V.M.; Chakraborty, P.; Chakraborty, T.; Tiwari, V. Application of copper as a pulse shaper in SHPB tests on brittle materials-experimental study, constitutive parameters identification, and numerical simulations. *Mech. Mater.* **2022**, *171*, 104336. [CrossRef]
25. Daoud, M.; Chatelain, J.F.; Bouzid, A. Effect of rake angle-based Johnson-Cook material constants on the prediction of residual stresses and temperatures induced in Al2024-T3 machining. *Int. J. Mech. Sci.* **2017**, *122*, 392–404. [CrossRef]
26. Chao, Z.L.; Jiang, L.T.; Chen, G.Q.; Zhang, Q.; Zhang, N.B.; Zhao, Q.Q.; Pang, B.J.; Wu, G.H. A modified Johnson-Cook model with damage degradation for B4Cp/Al composites. *Compos. Struct.* **2022**, *282*, 115029. [CrossRef]
27. Liu, Y.; Li, M.; Ren, X.W.; Xiao, Z.B.; Zhang, X.Y.; Huang, Y.C. Flow stress prediction of Hastelloy C-276 alloy using modified Zerilli-Armstrong, Johnson-Cook and Arrhenius-type constitutive models. *Trans. Nonferrous Met. Soc. China* **2020**, *30*, 3031–3042. [CrossRef]
28. Xie, G.; Yu, X.; Gao, Z.; Xue, W.; Zheng, L. The modified Johnson-Cook strain-stress constitutive model according to the deformation behaviors of a Ni-W-Co-C alloy. *J. Mater. Res. Technol.* **2022**, *20*, 1020–1027. [CrossRef]
29. Wei, G.; Zhang, W.; Deng, Y.F. Identification and validation of constitutive parameters of 45 Steel based on J-C model. *J. Vib. Shock.* **2019**, *38*, 173–178.
30. Zhu, Y. Research on Dynamic Constitutive Relationship of Q355B Steel Based on Johnson-Cook Model. Ph.D. Thesis, Harbin University of Science and Technology, Harbin, China, 2019.
31. Chen, J.L.; Shu, W.Y.; Li, J.W. Experimental study on dynamic mechanical property of Q235 steel at different strain rates. *J. Tongji Univ.* **2016**, *44*, 1071–1075.
32. Chen, X.W.; Wei, L.M.; Li, J.C. Experimental research on the long rod penetration of tungsten-fiber/Zr-based metallic glass matrix composite into Q235 steel target. *Int. J. Impact Eng.* **2015**, *79*, 102–116. [CrossRef]
33. Guo, Z.T.; Gao, B.; Guo, Z.; Zhang, W. Dynamic constitutive relation based on J-C model of Q235 steel. *Explos. Shock. Waves* **2018**, *38*, 804–810.
34. Shokry, A.; Gowid, S.; Mulki, H.; Kharmanda, G. On the Prediction of the Flow Behavior of Metals and Alloys at a Wide Range of Temperatures and Strain Rates Using Johnson–Cook and Modified Johnson–Cook-Based Models: A Review. *Materials* **2023**, *16*, 1574. [CrossRef]
35. Sirigiri VK, R.; Gudiga, V.Y.; Gattu, U.S.; Suneesh, G.; Buddaraju, K.M. A review on Johnson Cook material model. *Mater. Today Proc.* **2022**, *62*, 3450–3456. [CrossRef]
36. Carra, F.; Charrondiere, C.; Guinchard, M.; Sacristan de Frutos, O. Design and construction of an instrumentation system to capture the response of advanced materials impacted by intense proton pulses. *Shock. Vib.* **2021**, *2021*, 8855582. [CrossRef]

Journal of
Marine Science and Engineering

MDPI

Article

Theoretical Analysis of Plastic Behavior of Sandwich Beam with Metal Foam under Repeated Impacts

Kailing Guo [1,2], Mengying Mu [2], Wei Cai [3,*], Bofang Xu [4] and Ling Zhu [2,*]

1 Key Laboratory of High Performance Ship Technology, Wuhan University of Technology, Ministry of Education, Wuhan 430063, China; guokailing@whut.edu.cn
2 School of Naval Architecture, Ocean and Energy Power Engineering, Wuhan University of Technology, Wuhan 430063, China; mumengying30@163.com
3 State Key Laboratory of Materials Synthesis and Processing, Wuhan University of Technology, Wuhan 430070, China
4 Marine Design and Research Institute of China, Shanghai 200021, China; 15072477235@163.com
* Correspondence: caiwei199696@whut.edu.cn (W.C.); lingzhu@whut.edu.cn (L.Z.)

Abstract: The phenomenon of repeated impacts on engineering structures is very common, especially in naval and ocean engineering. When marine structures are subjected to repeated impact loadings, deformation and damage will accumulate as the impact number increases, resulting in the failure and damage of the structures, even causing serious accidents. Based on the rigid-plastic assumption, a theoretical model is established to analyze the plastic mechanical behavior of metal foam sandwich beams (MFSBs) suffering from repeated impacts, in which the membrane factor method (MFM) is applied to derive analytical solutions for the plastic responses of MFSBs. The theoretical predictions agree well with the results of impact tests and numerical simulations, indicating that the theoretical model is accurate and reliable. In addition, the dynamic responses of MFSBs are analyzed based on the MFM, and the effects of the core strength and the face thickness on the deflection responses are determined. The results show that the dimensionless permanent deflection of MFSBs is sensitive to the core strength ratio and the face thickness ratio, and as the core strength ratio or the face thickness ratio increases, the dimensionless permanent deflection decreases gradually in an exponential form. In addition, the influence of the core strength ratio and face thickness ratio becomes more significant as the impact number increases. The proposed theoretical method can provide a theoretical reference and technical support for the design of metal foam sandwich structures with improved impact resistance under repeated impact loadings.

Keywords: sandwich beam with metal foam; repeated impacts; plastic behavior; theoretical analysis; membrane factor method

Citation: Guo, K.; Mu, M.; Cai, W.; Xu, B.; Zhu, L. Theoretical Analysis of Plastic Behavior of Sandwich Beam with Metal Foam under Repeated Impacts. *J. Mar. Sci. Eng.* **2023**, *11*, 1974. https://doi.org/10.3390/jmse11101974

Academic Editor: Vincenzo Crupi

Received: 1 September 2023
Revised: 23 September 2023
Accepted: 8 October 2023
Published: 12 October 2023

1. Introduction

Marine structures are frequently subjected to repeated impact loadings, such as from supply ships, dropped objects, and floe ice. During navigation and operation, damage will accumulate, resulting in the failure of the structures and even causing serious accidents. Therefore, it is necessary to study the dynamic behavior of marine structures under repeated impact loadings.

In order to investigate the dynamic behavior of marine structures subjected to repeated impacts, theoretical analysis, numerical simulations, and impact tests have been performed by many academics. Zhu [1] conducted repeated collision tests on fully clamped rectangular plates and developed a numerical program based on the finite difference method to study the dynamic behavior of ship plates subjected to repeated loadings, and an expression of permanent deflection was given. In order to investigate the effect of low temperatures, Truong et al. [2,3] performed experimental and numerical investigations on steel beam and grillage structures subjected to repeated lateral impacts under low temperatures. The results showed that the permanent deflections at low temperatures were smaller than those at room temperature, and the influence was enhanced with the increase in the impact number. Zhu et al. [4] carried out repeated-impact tests on stiffened plates and proposed a theoretical method based on inscribing and circumscribing a yield surface, which provided analytical predictions of the permanent deflection. Zeng et al. [5] conducted experiments on circular mild-steel plates with surface cracks suffering from repeated impacts at low temperatures, and the effects of low temperatures as well as surface cracks on the peak impact force and permanent deflection were determined.

Recently, the pseudo-shakedown phenomenon of structures, caused by repeated impacts, has attracted more and more attention. Jones [6] applied a rigid-plastic method of analysis to plates subjected to repeated mass impacts and discussed the occurrence conditions of the pseudo-shakedown phenomenon. The results showed that plates subjected to repeated identical mass impact loadings would not achieve a pseudo-shakedown state for most impact cases, except in the special case when a small enough impact energy could be absorbed. He and Soares [7–10] performed experimental and numerical studies on the dynamic behavior of beams under repeated impacts, and the phenomenon of pseudo-shakedown was discussed. Cai et al. [11] employed experimental and numerical methods to determine the dynamic behavior of steel plates suffering from repeated ice impacts and discussed the conditions of the occurrence of pseudo-shakedown considering the energy consumption of ice during repeated impacts. The results showed that the pseudo-shakedown phenomenon occurred when the initial impact energy was smaller than the elastic energy of the structures.

Recently, increasing attention has been paid to the dynamic behavior of marine structures suffering from repeated impact loadings, and some new kinds of structures have been investigated, such as sandwich structures. Zhang et al. [12] performed repeated-impact experiments and numerical simulations on honeycomb sandwich plates, and the energy-absorption performance was analyzed. Zeng et al. [13] conducted experimental studies on aluminum corrugated-core sandwich structures subjected to repeated impacts, and the failure modes and energy absorption of the sandwich structures were explored.

As sandwich structures with lightweight metal cores exhibit excellent energy absorption and impact resistance, they have been widely used in the engineering field [14–18], resulting in an increasing number of investigations into the dynamic behavior of lightweight metal-core sandwich structures. Jing et al. [19] conducted impact tests on metal foam sandwich beams (MFSBs) subjected to aluminum foam projectiles, and the dynamic impact process was recorded using a high-speed camera system. The deformation and failure modes of the sandwich beam were analyzed. In order to investigate the dynamic responses of MFSBs suffering from blast loadings, Qin et al. [20] established a rigid-plastic model considering the combined effects of the member force and moment and derived analytical expressions of the dimensionless deflection. In addition, Qin et al. [21] performed a theoretical analysis of MFSBs subjected to low-velocity mass impacts, in which both quasi-static and dynamic methods were employed and the permanent deflections obtained from those methods were compared. In order to improve the accuracy of the theoretical solution, local denting of the front face sheet was taken into account by Qin et al. [22,23]. The results showed that local denting would absorb impact energy, resulting in the permanent deflections of MFSBs being decreased. Recently, increasing attention has been paid to sandwich structures with gradient cores [24–26]. Zhang et al. [27] investigated the stepwise gradient of the dynamic failure of composite sandwich beams with metal foam cores subjected to low-velocity impacts. Fang et al. [28] performed studies on the high-velocity impact resistance of stepwise gradient sandwich beams with metal foam cores. Zhou et al. [29] proposed an analytical model of fully clamped sandwich beams with layered-gradient foam cores subjected to low-velocity impact, and the dynamic solutions for the deflection responses of three-layer-graded core sandwich beams were attained. Recently, some studies have been performed on MFSBs subjected to repeated-impact loadings. Guo et al. [30,31] applied experimental and numerical methods to MFSBs subjected to repeated impact loadings, and the deformation modes and failure modes of the MFSBs were discussed. Meanwhile, the effects of the impact location and face thickness distribution on permanent deflections were evaluated.

There have been many investigations into the dynamic responses of marine structures subjected to repeated impacts and metal foam sandwich structures suffering from a single impact, in which some theoretical models have been established. However, there are few studies that provide a theoretical analysis of the dynamic mechanical behavior of metal foam sandwich structures under repeated-impact loadings. Therefore, establishing a theoretical model is essential to reveal the mechanism of deformation accumulation in metal foam sandwich structures suffering from repeated-impact loadings.

The objective of this work is to propose an efficient analytical method to predict the plastic responses of MFSBs subjected to repeated mass impacts, in which the theoretical model is established and the rebound effect of the sandwich beam is taken into account. Meanwhile, based on the maximum normal yield surface, the incorporation between the moment and membrane force is considered in the theoretical model, and the membrane factor method (MFM) is developed and applied to the solutions, the results of which are much more accurate than those of previous works. Moreover, the bound solutions based on the square yield surface are derived, and the dimensionless permanent deflections are given. In addition, the deflections obtained from the proposed theoretical model are compared with those of impact tests and numerical simulations, and the accuracy of the theoretical model is verified. At last, the effects of core strength and face thickness on permanent deflections are determined. Figure 1 presents a flowchart that summarizes the sequence of the developed analytical model.

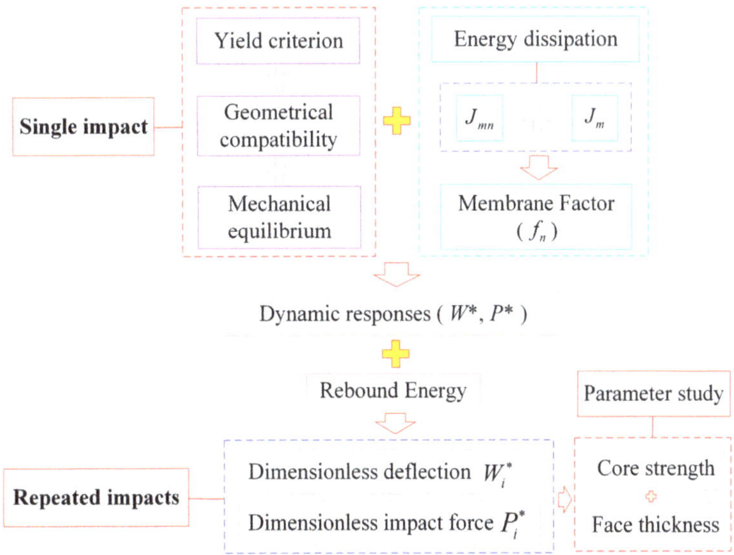

Figure 1. Flowchart of sequence of the developed analytical model.

The main content of this paper includes the following:

(1) Formula derivation for a unified yield criterion for sandwich structures.

(2) Theoretical derivation of analytical solutions based on the membrane factor method and square yield surface.

(3) Validation of the theoretical model by comparing the results of analytical solutions with those of impact tests and numerical simulations.

(4) Discussion of the dynamic responses of MFSBs, including dimensionless permanent deflection, dimensionless impact force, and the relationship between them and impact number.

(5) Investigation of parameter influences, including the effect of core strength and face thickness.

(6) Summary of the main findings and presentation of our main conclusions.

2. A Unified Yield Criterion for Sandwich Structures

When suffering from external loadings, the structures may appear deformed, including bending and axial tension. Therefore, when establishing the yield criteria, the interaction between the bending moment and the membrane force should be taken into account [32]. As for the sandwich beam, it is composed of three parts, including the front and back face sheets and the core layer, and its stress state is much more complex than that of the single-layer beam. According to the superposition principle of force, the complex stress state of a sandwich beam can be decomposed into pure bending and uniaxial tension. When the position of the neutral axis is different, the corresponding force state and direction are different.

It can be observed in reference [30] that, when MFSBs are subjected to repeated impacts, there is no delamination between the face sheets and the metal foam core. Thus, it is assumed that the face sheets are perfectly bonded to the core, and the face sheets and metal foam core obey the perfectly rigid plastic law. Meanwhile, the shear force is negligible compared to the bending moment and the axial force. Therefore, the slip interaction between the three layers is neglected in the theoretical model.

It is assumed that the thickness of the front and back face sheets is t, the thickness of the metal core is c, the width of the MFSB is b, and the distance from the bottom of the back face to the neutral axis is h. The yield strength of the face sheet and the metal foam is assumed to be σ_f and σ_c, respectively. Meanwhile, the moment and axial force of the MFSB are defined as M and N, respectively, and the plastic yield moment and plastic yield axial force are M_P and N_P. It is assumed that $\xi = h/(c+2t)$ is a dimensionless distance from the neutral axis; then, the stress state of the MFSB can be divided into two forms according to the position of the neutral axis, as shown in Figure 2.

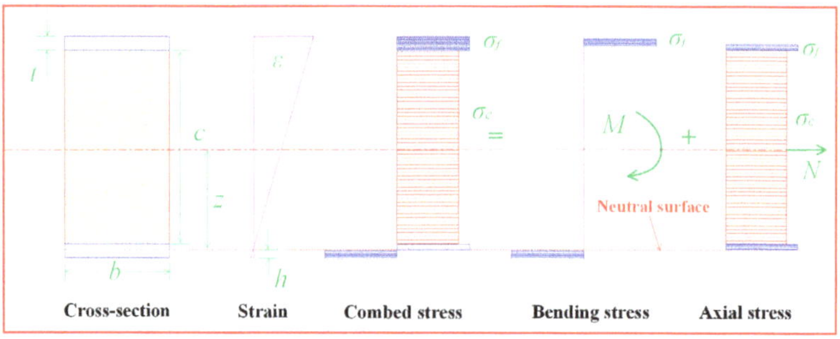

$$0 \le \xi \le \frac{t}{c+2t}$$

(a)

Figure 2. *Cont.*

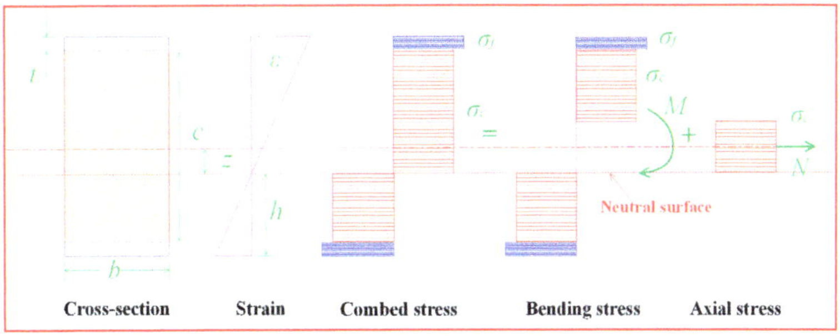

(b) $\frac{t}{c+2t} \leq \xi \leq \frac{1}{2}$

Figure 2. Distribution of the strain and stress in the sandwich cross-section.

If $0 \leq \xi \leq \frac{t}{c+2t}$,

$$N = \sigma_c bc + 2\sigma_f b[t - \xi(c + 2t)] \tag{1}$$

$$M = \sigma_f b(c + 2t)(1 - 2\xi) \tag{2}$$

Otherwise, if $\frac{t}{c+2t} \leq \xi \leq \frac{1}{2}$,

$$N = b \cdot \sigma_c \cdot (c + 2t)(1 - 2\xi) \tag{3}$$

$$M = \frac{b}{4}\left\{\sigma_c\left[c^2 - (c + 2t)^2(1 - 2\xi)^2\right] + \sigma_f\left[(c + 2t)^2 - c^2\right]\right\} \tag{4}$$

The plastic yield moment and plastic yield axial force can be expressed as

$$N_P = bc \cdot \sigma_c + 2b \cdot t \cdot \sigma_f \tag{5}$$

$$M_P = \frac{b}{4}c^2 \cdot \sigma_c + b \cdot t(c + t) \cdot \sigma_f \tag{6}$$

Dimensionless parameters can be introduced, i.e., $\bar{\sigma} = \sigma_c/\sigma_f$, $\bar{t} = t/c$. Then, combining Equations (5) and (6), we can obtain

$$\frac{N_P \cdot c}{M_P} = \frac{4(\bar{\sigma} + 2\bar{t})}{\bar{\sigma} + 4\bar{t}(1 + \bar{t})} \tag{7}$$

It can be derived from Equations (1)–(7):

$$\frac{N}{N_P} = \begin{cases} \frac{\bar{\sigma}+2\left[\bar{t}-\xi(1+2\bar{t})\right]}{\bar{\sigma}+2\bar{t}} & \left(0 \leq \xi \leq \frac{t}{c+2t}\right) \\ \frac{\bar{\sigma}(1+2\bar{t})(1-2\xi)}{\bar{\sigma}+2\bar{t}} & \left(\frac{t}{c+2t} \leq \xi \leq \frac{1}{2}\right) \end{cases} \tag{8}$$

$$\frac{M}{M_P} = \begin{cases} \frac{4(1+2\bar{t})^2(1-\xi)\xi}{\bar{\sigma}+2\bar{t}} & \left(0 \leq \xi \leq \frac{t}{c+2t}\right) \\ 1 - \frac{(1+2\bar{t})^2(1-2\xi)^2}{\bar{\sigma}+4\bar{t}(1+\bar{t})}\bar{\sigma} & \left(\frac{t}{c+2t} \leq \xi \leq \frac{1}{2}\right) \end{cases} \tag{9}$$

Introduce dimensionless moment and axial force, i.e., $m = M/M_P$ and $n = N/N_P$.

If $0 \le \xi \le \frac{t}{c+2t}$, i.e., $\frac{\overline{\sigma}}{\overline{\sigma}+2\overline{t}} \le |n| \le 1$, then $\xi = \frac{(1-|n|)(\overline{\sigma}+2\overline{t})}{2(1+2\overline{t})}$, substitute it into Equations (8) and (9):

$$|m| + \frac{[(\overline{\sigma}+2\overline{t})|n| + 2t - \sigma + 2](|n|-1)(\overline{\sigma}+2\overline{t})}{\overline{\sigma}+4\overline{t}(1+\overline{t})} = 0 \tag{10}$$

Otherwise, if $\frac{t}{c+2t} \le \xi \le 1$, i.e., $0 \le |n| \le \frac{\overline{\sigma}}{\overline{\sigma}+2\overline{t}}$, then $\xi = \frac{1}{2}\left[1 - \frac{|n|(\overline{\sigma}+2\overline{t})}{\overline{\sigma}(1+2\overline{t})}\right]$, substitute it into Equations (8) and (9):

$$|m| + \frac{(\overline{\sigma}+2\overline{t})^2}{\overline{\sigma}^2+4\overline{\sigma}\overline{t}(1+\overline{t})}n^2 = 1 \tag{11}$$

Combine Equations (10) and (11), and the yield criterion [20] can be obtained:

$$\begin{cases} |m| + \frac{(\overline{\sigma}+2\overline{t})^2}{\overline{\sigma}^2+4\overline{\sigma}\overline{t}(1+\overline{t})}n^2 = 1 & \left(0 \le |n| \le \frac{\overline{\sigma}}{\overline{\sigma}+2\overline{t}}\right) \\ |m| + \frac{[(\overline{\sigma}+2\overline{t})|n|+2t-\sigma+2](|n|-1)(\overline{\sigma}+2\overline{t})}{\overline{\sigma}+4\overline{t}(1+\overline{t})} = 0 & \left(\frac{\overline{\sigma}}{\overline{\sigma}+2\overline{t}} \le |n| \le 1\right) \end{cases} \tag{12}$$

When the sandwich beam reduces to a solid beam, namely, $\overline{\sigma} = 1$, $\overline{t} = 0$, then Equation (12) can be rewritten as $|m| + n^2 = 1$, which is the same as the yield criterion of the solid beam. The yield surface of metal foam sandwich structures is illustrated in Figure 3.

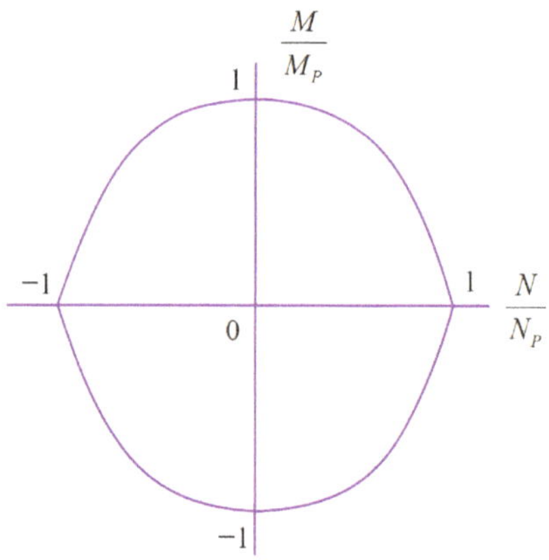

Figure 3. Yield surface of metal foam sandwich structures.

3. Analytical Solutions

3.1. Solutions Based on the Membrane Factor Method

In order to obtain much more accurate solutions and reduce the difficulty to solve the equation, the membrane factor method (MFM) was proposed and developed by Yu and Strong [33–35] based on maximal normal stress yield criteria. In this method, the effects of the bending moment and the axial force were both considered, and the axial force was connected with the bending moment by the membrane factor. Thus, in this paper, the MFM will be employed and developed to analyze the dynamic responses of MFSBs suffering from repeated wedge mass impacts.

As shown in Figure 4, the MFSB is subjected to a low-velocity impact from a rigid wedge striker. The length of the beam is $2L$, the width is b, and the thickness of the metal foam core is c. The mass of the wedge impactor is G_s, and the initial impact velocity is V_0.

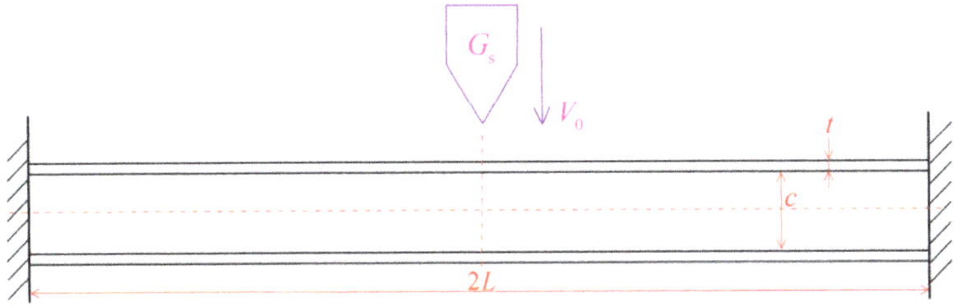

Figure 4. Sketch of fully clamped sandwich beam under impact from wedge mass.

According to associated flow rules, it can be expressed as

$$\frac{\dot{\varepsilon}}{\dot{\kappa}} = -\frac{M_P}{N_P}\frac{dm}{dn} \tag{13}$$

(i) If $0 \le |n| \le \frac{\overline{\sigma}}{\overline{\sigma}+2\overline{t}}$, we obtain

$$\frac{\dot{\varepsilon}}{\dot{\kappa}} = -\frac{M_P}{N_P}\cdot\frac{(\overline{\sigma}+2\overline{t})^2}{4\overline{\sigma}\overline{t}(1+\overline{t})+\sigma^2}\cdot 2|n| = \frac{\overline{\sigma}+2\overline{t}}{2\overline{\sigma}}|n|c \tag{14}$$

(ii) Otherwise, if $\frac{\overline{\sigma}}{\overline{\sigma}+2\overline{t}} \le |n| \le 1$, we have

$$\frac{\dot{\varepsilon}}{\dot{\kappa}} = -\frac{M_P}{N_P}\cdot\frac{(\overline{\sigma}+2\overline{t})\left[2|n|\cdot(\overline{\sigma}+2\overline{t})+2-2\overline{\sigma}\right]}{4\overline{\sigma}\overline{t}(1+\overline{t})+\sigma^2} = \frac{|n|(\overline{\sigma}+2\overline{t})+1-\overline{\sigma}}{2}c \tag{15}$$

As for a fully clamped MFSB, based on the results of the repeated-impact tests conducted by Guo [30], it can be assumed that the displacement of the sandwich beam is almost linearly distributed, as displayed in Figure 5.

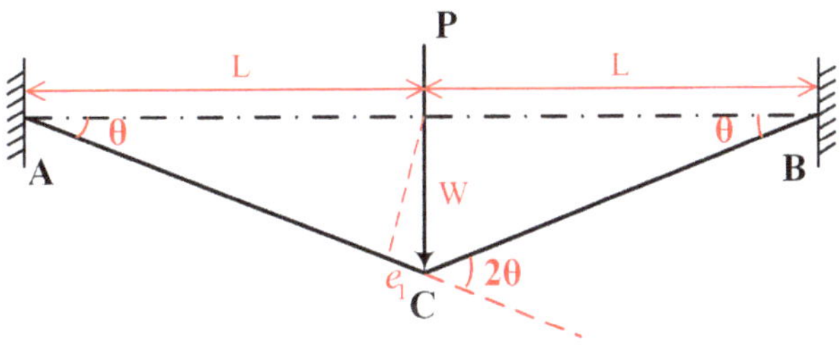

Figure 5. Deformation pattern of the neutral axis of the fully clamped MFSB.

From the geometrical compatibility equation, we obtain

$$\frac{\dot{\varepsilon}}{\dot{\kappa}} = \frac{W_0}{2} \tag{16}$$

Combining Equations (14)–(16), the dimensionless deflection can be given by

$$W^* = \frac{W_0}{c+2t} = \begin{cases} \frac{\overline{\sigma}+2\overline{t}}{\overline{\sigma}(1+2\overline{t})}|n| & \left(0 \le |n| \le \frac{\overline{\sigma}}{\overline{\sigma}+2\overline{t}}\right) \\ \frac{(\overline{\sigma}+2\overline{t})|n|-\overline{\sigma}+1}{1+2\overline{t}} & \left(\frac{\overline{\sigma}}{\overline{\sigma}+2\overline{t}} \le |n| \le 1\right) \end{cases} \tag{17}$$

Taking the interaction between the moment and the axial force, the energy dissipation of the MFSB can be expressed as follows:

$$J_{mn} = M_P\dot{\kappa}(m + n \cdot \frac{N_P}{M_P}\frac{\dot{\varepsilon}}{\dot{\kappa}}) \tag{18}$$

When only the moment is considered, the energy dissipation of the MFSB can be written as follows:

$$J_m = M_P\dot{\kappa} \tag{19}$$

From Equations (18) and (19), the membrane force factor f_n [34] can be obtained:

$$f_n = \frac{J_{mn}}{J_m} = m + n \cdot \frac{N_P}{M_P}\frac{W}{2} \tag{20}$$

Due to the sandwich beam being symmetrical, half the beam can be taken as an example to perform a force analysis, as presented in Figure 6.

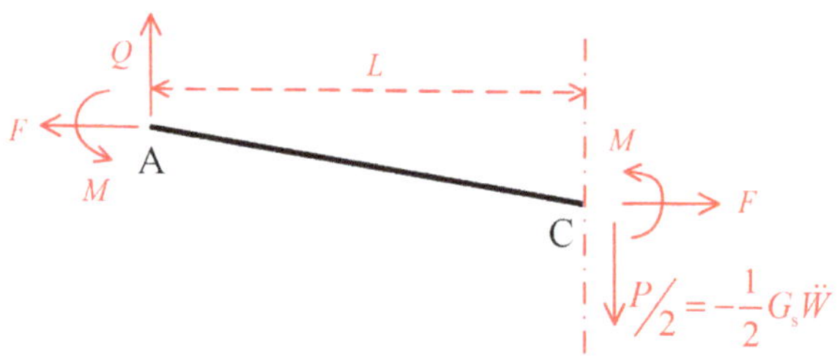

Figure 6. Sketch of the force state for the fully clamped MFSB.

The moment equilibrium equation of the MFSB is

$$2M + N \cdot W_0 - \frac{P}{2} \cdot L = 0 \tag{21}$$

It can be reduced to

$$P = \frac{4M_0}{L}\left(m + n \cdot \frac{N_0}{M_0} \cdot \frac{W_0}{2}\right) \tag{22}$$

The initial static collapse load is assumed to be

$$P_i = \frac{4M_0}{L} \tag{23}$$

Then, the dimensionless reaction force can be expressed as

$$P^* = \frac{P}{P_i} = m + n \cdot \frac{N_0}{M_0} \cdot \frac{W_0}{2} \tag{24}$$

Then, from Equations (20) and (24), the dimensionless reaction force can be reduced to

$$P^* = f_n \tag{25}$$

Combine Equations (5), (6) and (20) and we have

$$f_n = m + \frac{2(\bar{\sigma} + 2\bar{t})(1 + 2\bar{t})}{\bar{\sigma} + 4\bar{t}(1 + \bar{t})} n \cdot W^* \tag{26}$$

(i) If $0 \le |n| \le \frac{\bar{\sigma}}{\bar{\sigma}+2\bar{t}}$, then $0 \le W^* \le \frac{1}{1+2\bar{t}}$

From Equation (17), we obtain

$$n = \frac{\overline{\sigma}(1+2\overline{t})}{\overline{\sigma}+2\overline{t}}W^* \tag{27}$$

Substitute Equation (27) into Equation (12) and we have

$$m = 1 - \frac{\overline{\sigma}(1+2\overline{t})^2}{4\overline{t}(1+\overline{t})+\overline{\sigma}}W^{*2} \tag{28}$$

Substitute Equations (27) and (28) into Equation (26), and the membrane force factor can be expressed as

$$f_n = 1 + \frac{\overline{\sigma}(1+2\overline{t})^2}{\overline{\sigma}+4\overline{t}(1+\overline{t})}W^{*2} \tag{29}$$

(ii) If $\frac{\overline{\sigma}}{\overline{\sigma}+2\overline{t}} \le |n| \le 1$, then $\frac{1}{1+2\overline{t}} \le W^* \le 1$
According to Equation (17), we obtain

$$n = \frac{(1+2\overline{t})W^* + \overline{\sigma} - 1}{\overline{\sigma}+2\overline{t}} \tag{30}$$

Substitute the above equation into Equation (26) and we have

$$m = \frac{(1+2\overline{t})^2}{4\overline{t}(1+\overline{t})+\overline{\sigma}}(1-W^{*2}) \tag{31}$$

Substitute Equations (30) and (31) into Equation (26), and the membrane force factor can be rewritten as

$$f_n = \frac{(1+2\overline{t})\left[(1+2\overline{t})(W^{*2}+1)+2(\overline{\sigma}-1)W^*\right]}{\overline{\sigma}+4\overline{t}(1+\overline{t})} \tag{32}$$

(iii) If the deflection of the beam increases to a specific value, the beam will become like a string, which only suffers from axial force. At this moment, $n = 1$, $m = 0$, $W^* \ge 1$, and from Equation (26), we have

$$f_n = \frac{2(\overline{\sigma}+2\overline{t})(1+2\overline{t})}{\overline{\sigma}+4\overline{t}(1+\overline{t})}W^* \tag{33}$$

Combine Equations (29), (32) and (33), and the membrane force factor can be expressed as

$$f_n = \begin{cases} 1 + \frac{\overline{\sigma}(1+2\overline{t})^2}{\overline{\sigma}+4\overline{t}(1+\overline{t})}W^{*2} & \left(0 \le W^* \le \frac{1}{1+2\overline{t}}\right) \\ \frac{(1+2\overline{t})\left[(1+2\overline{t})(W^{*2}+1)+2(\overline{\sigma}-1)W^*\right]}{\overline{\sigma}+4\overline{t}(1+\overline{t})} & \left(\frac{1}{1+2\overline{t}} \le W^* \le 1\right) \\ \frac{2(\overline{\sigma}+2\overline{t})(1+2\overline{t})}{\overline{\sigma}+4\overline{t}(1+\overline{t})}W^* & (W^* \ge 1) \end{cases} \tag{34}$$

If only the moment works, the energy dissipation of the MFSB can be given by

$$D_m = \int_0^{\theta} 4M_P d\theta = \int_0^{W_0^*} \frac{4M_P}{L}(c+2t)dW^* \tag{35}$$

However, when both the moment and the axial force are considered, the energy dissipation of the MFSB can be written as

$$D_{mn} = \int_0^{\theta} 4M_P \cdot f_n \cdot d\theta = \int_0^{W_0^*} \frac{4M_P}{L}(c+2t)\cdot f_n \cdot dW^* = 4M_P \cdot \overline{c}(1+2\overline{t})\int_0^{W_0^*} f_n dW^* \tag{36}$$

Assuming that the mass of impactor is G_s and the initial impact velocity is V_0, then the initial impact energy is

$$E_K = \frac{1}{2} G_s V_0^2 \tag{37}$$

Define dimensionless kinetic energy and plastic energy as

$$E_K{}^* = \frac{E_K}{4M_P \cdot \bar{c}(1+2\bar{t})} \tag{38a}$$

$$D_{mn}{}^* = \frac{D_{mn}}{4M_P \cdot \bar{c}(1+2\bar{t})} \tag{38b}$$

Combine Equations (36)–(38) and we have

$$D_{mn}{}^* = \begin{cases} \frac{\bar{\sigma}(1+2\bar{t})^2}{3[\bar{\sigma}+4\bar{t}(1+\bar{t})]} W_0^{*3} + W_0^*, \left(0 \le W^* \le \frac{1}{1+2\bar{t}}\right) \\ \frac{(1+2\bar{t})^2}{3[\bar{\sigma}+4\bar{t}(1+\bar{t})]} W_0^{*3} + \frac{(1+2\bar{t})(\bar{\sigma}-1)}{\bar{\sigma}+4\bar{t}(1+\bar{t})} W_0^{*2} + \frac{(1+2\bar{t})^2}{\bar{\sigma}+4\bar{t}(1+\bar{t})} W_0^* \\ + \frac{\bar{\sigma}-1}{3[\bar{\sigma}+4\bar{t}(1+\bar{t})](1+2\bar{t})}, \left(\frac{1}{1+2\bar{t}} \le W^* \le 1\right) \\ \frac{(1+2\bar{t})(\bar{\sigma}+2\bar{t})}{\bar{\sigma}+4\bar{t}(1+\bar{t})} (W_0^{*2} - 1) + \frac{4(1+2\bar{t})^3 + 4(3\bar{t}^2+3\bar{t}+1)(\bar{\sigma}-1)}{3[\bar{\sigma}+4\bar{t}(1+\bar{t})](1+2\bar{t})}, (W^* \ge 1) \end{cases} \tag{39}$$

Define $W_0^* = \delta$, and introduce a, b, c, and d as coefficients of δ, namely

$a_1 = \frac{\bar{\sigma}(1+2\bar{t})^2}{3[\bar{\sigma}+4\bar{t}(1+\bar{t})]}, a_2 = \frac{(1+2\bar{t})^2}{3[\bar{\sigma}+4\bar{t}(1+\bar{t})]}, a_3 = \frac{(1+2\bar{t})(\bar{\sigma}+2\bar{t})}{\bar{\sigma}+4\bar{t}(1+\bar{t})}$

$b_2 = \frac{(1+2\bar{t})(\bar{\sigma}-1)}{\bar{\sigma}+4\bar{t}(1+\bar{t})}, c_2 = \frac{(1+2\bar{t})^2}{\bar{\sigma}+4\bar{t}(1+\bar{t})}$

$d_2 = \frac{\bar{\sigma}-1}{3[\bar{\sigma}+4\bar{t}(1+\bar{t})](1+2\bar{t})}, d_3 = \frac{4(1+2\bar{t})^3+4(3\bar{t}^2+3\bar{t}+1)(\bar{\sigma}-1)}{3[\bar{\sigma}+4\bar{t}(1+\bar{t})](1+2\bar{t})} - \frac{(1+2\bar{t})(\bar{\sigma}+2\bar{t})}{\bar{\sigma}+4\bar{t}(1+\bar{t})}$

Then, Equation (39) can be reduced to

$$D_{mn}{}^* = \begin{cases} a_1\delta^3 + \delta, \left(0 \le \delta \le \frac{1}{1+2\bar{t}}\right) \\ a_2\delta^3 + b_2\delta^2 + c_2\delta + d_2, \left(\frac{1}{1+2\bar{t}} \le \delta \le 1\right) \\ b_3\delta^2 + d_3, (\delta \ge 1) \end{cases} \tag{40}$$

Assume the impact number is i; then, according to the energy conservation theorem, we have

$$\begin{cases} a_1\delta_i{}^3 + \delta_i = \sum_0^i E_{Kn}^*, \left(0 \le \delta_i \le \frac{1}{1+2\bar{t}}\right) \\ a_2\delta_i{}^3 + b_2\delta_i{}^2 + c_2\delta_i + d_2 = \sum_0^i E_{Kn}^*, \left(\frac{1}{1+2\bar{t}} \le \delta_{i-1} \le 1\right) \\ a_3\delta_i{}^2 + d_3 = \sum_0^i E_{Kn}^*, (\delta_i \ge 1) \end{cases} \tag{41}$$

The analytical expression of permanent deflection of MFSBs in the ith impact number can be obtained from Equation (41), as follows:

(1) If $0 \le \delta_i \le \frac{1}{1+2\bar{t}}$

Assume that

$$\begin{cases} p_1 = \frac{1}{a_1} = \frac{3[\bar{\sigma}+4\bar{t}(1+\bar{t})]}{\bar{\sigma}(1+2\bar{t})^2} \\ q_1 = -\frac{i \cdot E_{K0}^*}{a_1} = -\frac{3[\bar{\sigma}+4\bar{t}(1+\bar{t})]}{\bar{\sigma}(1+2\bar{t})^2} \cdot \sum_0^i E_{Kn}^* \end{cases} \tag{42}$$

Equation (41) can be reduced

$$\delta_i{}^3 + p_1 \cdot \delta_i + q_1 = 0 \tag{43}$$

Define

$$\Delta_1 = \frac{{q_1}^2}{4} + \frac{{p_1}^3}{27} \tag{44}$$

Then, the expression of a dimensionless deflection can be given as

$$\delta_i = \sqrt[3]{-\frac{q_1}{2} + \sqrt{\Delta_1}} + \sqrt[3]{-\frac{q_1}{2} + \sqrt{\Delta_1}} \tag{45}$$

(2) If $\frac{1}{1+2\bar{t}} \leq \delta_{i-1} \leq 1$
Define

$$\begin{cases} \varphi = \delta_i + \frac{b_2}{3a_2} \\ p_2 = \frac{3a_2c_2 - b_2^2}{3a_2^2} \\ q_2 = \frac{2b_2^3 - 9a_2b_2c_2 + 27a_2^2 d'_2}{27a_2^3} \\ d'_2 = d_2 - \sum\limits_0^i \mathrm{E}^*_{Kn} \end{cases} \tag{46}$$

Then, we have

$$\varphi^3 + p_2 \cdot \varphi + q_2 = 0 \tag{47}$$

$$\Delta_2 = \frac{{q_2}^2}{4} + \frac{{p_2}^3}{27} \tag{48}$$

$$\varphi = \sqrt[3]{-\frac{q_2}{2} + \sqrt{\Delta_2}} + \sqrt[3]{-\frac{q_2}{2} + \sqrt{\Delta_2}} \tag{49}$$

Then, we can obtain the following:

$$\delta_i = \sqrt[3]{-\frac{q_2}{2} + \sqrt{\Delta_2}} + \sqrt[3]{-\frac{q_2}{2} + \sqrt{\Delta_2}} - \frac{b_2}{3a_2} \tag{50}$$

(3) If $\delta_i \geq 1$

$$\delta_i{}^2 + \frac{\mathrm{d}_3}{a_3} - \frac{1}{a_3}\sum_0^i \mathrm{E}^*_{Kn} = 0 \tag{51}$$

$$\delta_i = \sqrt{\frac{\sum\limits_0^i \mathrm{E}^*_{Kn} - d_3}{b_3}} \tag{52}$$

3.2. Solutions Based on Square Yield Surface

Similar to solid structures, the square yield surface can be employed to solve the plastic responses of sandwich beams subjected to repeated-impact loadings, and the corresponding bounds of solutions can be obtained.

As for circumscribing yield surface, $|m| = |n| = 1$, i.e., $M = M_P$, $N = N_P$

On the other hand, for inscribing yield surface, it is assumed that $|N| = \theta \cdot N_P$ and $|M| = \theta \cdot M_P$, namely,

$$|m| = |n| = \theta \tag{53}$$

(i) If $\theta \leq \frac{\bar{\sigma}}{\bar{\sigma}+2\bar{t}}$
Substituting Equation (53) into Equation (12), we obtain

$$\theta + \frac{(\bar{\sigma} + 2\bar{t})^2}{4\bar{\sigma}\bar{t}(1+\bar{t}) + \bar{\sigma}^2}\theta^2 = 1 \tag{54}$$

Assume $k_0 = \frac{(\bar{\sigma}+2\bar{t})^2}{4\bar{\sigma}\bar{t}(1+\bar{t})+\bar{\sigma}^2}$

Then, Equation (54) can be reduced to

$$\theta = \frac{\sqrt{1+4k_0}-1}{2k_0} \tag{55}$$

Meanwhile, $\theta \leq \frac{\overline{\sigma}}{\overline{\sigma}+2\overline{t}}$; then, from Equation (55), we obtain

$$8\overline{t}(1+\overline{t}) - \overline{\sigma}^2 \leq 0 \tag{56}$$

(ii) If $\theta \geq \frac{\overline{\sigma}}{\overline{\sigma}+2\overline{t}}$
Substitute Equation (53) into Equation (12) and we obtain

$$\theta^2 + \left[\frac{(1-\overline{\sigma})(3\overline{\sigma}+8\overline{t})}{(\overline{\sigma}+2\overline{t})^2}+1\right]\theta - \left[\frac{2(1-\overline{\sigma})}{\overline{\sigma}+2\overline{t}}+1\right] = 0 \tag{57}$$

Assume $k_1 = \frac{(1-\overline{\sigma})(3\overline{\sigma}+8\overline{t})}{(\overline{\sigma}+2\overline{t})^2}+1$, $k_2 = \frac{2(1-\overline{\sigma})}{\overline{\sigma}+2\overline{t}}+1$
Then, Equation (57) can be reduced to

$$\theta = \frac{\sqrt{k_1^2+4k_2}-k_1}{2} \tag{58}$$

And we obtain

$$8\overline{t}(1+\overline{t}) - \overline{\sigma}^2 \geq 0 \tag{59}$$

Then, the square yield surface of sandwich structures can be obtained, as presented in Figure 7.

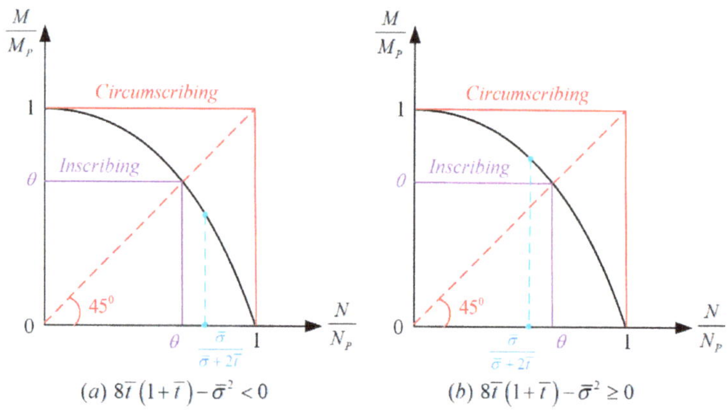

Figure 7. Square yield surface of sandwich structures.

For a fully clamped sandwich beam, when only the plastic yield moment works, the plastic energy dissipation rate can be expressed as

$$\dot{D}_m = 4M_P \cdot \theta_m \tag{60}$$

While both the moment and the axial force are considered for the plastic energy dissipation, the energy dissipation rate can be written as

$$\dot{D}_{mn} = 4M_P \cdot \left(1+\frac{N_P}{M_P}\frac{W}{2}\right)\theta_m \tag{61}$$

Integrate the above equation and we can obtain

$$D_{mn} = \int_0^{\theta_m} 4M_P \cdot \left(1 + \frac{N_P}{M_P}\frac{W}{2}\right)\theta_m d\theta = \int_0^{W^*} 4M_P \cdot \bar{c}(1+2\bar{t}) \cdot \left(1 + \frac{N_P}{M_P}\frac{W}{2}\right) dW^* \tag{62}$$

Based on energy conservation $D_{mn} = E_{K0}$, we obtain

$$E^*_{K0} = \frac{(\bar{\sigma}+2\bar{t})(1+2\bar{t})}{\bar{\sigma}+4\bar{t}(1+\bar{t})} W^{*2} + W^* \tag{63}$$

Solve Equation (63), and the dimensionless deflection of the MFSB based on circumscribing the yield surface can be derived as

$$W^*_{0c} = \frac{\bar{\sigma}+4\bar{t}(1+\bar{t})}{2(\bar{\sigma}+2\bar{t})(1+2\bar{t})}\left[\sqrt{1+\frac{4(\bar{\sigma}+2\bar{t})(1+2\bar{t})}{\bar{\sigma}+4\bar{t}(1+\bar{t})}E^*_{K0}} - 1\right] \tag{64}$$

Similarly, the dimensionless deflection of the MFSB based on inscribing yield surface can be given as

$$W^*_{0i} = \frac{\bar{\sigma}+4\bar{t}(1+\bar{t})}{2(\bar{\sigma}+2\bar{t})(1+2\bar{t})}\left[\sqrt{1+\frac{4(\bar{\sigma}+2\bar{t})(1+2\bar{t})}{\bar{\sigma}+4\bar{t}(1+\bar{t})}\frac{E^*_{K0}}{\theta}} - 1\right] \tag{65}$$

As for repeated impacts, employing the principle of energy and deformation accumulation, the dimensionless deflection for the ith impact can be derived as follows:

$$W^*_{0ci} = \frac{\bar{\sigma}+4\bar{t}(1+\bar{t})}{2(\bar{\sigma}+2\bar{t})(1+2\bar{t})}\left[\sqrt{1+\frac{4(\bar{\sigma}+2\bar{t})(1+2\bar{t})}{\bar{\sigma}+4\bar{t}(1+\bar{t})}iE^*_{K0}} - 1\right] \tag{66a}$$

$$W^*_{0ii} = \frac{\bar{\sigma}+4\bar{t}(1+\bar{t})}{2(\bar{\sigma}+2\bar{t})(1+2\bar{t})}\left[\sqrt{1+\frac{4(\bar{\sigma}+2\bar{t})(1+2\bar{t})}{\bar{\sigma}+4\bar{t}(1+\bar{t})}\frac{iE^*_{K0}}{\theta}} - 1\right] \tag{66b}$$

4. Results and Discussion

4.1. Validation of Theoretical Model

In order to verify the accuracy of the theoretical model, the permanent deflections predicted by the theoretical method are compared with those of repeated-impact tests [30] and numerical simulations [31].

The metal foam sandwich beam was composed of a front face sheet, a back face sheet, and an aluminum foam core. Epoxy resin was used to glue the face sheets and the core. The material of the face sheets was mild steel, and the core material was closed-cell aluminum foam with a density of 0.5 g/cm^3. The thicknesses of the face sheets and the core were t = 1 mm and c = 10 mm, respectively. The total length and the span length of the MFSB were L_B = 250 mm and L_S = 150 mm, respectively, and the width of the beam was B = 30 mm. The impactor mass was 7.884 kg, and the impact velocity was 2.12 m/s, i.e., the impact energy was 17.8 J. The apparatus used for repeated-impact tests is presented in Figure 8.

Dynamic responses of MFSBs obtained from repeated-impact tests are illustrated in Table 1, comprising permanent deflections, rebound energy, and absorbed energy.

The numerical model is created by ABAQUS. In the numerical model, the length of the sandwich beam is 150 mm, the width of the beam is 30 mm, and the thickness of the core and the face sheet is 10 mm and 1 mm, respectively. The width of the wedge impactor is 40 mm, the included angle of the wedge is 60°, and the fillet radius is 1.5 mm. In the numerical model, the face sheet is specified as an elastic/plastic material, and the elastic part is defined by the following parameters: Young's modulus and Poisson's ratio, and the hardening behavior is defined using the true plastic stress–strain curve. As for the metal foam core, the material model is defined as crushable foam. As for aluminum foam, the elastic and plastic Poisson's ratios are 0.3 and 0, respectively, and the plastic stress ratio is 1.73. The plastic stress–strain curves of mild steel and aluminum foam obtained from material tests are presented in Figure 9.

Table 1. Dynamic responses of MFSBs in repeated-impact tests.

Impact Number	Permanent Deflection (mm)	Rebound Energy (J)	Absorbed Energy (J)
1	3.85	0.41	17.39
2	6.86	0.53	17.27
3	9.46	0.68	17.12
4	11.78	1.03	16.77
5	13.85	1.41	16.39
6	15.61	1.50	16.30
7	17.04	1.69	16.11
8	18.35	2.12	15.68
9	19.30	2.28	15.52
10	20.05	2.60	15.20

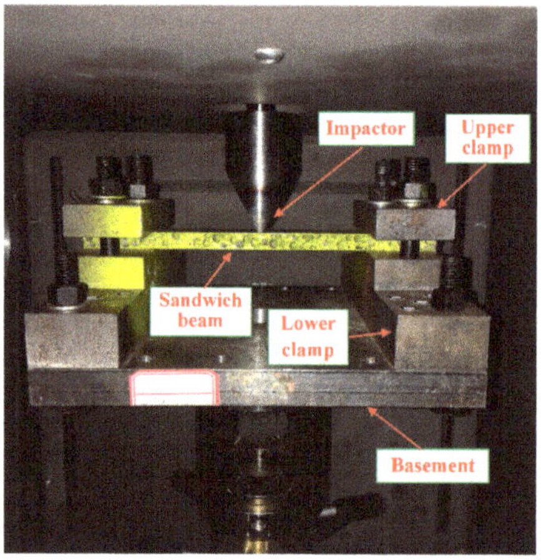

Figure 8. Apparatus of repeated-impact test [30].

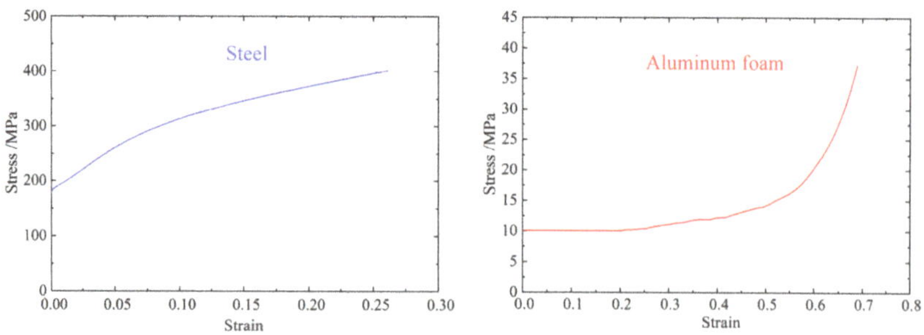

(**a**) Mild steel (**b**) Aluminum foam

Figure 9. Plastic stress–strain curve.

As for the mesh, the width of the refined area is 60 mm, and the mesh sizes of refined and non-refined areas are 1 mm and 2.5 mm, respectively. The mesh sizes of the beam along the direction of thickness and width are 1 mm and 1.5 mm, respectively. The mesh convergence was analyzed in our previous study [31], and the mesh size for face sheets and foam core in the present numerical model was satisfied with convergence requirements. The wedge impactor is defined as a discrete rigid body, and a quadrilateral shell element (R3D4) is chosen. Meanwhile, a core layer and face sheets adopt linear reduction integral hexahedral element (C3DR8) and quadrilateral shell element (S4R), respectively. The finite-element model of numerical simulations is illustrated in Figure 10.

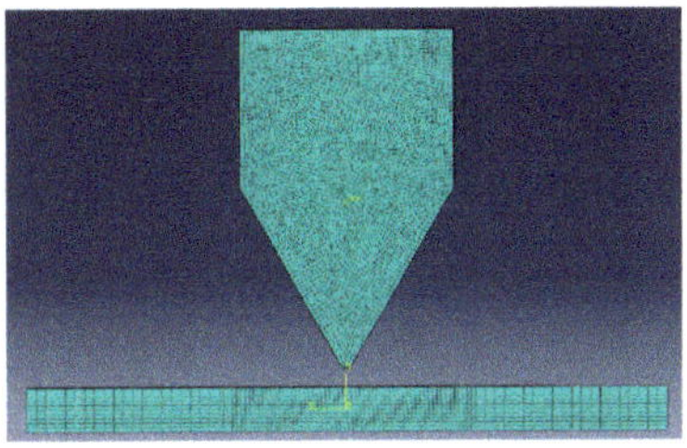

Figure 10. Finite-element model of numerical simulations [31].

The rebound effect of the MFSB is considered to be a form of the energy-absorption coefficient obtained from impact tests. E_k and E_{ir} are defined as initial impact energy and rebound energy, respectively. Meanwhile, V_0 and V_{ir} are defined as initial impact velocity and rebound velocity, respectively. When considering the influence of rebound of the impactor, the elastic energy should be subtracted from the total energy, that is, the plastic dissipation energy of the MFSB can be expressed as

$$D_{mn} = \sum_{1}^{i}(E_K - E_{ir}) = \sum_{1}^{i}\frac{1}{2}m\left(V_0^2 - V_{ir}^2\right) \tag{67}$$

The relationship between absorbed energy and the impact number (N) can be obtained from repeated-impact tests [30], as presented in Table 1. Meanwhile, the expression of the energy-absorption coefficient in the form of an impact number (N) can be obtained, as illustrated in Equation (69).

$$\frac{E_K - E_{ir}}{E_K} = 0.997 - 0.014N \tag{68}$$

As for mild steel, the yield stress in the theoretical model is assumed to be the average value between the initial yield stress and the ultimate strength, as follows:

$$\sigma_f = \frac{\sigma_y + \sigma_u}{2} \tag{69}$$

For the metal foam core, the yield stress can be assumed to be equal to plateau stress, i.e.,

$$\sigma_c = \sigma_p \tag{70}$$

In the theoretical model, the material properties and geometric parameters are the same as in the impact tests. Meanwhile, the boundary condition, impact mass, and impact velocity also stay the same as in the impact tests. In the repeated-impact tests, the impact mass is 7.884 kg, the impact velocity is 2.12 m/s, and the impact energy is 17.8 J. The geometric parameters and material properties as well as the dimensionless parameters of the MFSB are presented in Table 2.

Table 2. Parameters employed in repeated-impact tests and theoretical model.

	Parameter	Symbol	Unit	Value
Geometric Parameters	Beam Length	$2L$	mm	150
	Face Thickness	t	mm	1
	Core Thickness	c	mm	10
	Beam Width	B	mm	30
Material Properties	Face Density	u_1	kg/mm^3	7.8×10^{-6}
	Core Density	u_2	kg/mm^3	5.0×10^{-7}
	Line Density of the Beam	u	kg/mm	2.06×10^{-6}
	Face Yield Stress	σ_f	GPa	0.25
	Core Yield Stress	σ_c	GPa	0.01
	Face Young's Moduli	E_f	GPa	201
	Core Young's Moduli	E_c	GPa	0.42
	Fully Plastic Moment	M_P	kg.mm^2/ms^2	3.00
	Fully Plastic Axial Force	N_P	kg.mm/ms^2	0.60
Energy	Beam Mass	G_B	kg	9.27×10^{-3}
	Impactor Mass	G_S	kg	7.884
	Mass Ratio	$G^* = G_S/G_S$	/	850
	Impact Velocity	V	m/s	2.12
	Impact Energy	E_{K0}	J	17.8
Dimensionless parameters	Yield Stress	$\overline{\sigma}$	/	0.04
	Thickness	$\overline{t}$	/	0.10
	Thickness to length	$\overline{c}$	/	0.133
	Kinetic energy	$E_{K0}{}^*$	/	0.25664

The correlation curves between dimensionless deflections with the impact number obtained from the membrane factor method, the square yield surface, the repeated-impact tests, and numerical simulations are illustrated in Table 3 and Figure 11. It can be observed that the dimensionless deflections predicted by the membrane factor method are very close to those of the impact tests and numerical simulations, and the discrepancies between the above three different methods are very small, which all lie between the results of inscribing and circumscribing yield surface. When the impact number is small, the differences in dimensionless deflections between different methods are relatively small. As the impact number increases, however, the discrepancy between dimensionless deflections obtained by the membrane force method, the square yield surface, repeated-impact tests, and numerical simulations increase gradually.

Table 3. Dimensionless permanent deflections of MFSB.

Impact Number	MFM	Circumscribing	Inscribing	Numerical [31]	Test [30]
1	0.268	0.236	0.405	0.332	0.321
2	0.530	0.425	0.703	0.613	0.572
3	0.781	0.586	0.947	0.845	0.788
4	1.014	0.728	1.157	1.035	0.982
5	1.208	0.856	1.344	1.195	1.154
6	1.374	0.972	1.513	1.333	1.301
7	1.520	1.080	1.667	1.456	1.420
8	1.652	1.179	1.809	1.567	1.529
9	1.773	1.272	1.941	1.669	1.608
10	1.886	1.359	2.065	1.765	1.671

Some assumptions have been made in the theoretical model, resulting in differences between the results of the theoretical analysis and those of the impact tests. On the one hand, the influence of friction has not been considered in the theoretical model, which would consume energy in the impact tests. On the other hand, in the theoretical model, the yield stress of the face sheet is assumed to be the average value of the initial yield stress and ultimate stress. While in the impact tests, the yield stress of the face sheet would change as the strain increases. Thus, when the deflection is small, the yield stress of the theoretical model would be larger than impact tests, but it would be smaller when the deflection is larger. Therefore, when the impact number is small, the permanent deflections obtained by the membrane factor method are smaller than those of impact tests and numerical simulations. By contrast, when the impact number is larger, the permanent deflections predicted by the membrane factor method are larger than those of impact tests and numerical simulations. The deformation profiles of the MFSB for the first, fifth, and tenth impact are presented in Figure 12. The deformation profiles obtained from MFM lie between those of circumscribing yield surface and inscribing yield surface, which are very close to those of impact tests and numerical simulations.

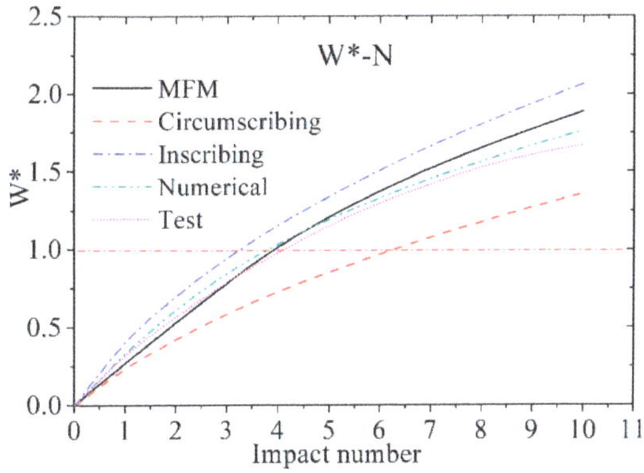

Figure 11. Correlation curves between dimensionless deflections with impact number.

Based on the above comparisons and analyses, it is confirmed that the predictions of the membrane factor method agree well with those of the corresponding impact tests and numerical simulations, indicating that the theoretical method proposed in this paper has high accuracy in analyzing the plastic responses of MFSBs suffering from repeated impacts.

4.2. Dynamic Responses of MFSBs

The correlation between dimensionless permanent deflections and the impact number is shown in Figure 13. As the impact number increases, the dimensionless deflection increases gradually, while the increment decreases. Meanwhile, the deflections derived from the MFM are between the upper and lower solutions obtained by the method based on the square yield surface. When a deflection is smaller than the total thickness of the sandwich beam ($W^* \leq 1$), both the moment and the membrane force work and consume energy. Therefore, the results obtained by the MFM are much closer to those of the inscribing yield surface. When deflection exceeds the total thickness of the sandwich beams ($W^* > 1$), the sandwich beams are in the membrane state and behave like a stretched plastic sting, and only the membrane force works. Thus, the impact number increases, and the results obtained by the MFM become much closer to those of circumscribing yield surface. The correlation between the dimensionless impact force and the impact number is shown in Figure 14. The dimensionless impact force predicted by the membrane factor method is between the upper and lower solutions based on the square yield surface. The initial collapse force solved by the membrane factor method is equal to that of the method based on circumscribing the yield surface. With the increase in the impact number, the dimensionless impact force predicted by the membrane factor method approaches the result obtained by the circumscribing yield surface.

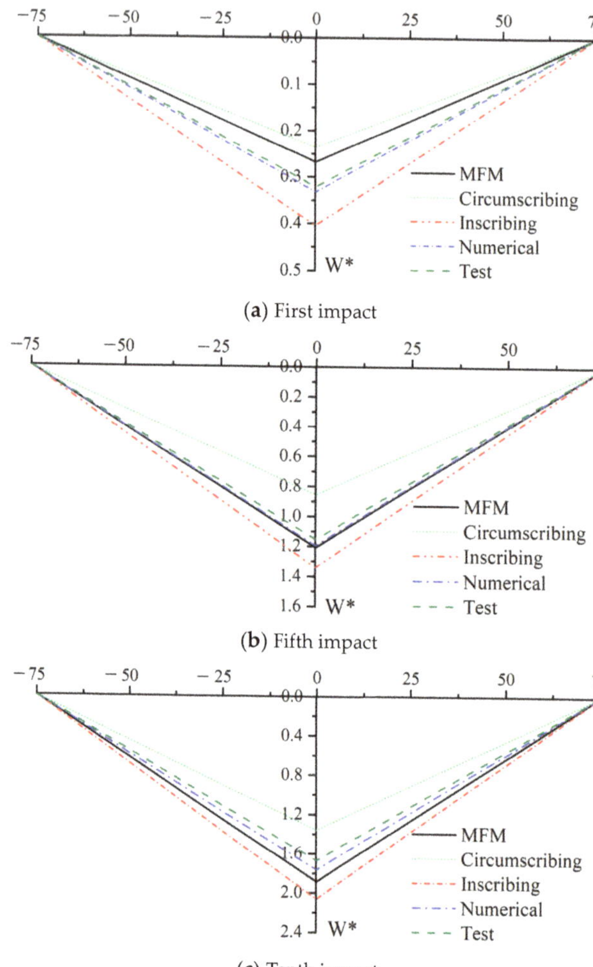

(**a**) First impact

(**b**) Fifth impact

(**c**) Tenth impact

Figure 12. Deformation profile of the MFSB for different impact numbers.

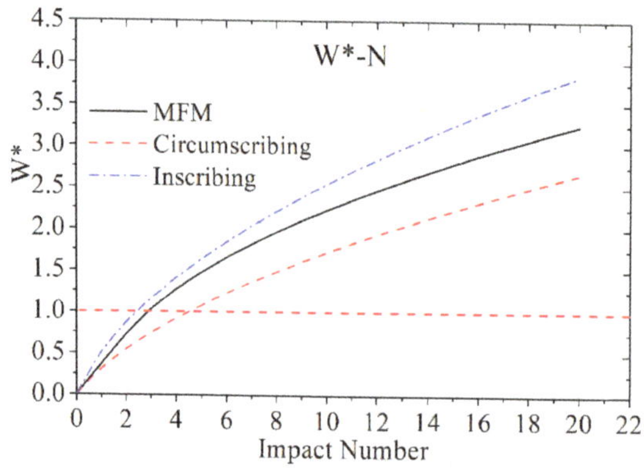

Figure 13. Dimensionless permanent deflection vs. impact number.

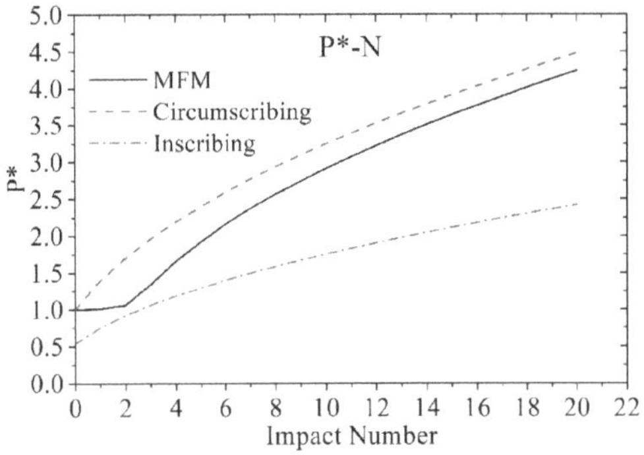

Figure 14. Dimensionless impact force vs. impact number.

The correlations between the dimensionless impact force and dimensionless deflection are presented in Figure 15. As for the results obtained from the square yield surface, the dimensionless impact force increases linearly with the increase in dimensionless deflection. Meanwhile, the results derived from the MFM lie between those of the inscribing and circumscribing yield surface. When $W^* \geq 1$, the slopes of the membrane factor method solutions are equal to those of the square yield surface solutions. The correlation between dimensionless deflection and dimensionless impact energy is displayed in Figure 16. The results obtained from the MFM are between the solutions of the upper and lower yield surfaces, and as impact energy increases, the results of the MFM become closer to those of the circumscribing yield surface.

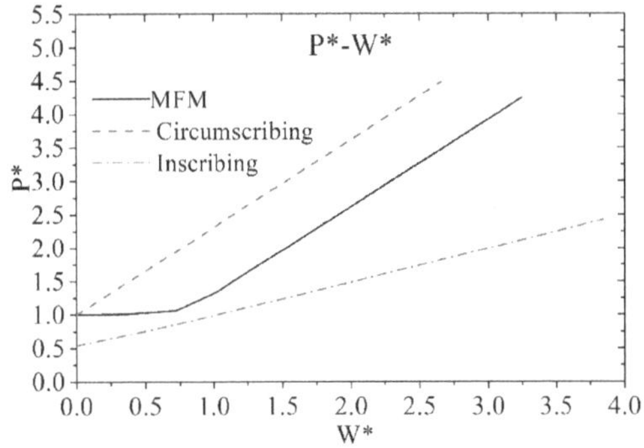

Figure 15. Dimensionless impact force vs. deflection.

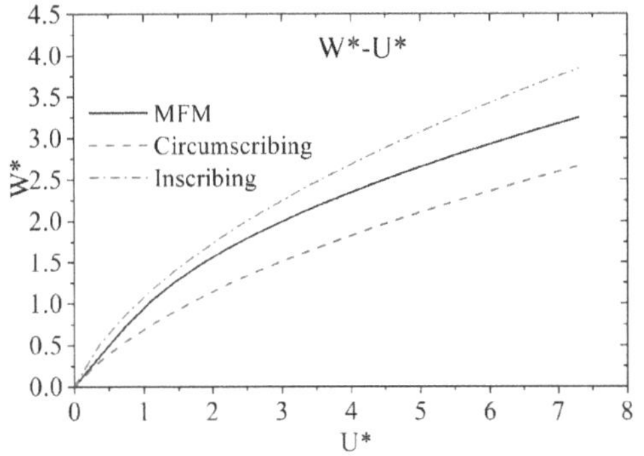

Figure 16. Dimensionless deflection vs. impact energy.

Through the above analysis, it can be seen that the results of the membrane factor method are within the upper and lower limits obtained by the square yield. When the impact number is small, the deflection obtained by the membrane factor method is close to the result of the inscribing yield surface solution. However, when the impact number is large, the deflection value obtained by the membrane factor method is close to the result of the circumscribing yield surface solution.

4.3. Effect of Core Strength

The core strength ratio (CSR) is defined as variation that can be used to study the effect of core strength on the dynamic responses of MFSBs, and seven cases are selected, as illustrated in Table 4, namely, CSR-0.01, CSR-0.05, CSR-0.1, CSR-0.2, CSR-0.3, CSR-0.5, and CSR-1. Most of the geometric parameters and energy parameters remain the same, as listed in Table 1, and only some parameters related to core yield strength are changed accordingly.

The effect of core strength on the dynamic responses of MFSBs is shown in Figure 15. Core strength has a significant effect on the permanent deflections of MFSBs. It can be seen from Figure 17a, as for different CSRs, that the trends of the W^*-N curves are almost the same, i.e., W^* increases gradually with the increasing impact number while the increment slope decreases. As the impact number increases, the total energy absorbed by the MFSB increases, and the deformation of the MFSB accumulates gradually, resulting in the effect of the core strength ratio becoming more significant with the increase in impact number. The relationship between dimensionless deflection and core strength ratio is presented in Figure 17b. It can be seen, both for different impact numbers, that the dimensionless permanent deflection decreases with the increase in the core strength ratio in exponential form. Meanwhile, the core strength ratio is smaller, and the effect on dimensionless deflection is much more significant. When the geometric parameters and impact energy as well as face strength remain unchanged, if the core strength ratio increases, both M_P and N_P increase, meaning the load-carrying capacity of the MFSB improves, resulting in a decrease in the dimensionless permanent deflection with the increase in the core strength ratio.

Table 4. Core strength ratio cases (unit system: kg/mm/ms).

Case	σ_c	σ_f	CSR
CSR-0.01	0.0025	0.25	0.01
CSR-0.05	0.0125	0.25	0.05
CSR-0.1	0.025	0.25	0.1
CSR-0.2	0.050	0.25	0.2
CSR-0.3	0.075	0.25	0.3
CSR-0.5	0.125	0.25	0.5
CSR-1	0.250	0.25	1

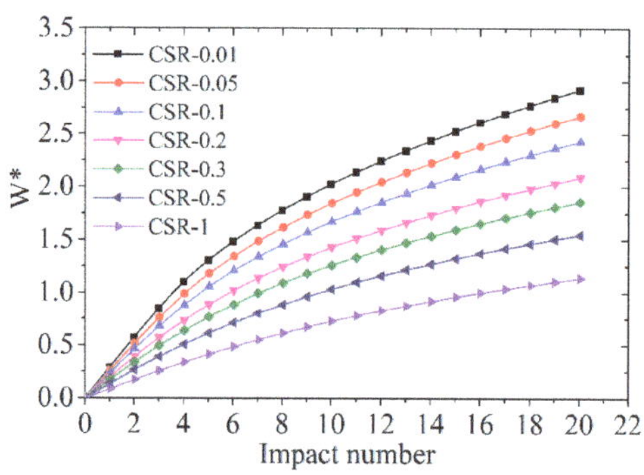

(**a**) Correlation between dimensionless deflection and impact number

Figure 17. *Cont.*

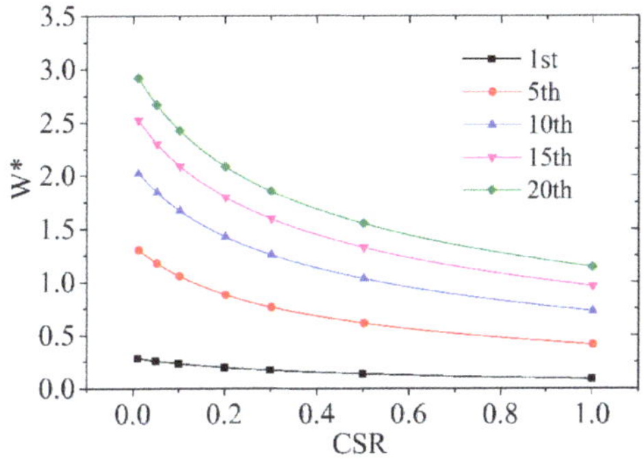

(**b**) Correlation between dimensionless deflection and core strength ratio

Figure 17. Effect of core strength on dynamic responses of MFSB.

In conclusion, the dimensionless permanent deflection of the MFSB is sensitive to the core strength ratio, and as the impact number increases, the dimensionless permanent deflection decreases gradually. Meanwhile, the influence of core strength becomes more significant as the impact number increases.

4.4. Effect of Face Thickness

In order to determine the effect of face thickness on the dynamic responses of MFSB, nine cases are selected, as illustrated in Table 5, and the face thickness ratio (FTR) is defined as the variation. Most of the geometric parameters and energy parameters remain the same as those listed in Table 1; only some parameters related to face thickness are changed accordingly.

Table 5. Face thickness ratio cases (unit system: kg/mm/ms).

Case	t	c	CSR
FTR-0.01	0.1	10	0.01
FTR-0.02	0.2	10	0.02
FTR-0.05	0.5	10	0.05
FTR-0.08	0.8	10	0.08
FTR-0.1	1.0	10	0.1
FTR-0.12	1.2	10	0.12
FTR-0.15	1.5	10	0.15
FTR-0.18	1.8	10	0.18
FTR-0.2	2.0	10	0.2

The effect of the face thickness ratio on the dynamic responses of the MFSB is shown in Figure 16. From Figure 18a, it can be observed that the dimensionless permanent deflection increases with impact number, while the increment decreases, and for different cases, the trends are very similar. Meanwhile, discrepancies between each case grow as the impact number increases. The correlation between dimensionless deflection and the face thickness ratio is displayed in Figure 18b. As the face thickness ratio increases, the dimensionless permanent deflection decreases gradually in exponential form. Meanwhile, when the face thickness ratio becomes smaller, the effect of face thickness becomes more significant. When the face thickness increases, the load-carrying capacity of the MFSB increases; thus, once subjected to identical impact energy, the permanent deflection decreases.

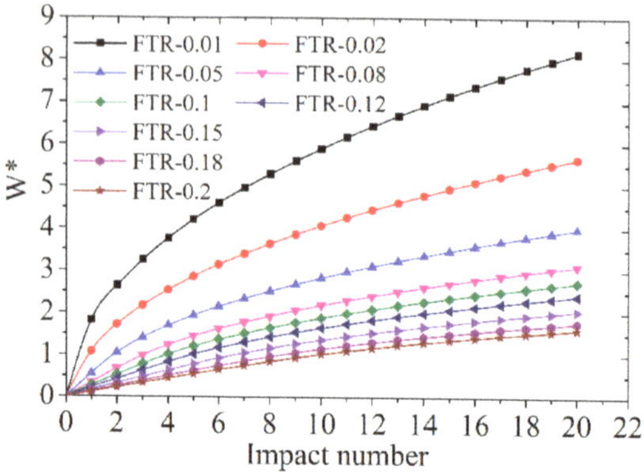

(**a**) Correlation between dimensionless deflection and impact number

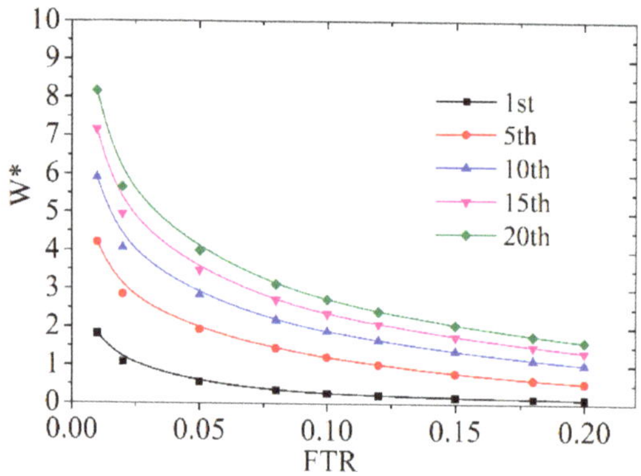

(**b**) Correlation between dimensionless deflection and face thickness ratio

Figure 18. Effect of face thickness on dynamic responses of MFSB.

5. Conclusions

In this paper, a theoretical model based on a perfect rigid-plasticity assumption is established, and the membrane factor method is employed to derive the dynamic plastic responses of MFSBs under repeated mass impact. Meanwhile, the rebound effect of MFSBs is taken into account based on the results of impact tests. In addition, the yield criterion based on the square yield surface is employed to obtain upper and lower bounds for plastic responses. In addition, the results obtained from the membrane factor method are compared with those from impact tests and numerical simulations to verify the accuracy of the proposed theoretical model. In addition, the effects of core strength ratio and face thickness ratio on the deflection responses of MFSBs are analyzed based on the MFM. Thus, the following conclusions can be drawn:

(1) The dimensionless permanent deflections derived from the membrane factor method agree well with those of repeated-impact tests and numerical simulations, indicating that the proposed method is of high accuracy for predicting the plastic responses of MFSBs suffering from repeated impacts.

(2) The process of solving the equation based on the square yield surface is much simpler, but the result of the membrane factor method is more accurate, which is within the bound solution range of the square yield surface.

(3) The dimensionless permanent deflection of MFSBs is sensitive to the core strength ratio, and as the core strength ratio increases, the dimensionless permanent deflection

decreases gradually in exponential form. Meanwhile, the influence of core strength becomes more significant as the impact number increases.

(4) The face thickness ratio has a visible effect on the dimensionless permanent deflection of MFSBs, which is more significant in the case of small face thickness ratios. As the face thickness ratio increases, the dimensionless permanent deflection decreases gradually in exponential form. Meanwhile, the influence of the face thickness ratio becomes more significant as the impact number increases.

Though the proposed theoretical method based on the MFM can predict the plastic responses of MFSBs subjected to repeated impacts with high accuracy, the corresponding theoretical model for metal foam sandwich plates (MFSPs) has not been established due to the complex structural forms and force states of the MFSPs. Thus, future investigations must establish the theoretical model of MFSPs suffering from repeated impacts and employ the MFM to obtain accurate solutions for plastic responses.

Author Contributions: Conceptualization, K.G.; methodology, L.Z.; formal analysis and investigation, B.X. and M.M.; writing—original draft preparation, K.G.; writing—review and editing, W.C. All authors have read and agreed to the published version of the manuscript.

Funding: This work was financially supported by the project of National Natural Science Foundation of China (Grant number: 12202328, 12172265), and the project of Key Laboratory of Impact and Safety Engineering (Ningbo University), Ministry of Education (Grant number: CJ202203).

Institutional Review Board Statement: Not applicable.

Informed Consent Statement: Not applicable.

Data Availability Statement: Not applicable.

Acknowledgments: The authors gratefully acknowledge the support from the project of National Natural Science Foundation of China (Grant number: 12202328, 12172265), and the project of Key Laboratory of Impact and Safety Engineering (Ningbo University), Ministry of Education (Grant number: CJ202203).

Conflicts of Interest: The authors declare no conflict of interest.

References

1. Zhu, L.; Faulkner, D. Damage estimate for plating of ships and platforms under repeated impacts. *Mar. Struct.* **1996**, *9*, 697–720. [CrossRef]
2. Truong, D.D.; Jung, H.J.; Shin, H.K.; Cho, S.R. Response of low-temperature steel beams subjected to single and repeated lateral impacts. *Int. J. Nav. Arch. Ocean* **2018**, *10*, 670–682. [CrossRef]
3. Truong, D.D.; Shin, H.K.; Cho, S.R. Repeated lateral impacts on steel grillage structures at room and sub-zero temperatures. *Int. J. Impact Eng.* **2018**, *113*, 40–53. [CrossRef]
4. Zhu, L.; Shi, S.; Jones, N. Dynamic response of stiffened plates under repeated impacts. *Int. J. Impact Eng.* **2018**, *117*, 113–122. [CrossRef]
5. Zeng, Y.; Chen, H.; Yu, R.; Shen, Z.; Yu, Z.; Liu, J. Experimental research on dynamic behavior of circular mild steel plates with surface cracks subjected to repeated impacts in low temperature. *Shock Vib.* **2020**, *2020*, 3713709. [CrossRef]
6. Jones, N. Pseudo-shakedown phenomenon for the mass impact loading of plating. *Int. J. Impact Eng.* **2014**, *65*, 33–39. [CrossRef]
7. He, X.; Soares, C.G. Experimental study on the dynamic behavior of beams under repeated impacts. *Int. J. Impact Eng.* **2021**, *147*, 103724. [CrossRef]
8. He, X.; Soares, C.G. Numerical study on the pseudo-shakedown of beams under repeated impacts. *Ocean Eng.* **2021**, *242*, 110137. [CrossRef]
9. He, X.; Garbatov, Y.; Soares, C.G. Analysis of pseudo-shakedown of rectangular plates under repeated impacts. *Ocean Eng.* **2022**, *265*, 112609. [CrossRef]
10. He, X.; Garbatov, Y.; Soares, C.G. Pseudo-shakedown of rectangular plates under repeated impacts. *Mar. Struct.* **2022**, *85*, 103258. [CrossRef]
11. Cai, W.; Zhu, L.; Qian, X. Dynamic responses of steel plates under repeated ice impacts. *Int. J. Impact Eng.* **2022**, *162*, 104129. [CrossRef]
12. Zhang, Y.; Li, Y.G.; Guo, K.L.; Zhu, L. Dynamic mechanical behaviour and energy absorption of aluminium honeycomb sandwich panels under repeated impact loads. *Ocean Eng.* **2021**, *209*, 108344. [CrossRef]
13. Zeng, Y.; Liu, J.X.; Zhao, Y.J. Experimental research on the repeated impacts behavior of aluminium corrugated-core sandwich structures. *Shock Vib.* **2022**, *2022*, 4644029.

14. Palomba, G.; Epasto, G.; Sutherland, L.; Crupi, V. Aluminium honeycomb sandwich as a design alternative for lightweight marine structures. *Ships Offshore Struct.* **2022**, *17*, 2355–2366. [CrossRef]
15. Mozafari, H.; Distefano, F.; Epasto, G.; Gu, L.; Linul, E.; Crupi, V. Design of an Innovative Hybrid Sandwich Protective Device for Offshore Structures. *J. Mar. Sci. Eng.* **2022**, *10*, 1385. [CrossRef]
16. Lin, H.; Han, C.; Yang, L.; Karampour, H.; Luan, H.; Han, P.; Xu, H.; Zhang, S. Dynamic Performance and Crashworthiness Assessment of Honeycomb Reinforced Tubular Pipe in the Jacket Platform under Ship Collision. *J. Mar. Sci. Eng.* **2022**, *10*, 1194. [CrossRef]
17. Corigliano, P.; Palomba, G.; Crupi, V.; Garbatov, Y. Stress–Strain Assessment of Honeycomb Sandwich Panel Subjected to Uniaxial Compressive Load. *J. Mar. Sci. Eng.* **2023**, *11*, 365. [CrossRef]
18. Guo, H.Y.; Zhang, J.X. Expansion of sandwich tubes with metal foam core under axial compression. *J. Appl. Mech.* **2023**, *90*, 051008. [CrossRef]
19. Jing, L.; Wang, Z.; Ning, J.; Zhao, L.M. The dynamic response of sandwich beams with open-cell metal foam cores. *Compos. Part B-Eng.* **2011**, *42*, 1–10. [CrossRef]
20. Qin, Q.H.; Wang, T.J. A theoretical analysis of the dynamic response of metallic sandwich beam under impulsive loading. *Eur. J. Mech. A-Solid* **2009**, *28*, 1014–1025. [CrossRef]
21. Qin, Q.H.; Wang, T.J. Low-velocity heavy-mass impact response of slender metal foam core sandwich beam. *Compos. Struct.* **2011**, *93*, 1526–1537. [CrossRef]
22. Qin, Q.H.; Wang, T.J. Plastic Analysis of Metal Foam Core Sandwich Beam Transversely Loaded by a Flat Punch: Combined Local Denting and Overall Deformation. *Int. J. Appl. Mech.* **2012**, *79*, 041010. [CrossRef]
23. Qin, Q.H.; Wang, T.J. Low-velocity impact response of fully clamped metal foam core sandwich beam incorporating local denting effect. *Compos. Struct.* **2013**, *96*, 346–356. [CrossRef]
24. Asemi, K.; Salami, S.J. A study on low velocity impact response of FGM rectangular plates with 3D elasticity based graded finite element modeling. *J. Theor. Appl. Mech.* **2015**, *53*, 859–872. [CrossRef]
25. Zafarmand, H.; Salehi, M.; Asemi, K. Three dimensional free vibration and transient analysis of two directional functionally graded thick cylindrical panels under impact loading. *Lat. Am. J. Solids Struct.* **2015**, *12*, 205–225. [CrossRef]
26. Khatounabadia, M.; Jafariaand, M.; Asemi, K. Low-velocity impact analysis of functionally graded porous circular plate reinforced with graphene platelets. *Wave Random Complex Media* **2022**, *6*, 2091182. [CrossRef]
27. Zhang, W.; Qin, Q.H.; Li, K.; Li, J.F.; Wang, Q. Effect of stepwise gradient on dynamic failure of composite sandwich beams with metal foam core subject to low-velocity impact. *Int. J. Solids Struct.* **2021**, *228*, 111125. [CrossRef]
28. Fang, B.; Huang, W.; Xu, H.; Jiang, C.; Liu, J. High-velocity impact resistance of stepwise gradient sandwich beams with metal foam cores. *Thin Wall Struct.* **2022**, *181*, 110054. [CrossRef]
29. Zhou, X.F.; Jing, L. Large deflection response of sandwich beams with layered-gradient foam cores subjected to low-velocity impact. *Int. J. Impact Eng.* **2023**, *172*, 104429. [CrossRef]
30. Guo, K.L.; Mu, M.Y.; Zhou, S. Investigation on the dynamic behaviors of aluminum foam sandwich beams subjected to repeated low-velocity impacts. *Metals* **2023**, *13*, 1115. [CrossRef]
31. Guo, K.L.; Mu, M.Y.; Zhou, S.; Zhang, Y.J. Dynamic responses of metal foam sandwich beam under repeated impacts considering impact location and face thickness distribution. *Compos. Part C-Open* **2023**, *11*, 100372. [CrossRef]
32. Qin, Q.H.; Wang, T.J. An analytical solution for the large deflections of a slender sandwich beam with a metallic foam core under transverse loading by a flat punch. *Compos. Struct.* **2009**, *88*, 509–518. [CrossRef]
33. Yu, T.X.; Stronge, W.J. Large deflections of a rigid-plastic beam-on-foundation from impact. *Int. J. Impact Eng.* **1990**, *9*, 115–126. [CrossRef]
34. Yu, T.X.; Chen, F.L. Analysis of large deflection dynamic response of rigid-plastic beams. *J. Eng. Mech.* **1993**, *119*, 1293–1301.
35. Zhu, L.; Tian, L.R.; Chen, F.L.; Yu, T.X. A new equivalent method for complex- shaped pulse loading based on saturation analysis and membrane factor method. *Int. J. Impact Eng.* **2021**, *158*, 104018. [CrossRef]

Journal of
Marine Science and Engineering

MDPI

Article

Study on the Effect of Nodal Configuration on the Mechanical Properties of Hexa-Ligamentous Chiral Honeycombs

Zhiping Wang [1,2], Gang Chen [1,2], Xiaofei Cao [1,2], Wei Chen [3,*], Chun Bao Li [3] and Xiaobin Li [3]

[1] Hubei Key Laboratory of Theory and Application of Advanced Materials Mechanics, Department of Mechanics and Engineering Structure, Wuhan University of Technology, Wuhan 430070, China; xiaofei_cao@126.com (X.C.)
[2] School of Science, Wuhan University of Technology, Wuhan 430063, China
[3] School of Naval Architecture, Ocean and Energy Power Engineering, Wuhan University of Technology, Wuhan 430063, China
* Correspondence: whutcw01@126.com

Abstract: To investigate the effect of nodal configuration on the mechanical properties of hexachiral honeycombs, three specimens, namely, a standard honeycomb, a thickened-node honeycomb, and a filled-node honeycomb, were prepared using 3D-printing technology. Several quasi-static compression tests were performed, which revealed that nodal reinforcement can inhibit nodal aberrations during ligament winding, thus facilitating the "rotational" mechanism and improving the negative Poisson's ratio properties of the honeycomb. Experiments performed using the finite element method showed that nodal reinforcement mainly played a role in the stage of stress rise, and the role of nodal filling was more significant than that of nodal thickening. Low-strain standard honeycombs presented the highest specific absorption energy. However, the specific absorption energy of the filled-node honeycomb and the thickened-node honeycomb exceeded that of the standard honeycomb at a strain of 0.71. The conclusions presented herein can aid in the optimal design of honeycombs and contribute to the design of protective structures.

Keywords: node configuration; platform stress; transformation mode; negative Poisson's ratio

Citation: Wang, Z.; Chen, G.; Cao, X.; Chen, W.; Li, C.B.; Li, X. Study on the Effect of Nodal Configuration on the Mechanical Properties of Hexa-Ligamentous Chiral Honeycombs. *J. Mar. Sci. Eng.* **2023**, *11*, 1692. https://doi.org/10.3390/jmse11091692

Academic Editor: Joško Parunov

Received: 25 July 2023
Revised: 23 August 2023
Accepted: 25 August 2023
Published: 27 August 2023

1. Introduction

Protective structures are crucial for naval ships as they can safeguard the ships against explosions and impact loads. Modern ships are generally equipped with a combination of gas and liquid tank protection on the sides of the ship. How to endow these protective structures with special features, such as light weight, is a popular research topic. Porous materials, such as foam and honeycombs, can be engineered to possess these properties. Yang et al. [1] proposed a macroscopic honeycomb-side protection structure and analyzed the simulation results, concluding that the honeycomb protection structure had good impact resistance. An auxetic honeycomb is a structure that has a negative Poisson's ratio due to its special deformation mechanism [2]. Conventional honeycombs exhibit lateral expansion or contraction behavior under uniaxial compression or tension, while an auxetic honeycomb exhibits a negative Poisson's ratio (NPR). In recent years, research on the mechanical properties and energy absorption characteristics of negative Poisson's ratio materials has been very popular. Such research mainly involves improving energy absorption [3–5] and mechanical properties [6–9].

With the development of manufacturing technology, more and more new honeycomb structures are being used in a wide range of applications, such as for deformed wing structures [10], smart fasteners, replacement vessels, and curved door panels [11]. The cellular configuration of a honeycomb can significantly affect its mechanical properties [12,13], so investigating the relationship between the cellular configuration and the macroscopic mechanical behavior of a honeycomb is of great interest. A large number of cytosolic configurations have been studied by scholars [14–20], such as hexagonal, triangular, square, circular, concave-angled hexagonal, chiral cytosolic, and star-shaped cytosolic configurations. However, these traditional cytosolic configurations cannot fully meet the increasingly complex functional requirements in this field, and there is an urgent need for materials with better mechanical properties, better energy absorption characteristics, and a wider range of Poisson's ratios for specific application scenarios [21]. Therefore, optimized designs based on classical honeycomb structures are constantly being proposed.

The negative Poisson's ratio properties of honeycomb structures are induced by two main mechanisms, namely, an inner concave mechanism and a rotational mechanism [22]. Due to the unique mechanical properties of negative Poisson's ratio honeycombs, they have been the focus of a great deal of research. Prall et al. [23] investigated the in-plane mechanical properties of six-ligament chiral honeycombs using both theoretical analysis and experimental studies. Qi et al. [24] investigated the in-plane response of a four-ligament chiral honeycomb under quasi-static and dynamic loading conditions and derived theoretical equations for platform stresses applicable under quasi-static and dynamic loads. Dong et al. [25] experimentally investigated the effect of the thickness of a concave-angled hexagonal honeycomb on deformation and found that the deformation patterns of thin-walled honeycombs differed significantly from those of thick-walled honeycombs. Zhu et al. [26] established a chiral honeycomb model filled with cell elements in nodes and studied it using simulation methods. The results showed that this configuration improved the honeycomb's overall mechanical properties. Evidently, increasing the thickness of the nodes and filling them had an impact on their overall mechanical behavior. However, research on the effects of node thickening and node filling on mechanical properties is limited and thus deserves more attention.

The aim of this study was to investigate the effect of node thickening and node filling on mechanical properties and to compare them using a standard six-line chiral honeycomb. Quasi-static compression tests were carried out on three honeycombs with different nodes in order to investigate the similarities and differences in their deformation patterns, mechanical properties, and energy absorption characteristics.

2. Specimen Preparation and Testing

2.1. Geometry Description and Fabrication

Three different configurations of six-ligament chiral honeycombs were developed: (a) standard fovea, (b) thickened node, and (c) filled node honeycombs. The standard honeycomb consists of a node connected to six ligaments in a periodic arrangement. It has 10 columns (9 columns in even rows) in the transverse direction and 11 rows in the longitudinal direction, with transverse and vertical lengths of 211.2 mm and 203.6 mm, respectively [27]. The node-thickened honeycomb was developed by thickening the nodes of a standard honeycomb, thus effectively improving the honeycomb's overall impact resistance. The node-filled honeycomb is a standard honeycomb whose nodes are filled with small chiral cells. The six ligaments of the cell are connected to the ligaments at the nodes for transferring the load.

The above three honeycombs were prepared using 3D-printing technology for quasi-static compression tests with a thickness of 25 mm. The honeycomb material employed was 316L stainless steel, and its material parameters are shown in Table 1. The honeycomb specimens were fabricated using selective-laser-melting (SLM) technology, which was widely applied alongside 3D-printing technology because of the wide range of alloys and rapid fabrication enabled by this technique. During the manufacturing process, a high-power laser was used to completely melt the surface layer of the metal powder and process it layer-by-layer until the sample was completed. To avoid defects in the material structure and maximize the performance potential of the material, the specimens were surface-ground. All the fabricated specimens are shown in Figure 1, which shows their good molding quality.

Table 1. Material parameters.

Materials	Density (kg/m^3)	Young's Modulus (GPa)	Poisson's Ratio
316L steel	7980	205	0.3
Yield strength (MPa)	Tensile strength (MPa)		Failure strain
742	1042		0.33

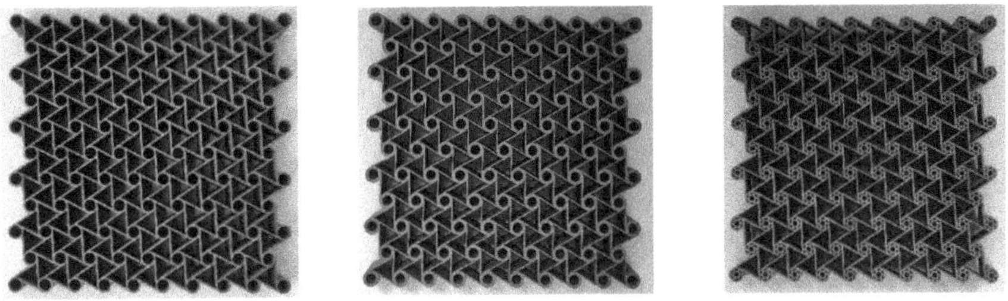

(a) (b) (c)

Figure 1. Honeycomb specimens: (**a**) standard honeycomb; (**b**) thickened-node honeycomb; (**c**) filled-node honeycomb.

The detailed geometric parameters of the honeycombs are shown in Figure 2 and Table 2, where r_1 is the node radius, r_2 is the filled node radius, t_1 is the ligament thickness, t_2 is the node thickness, l_1 is the ligament length, and l_2 is the filled ligament length.

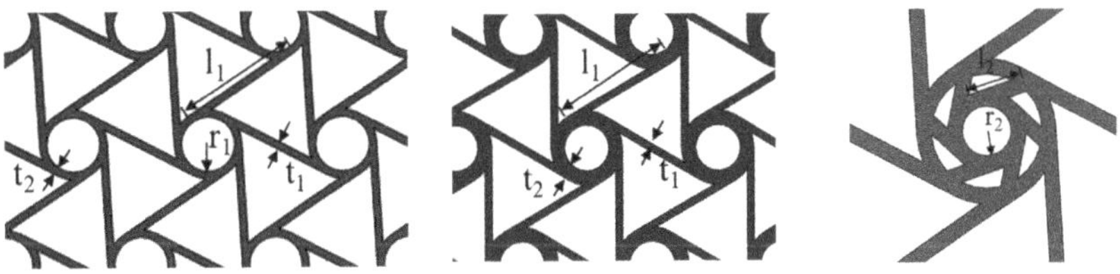

(a) (b) (c)

Figure 2. Honeycomb geometry: (**a**) standard honeycomb; (**b**) thickened-node honeycomb; (**c**) filled-node honeycomb.

Table 2. Honeycombs' geometric parameters.

Honeycomb Configuration	r_1 (mm)	r_2 (mm)	t_1 (mm)	t_2 (mm)	l_1 (mm)	l_2 (mm)
Standard honeycomb	5		1.2	1.2	20	
Node-thickened	5		1.2	2.4	20	
Node-filled	5	2.5	1.2		20	$2.5\sqrt{3}$

2.2. *Quasi-Static Compression Test*

The compressor selected for the test was a WDW-300 (see Figure 3) microcomputer-controlled electronic testing machine, with a maximum test force of 300 kN and a measurement accuracy of ±5%. This testing machine is mainly composed of two parts: an operating system and a testing machine. The operating system controls the compression speed of the testing machine and stores the relevant data during a test, and the testing machine is responsible for the compression or tension at a given time. To investigate the effect of the nodal configurations on the mechanical properties of the six-ligament chiral honeycombs, constant velocity compression tests at 1 mm/min were carried out on the three honeycombs. As a result, the compressive displacement, reaction force, and deformation behavior observed during the tests were recorded.

Figure 3. WDW-300 microcomputer-controlled electronic universal testing machine.

3. Experimental Results

3.1. Macroscopic Compression Deformation

The macroscopic compression deformation processes applied to the three different honeycomb configurations are shown in Figure 4. It can be observed that the three honeycombs exhibit similar overall deformation patterns under quasi-static compression, with each of them exhibiting a double V-shaped necking deformation pattern. This is because the applied stress on the six-ligament chiral honeycomb exceeds the elastic strength limit, causing the individual ligament to wrap around the node and rotate clockwise, with the ligaments on the left and right sides wrapping first and eventually spreading toward the center until all nodes are in full contact.

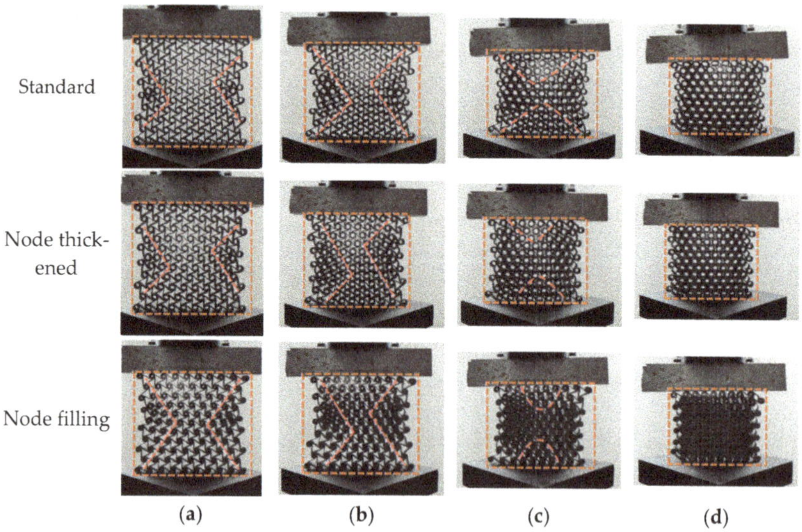

Figure 4. Compression deformation modes of three different honeycomb configurations. (**a**) $\varepsilon = 0.098$. (**b**) $\varepsilon = 0.196$. (**c**) $\varepsilon = 0.295$. (**d**) $\varepsilon = 0.393$.

All three honeycombs showed significant deformation at a nominal strain of 0.098; most of this strain was observed in the lower left and upper right sections of the honeycombs. However, the double V-shaped deformation zone in the standard honeycomb was slightly smaller than that of the other two honeycombs. When the nominal strain reached 0.196, the double V-shaped deformation zone increased significantly in all three honeycombs. Although the deformation area expanded from both sides toward the middle, the increase in the deformation area in the standard honeycomb was not as obvious as that of the other two honeycombs. When the nominal strain reached 0.295, as the ligament continued to expand, the force from the double-V deformation was transmitted from both sides. The standard honeycomb possessed the largest unconsolidated zone and the lowest degree of ligament entanglement. When the specimens were compressed further up to a strain of 0.393, the ligaments of all three honeycombs were fully wound, and the subsequent deformation mode was nodal extrusion. Due to the lower node stiffness of the standard honeycomb, the deformation speed of both the thickened-node honeycomb and the node-filled honeycomb was greater than that of the standard honeycomb. Nodal buckling and a slowed ligament-winding process were observed during compression for the standard honeycomb. The increase in nodal stiffness due to nodal strengthening accelerates the ligament entanglement process under certain circumstances. Moreover, the deformation processes of the honeycomb are closely related to the forms of nodal strengthening.

3.2. Mechanical Properties

The stress–strain curves of the standard honeycomb under quasi-static compression are presented in Figure 5a, which shows three different stages, including the initial linear elasticity stage, the plateau stage, and the stress elevation stage. In the early stage of compression, the whole honeycomb exhibits linear elastic deformation. The ligaments and nodes do not deform significantly, and the stress grows linearly with the strain, which is manifested as the initial linear elasticity stage in the stress–strain curves. As compression proceeds, the ligaments on both sides yield, and plastic hinges form at the node connections, causing the nodes to rotate. They intertwine with each other and propagate this deformation behavior uniformly to neighboring cytosolic elements. In this phase, stress is kept nearly constant to form a platform, which manifests as the platform stage in the curve. A curl-winding behavior of the cellular element leads to the "necking" phenomenon, which is a typical negative Poisson's ratio behavior. As the specimen is compressed further, the ligament is almost completely curled, and the nodes are in close contact with each other. The whole structure shows overall rigidity, for which the stress increases sharply in the curve.

A comparison of the stress–strain curves for the standard honeycomb, the thickened-node honeycomb, and the filled-node honeycomb is given in Figure 5b. It can be seen that the overall trend is consistent for all the structures and composed of three phases. However, the plateau stress and densification strain of the standard honeycomb are significantly greater than those of the thickened-node and filled-node honeycombs. The platform stress of the thickened-node honeycomb was 73.2% that of the standard honeycomb, and its platform strain was 96.1% that of the standard honeycomb; the platform stress of the filled-node honeycomb was 75.4% that of the standard honeycomb, and its platform strain was 91.5% that of the standard honeycomb. The nodes of the standard honeycomb deformed during the ligament entanglement phase to absorb energy, whereas little deformation occurred for the nodes of the thickened-node honeycomb and filled-node honeycomb, resulting in a larger plateau stress in the standard honeycomb. The nodal deformation of the standard honeycomb slowed down the degree of ligament winding; thus, the ligament bending deformation of the standard honeycomb was significantly less than that of the thickened-node and filled-node honeycombs, resulting in the standard honeycomb having the largest plateau strain. Moreover, the difference in densification strain between the thickened-node honeycomb and the filled-node honeycomb is small, which is the reason why the deformation mode for both structures was only ligament curling.

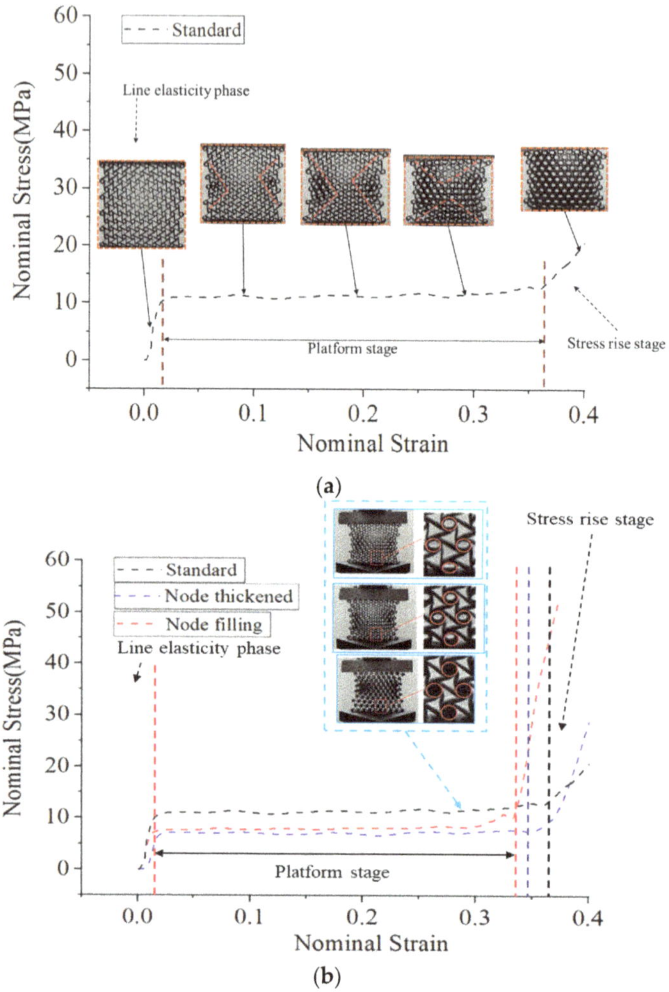

Figure 5. Stress–strain curves of honeycomb structures: (**a**) Standard honeycomb stress–strain curve; (**b**) stress–strain comparison.

3.3. Poisson's Ratio

Poisson's ratio is one of the most important indicators of the mechanical behavior of cellular materials. It describes the deformation characteristics of a material when it is subjected to external loads. The negative Poisson's ratio characteristic of the six-ligament chiral honeycomb mainly occurs in the platform stage, which is mainly investigated in this paper via the experimental results. The cross-sectional strain was calculated by dividing the average value of the transverse deformation displacement at the nodal edges on both sides of the honeycomb by the original transverse length value [28]. The transverse strain ε_x, longitudinal strain ε_y, and Poisson's ratio ν of the honeycomb were calculated as follows. Figure 6 shows the measurement locations of the points for calculating Poisson's ratio.

$$\Delta\overline{x} = \frac{\sum_{i=1}^{6}(A_{ix} + B_{ix})}{6} \tag{1}$$

$$\varepsilon_x = \frac{\Delta \overline{x}}{X} \tag{2}$$

$$\varepsilon_y = \frac{\Delta y}{Y} \tag{3}$$

$$\nu = -\frac{\varepsilon_x}{\varepsilon_y} \tag{4}$$

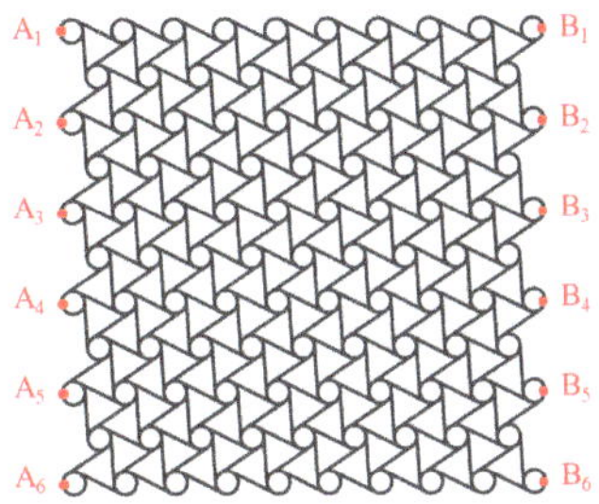

Figure 6. Boundary measurement points for calculating Poisson's ratio.

Figure 7 illustrates the variation in the Poisson's ratio values during the compression process. It can be seen that significant differences exist in the Poisson's ratio values of the three kinds of honeycombs. Overall, the trend was the same for all three structures. Poisson's ratio increases with increasing nominal strain. The negative Poisson's ratio effect was more evident for filled-node honeycombs than thickened-node honeycombs and was weakest for standard honeycombs. Moreover, as shown in Figure 7, the lateral shrinkage of the thickened honeycomb under compression is significantly greater than that of the standard honeycomb (shaded in blue in the figure). It should be noted that the deformation of the filled-node honeycomb was similar to that of the thickened-node honeycomb, so only the thickened-node honeycomb was used in this section for image overlap comparison. The compression displacement of the thickened honeycomb at its nodes was completely caused by the winding deformation of the ligaments, while the compression displacement of the standard honeycomb was induced by both ligament deformation and nodal yielding, thus preserving a certain deformation margin for the ligaments, while the other two honeycombs did not undergo nodal yielding due to their nodal stiffness. The other two types of honeycombs exhibited no nodal yielding due to their increased nodal stiffness. The main reason for this phenomenon is that nodal strengthening increases the stiffness of a node and enhances the 'rotational' deformation mechanism; thus, an increasing negative Poisson's ratio effect is achieved.

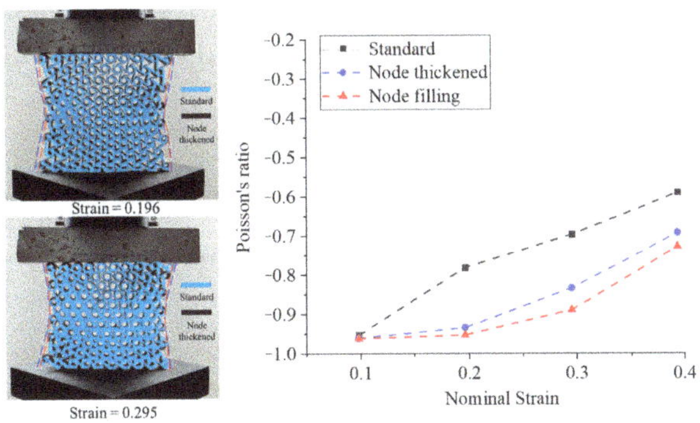

Figure 7. Variation of the Poisson's ratio during the compression of the honeycomb structures.

4. Numerical Simulation

4.1. Numerical Method and Validation

With the development of computer technology and computational mechanics, the use of simulation software for numerical studies has gradually become an accepted research method by mainstream scholars. Due to the high strength of the honeycombs employed in this study, it was difficult to fully compress the honeycombs until densification. Thus, a simulation was conducted, and its results are presented in this section. In this paper, the finite element software ABAQUS was selected to carry out in-plane compression numerical simulations for the standard honeycomb, the thickened-node honeycomb, and the filled-node honeycomb. The material parameters used in the finite element model were an ideal elastic–plastic principal structure with a density of 7980 kg/m^3, a Young's modulus of 205 Gpa, a Poisson's ratio of 0.3, and a yield strength of 742 MPa. The detailed settings of the computational model are shown in Figure 8. Two discrete rigid plates were arranged at the top and bottom positions, with the lower plate fixed and the upper plate compressed vertically at a constant speed of 0.001 m/s. A general degree of contact was defined between the honeycomb itself and the upper and lower rigid plates with a friction coefficient of 0.20. The S4R shell element type was selected. When the cell size is reduced, the equivalent force tends to converge; at the same time, for the consideration of accuracy and computational efficiency, we chose a 0.5 mm mesh for finite element calculation. The overall mesh size of the structure was 0.5 mm, and 20,832 elements were generated. Mesh sensitivity analysis was conducted to ensure the selected mesh size, which is appropriate for calculation, was employed. The out-of-plane displacement of the model was constrained to ensure its in-plane compression state.

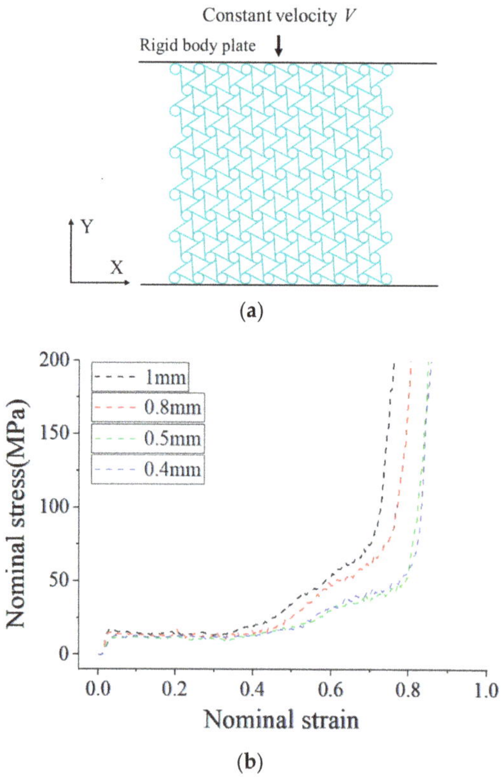

Figure 8. Quasi-static compression finite element model. (**a**) Simulation settings. (**b**) Comparison of simulation results for different mesh sizes.

In Figure 9, the simulation stress–strain curves as well as the compression processes are compared with those of the experimental results. It can be seen that both structures show a similar double-V deformation pattern. Moreover, the stress–strain curves are also almost the same, showing good consistency. The comparison results validate the reliability of the numerical simulation method used in this study.

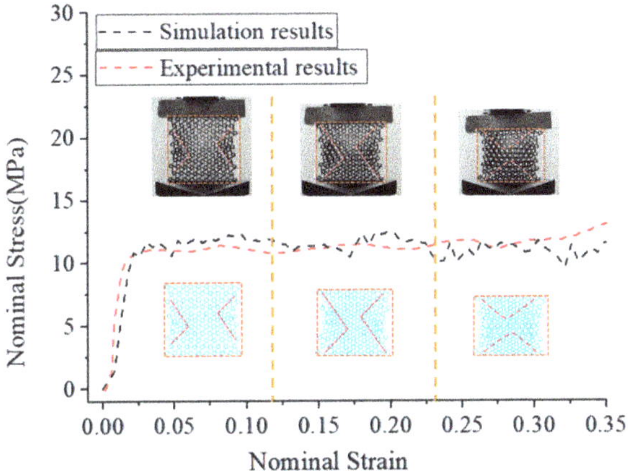

Figure 9. Comparison of test results with simulation results.

4.2. Analysis of Deformation Process

The simulated deformation processes of the standard, thickened-node, and filled-node honeycombs are illustrated in Figures 10, 11, and 12, respectively. The deformation of all three kinds of honeycombs was consistent with the deformation presented in the experimental results. The main structural difference between the three honeycombs pertained to the nodes, and a change in node configuration will affect the deformation of the ligaments when winding. After a certain amount of ligament winding, a node collapse behavior will appear. At this stage, the influence of node configuration is more obvious, where the deformation of the thickened-node honeycomb and filled-node honeycomb is more uniform when compared with that of the standard honeycomb. Moreover, no localized node collapse occurred for the two-node configurations.

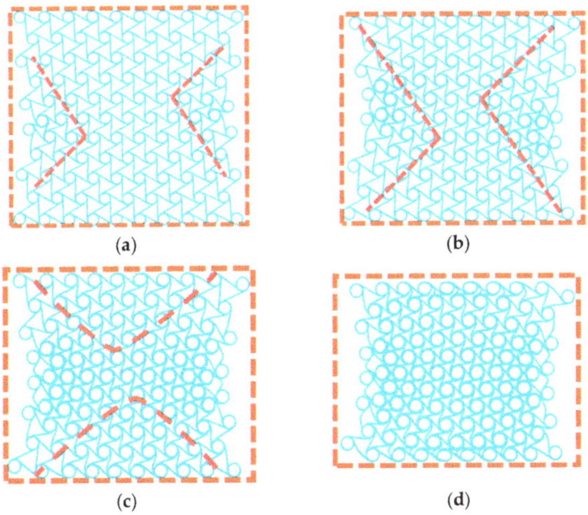

Figure 10. *Cont.*

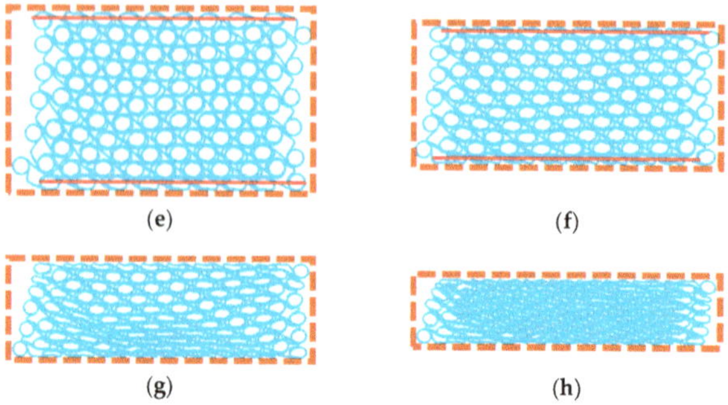

Figure 10. Deformation process of the standard honeycomb under different compression strains: (**a**) ε = 0.098; (**b**) ε = 0.196; (**c**) ε = 0.295; (**d**) ε = 0.393; (**e**) ε = 0.491; (**f**) ε = 0.589; (**g**) ε = 0.688; (**h**) ε = 0.786.

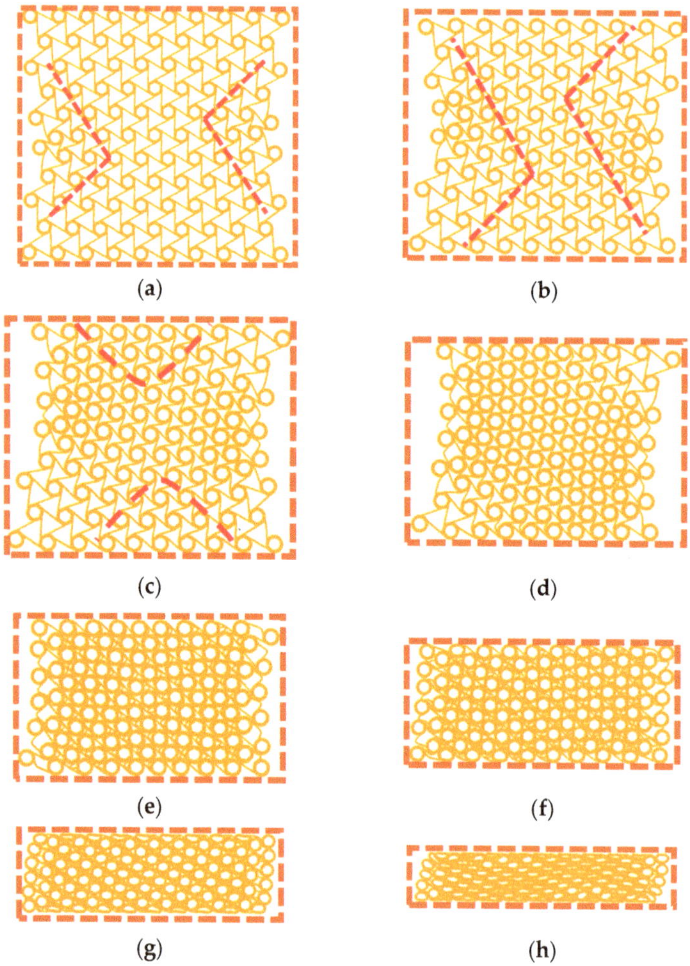

Figure 11. Deformation process of the thickened-node honeycomb under different compression strains: (**a**) ε = 0.098; (**b**) ε = 0.196; (**c**) ε = 0.295; (**d**) ε = 0.393; (**e**) ε = 0.491; (**f**) ε = 0.589; (**g**) ε = 0.688; (**h**) ε = 0.786.

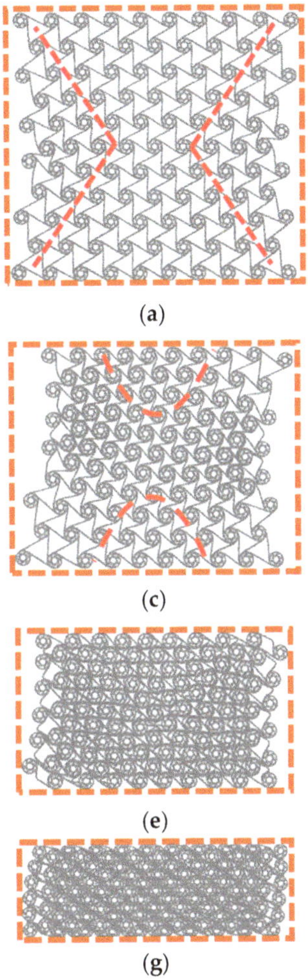

Figure 12. Deformation process of the filled-node honeycomb under different compression strains: (**a**) $\varepsilon = 0.098$; (**b**) $\varepsilon = 0.196$; (**c**) $\varepsilon = 0.295$; (**d**) $\varepsilon = 0.393$; (**e**) $\varepsilon = 0.491$; (**f**) $\varepsilon = 0.589$; (**g**) $\varepsilon = 0.688$; (**h**) $\varepsilon = 0.786$.

When the compression strain was 0.491, the stress of the two node configurations was higher than that of the standard honeycomb. At this stage, while the difference between these two types of honeycombs was not significant, the nodes at the top and bottom of the standard honeycomb were significantly deformed. This was due to the greater nodal stiffness of these two types of honeycomb node configurations, allowing them to resist compression. When the compression strain reached 0.589, the stress of the filled-node honeycomb was higher than that of the thickened-node honeycomb despite the deformation patterns being similar. When the compression strain reached 0.688, the stress of the filled-node honeycomb was significantly greater than that of the thickened-node honeycomb. The nodes of the node-filled honeycomb are almost invisible, while significant deformation can be observed in the nodes at the top and bottom ends of the thickened-node honeycomb. When the specimens were compressed further to a strain of 0.786, all three types of honeycombs were densified, with the filled-node honeycomb presenting the highest stress, the thickened-node honeycomb presenting the next highest stress, and the standard honeycomb presenting the lowest stress. This result indicates that node stiffness can have a significant impact on the compressive performance of cellular structures and that the deformation process varies under different reinforcement methods.

4.3. Mechanical Performance

Figure 13a presents the simulated results regarding quasi-static compression for the standard honeycomb. The curve characteristics of the initial linear elasticity phase and the plateau phase are consistent with the experimental results. In the stress rise stage, the ligament is almost completely curled, and the nodes are in close contact. As the specimen is compressed further, the nodes of the upper and lower sides first undergo compression collapse; then, the overall nodes yield until they are destroyed.

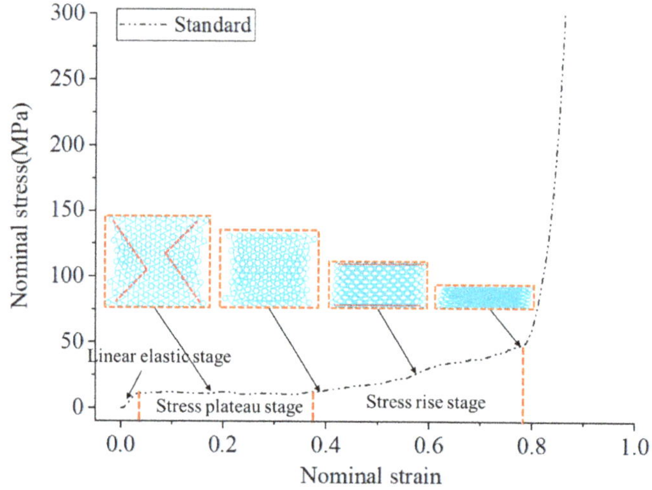

(**a**)

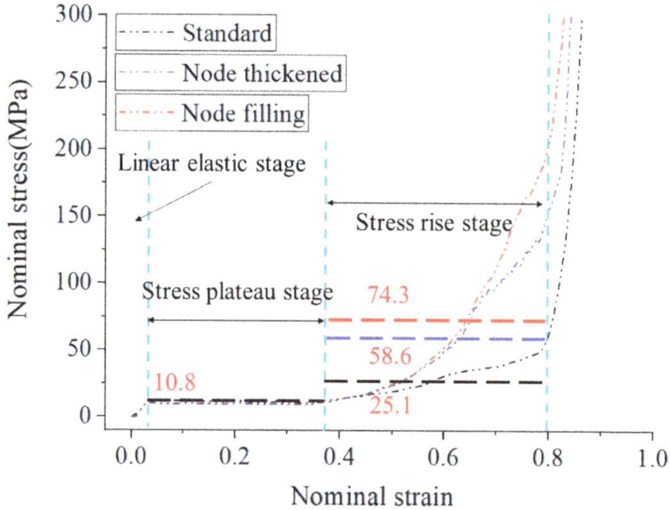

(**b**)

Figure 13. Simulated compression stress–strain curves. (**a**) Simulated compression stress–strain curve for the standard honeycomb. (**b**) Comparison of simulated compression stress–strain curves.

Figure 13b compares the simulated compression stress–strain curves of the three kinds of honeycomb structures. It can be seen that the thickened-node honeycomb mainly starts to exhibit affected mechanical properties after the plateau stage, for which the difference is not significant. In the stress enhancement stage, the average stress of the standard honeycomb is 25.1 MPa, and the average stress of the thickened-node honeycomb is 58.6 MPa. A 133.5% increase in stress was obtained because the thickened nodes are more resistant to yielding during mutual extrusion, and the overall stiffness of the thickened-node honeycomb is higher than that of the standard honeycomb at this stage. The average stress of the filled-node honeycomb was 74.3 MPa, which is 196.0% higher than that of the standard honeycomb. This is because the filled cells inside the nodes play a supporting role, significantly increasing the node stiffness of the filled-node honeycomb.

5. Discussion

5.1. Energy Absorption

Energy absorption characteristics are important properties of honeycomb materials, and they are critical for the design and application of impact-resistant honeycomb structures. Based on analyzing the compression deformation and mechanical properties of the honeycombs, the energy absorption characteristics of standard honeycomb materials, thickened-node honeycomb materials, and filled-node honeycomb materials were compared, and the specific energy absorptions (*SEA*) of the three kinds of honeycombs were investigated. Specific energy absorption is defined as follows

$$SEA = \frac{U}{m} = \frac{\int \sigma(\varepsilon) d\varepsilon}{\rho \overline{\rho}} \quad (5)$$

where U is the total absorbed energy of the honeycomb, m is the mass of the honeycomb, $\sigma(\varepsilon)$ is the nominal stress of the honeycomb, ρ is the density of the base material, and $\overline{\rho}$ is the relative density of the honeycomb.

As shown in Figure 14, the specific absorption energy of the three honeycombs increases linearly during the initial linear elasticity and plateau phases, with the standard honeycomb outperforming the thickened-node and filled-node honeycombs. This is because the relative density of the standard honeycomb is less than that of the thickened-node and filled-node honeycombs and the reinforcement of the node is not involved in deformation energy absorption at this time. However, there are both localized nodal-crushing and ligament-wrapping mechanisms in the standard honeycomb and, essentially, only one ligament wrap mechanism in the thickened-node and filled-node honeycombs. The combination of the above factors leads to the fact that nodal strengthening of the honeycomb node reduces the efficiency of energy absorption in the early stage. After the first platform stage, the effect of nodal reinforcement begins to manifest itself gradually, and the differences between the two node-reinforced honeycombs and the standard honeycomb gradually decrease. Particularly, when the nominal strain is 0.71, equivalent specific energy absorption can be obtained. Afterwards, the *SEA* of the two nodal-reinforced honeycombs gradually exceeds that of the standard honeycomb, and the role of the node starts to become prominent. Moreover, as the nodal stiffness of the node-filled honeycomb is greater than that of the thickened-node honeycomb, the specific energy absorption of the nodal-filled honeycomb under high strain is higher than that of the thickened-node honeycomb.

5.2. Comparison of Node Enhancement Effects

In this paper, to analyze the internal mechanisms that induce the differences between the modified honeycombs and the standard honeycomb, two different nodal reinforcement configurations were designed. The main effect of the strengthening of the nodes is an increase in the stress level in the extrusion phase of the nodes and an enhancement of the negative Poisson's ratio effect of the whole structure.

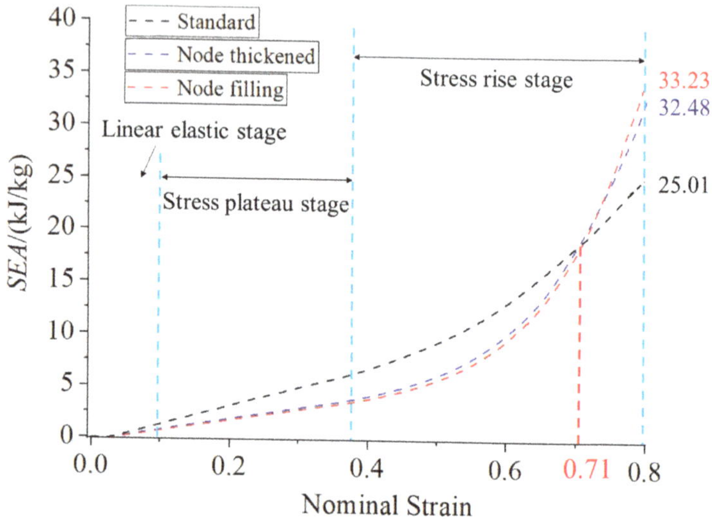

Figure 14. Specific energy absorption curves of the three different kinds of honeycombs.

Zhu et al. [26] designed a honeycomb configuration based on the definition of a multilevel honeycomb (see Figure 15), whose main feature was that the cytokinetic elements in the nodes of the base structure were filled. The difference in the mechanical properties between the proposed structure and the base structure under quasi-static compression was investigated using numerical simulation methods. It was observed that the filling of the nodes with cytosolic elements can reduce the distortion of the nodes during the ligament-winding process, which can help to promote the "rotation" mechanism and strengthen the overall negative Poisson's ratio effect.

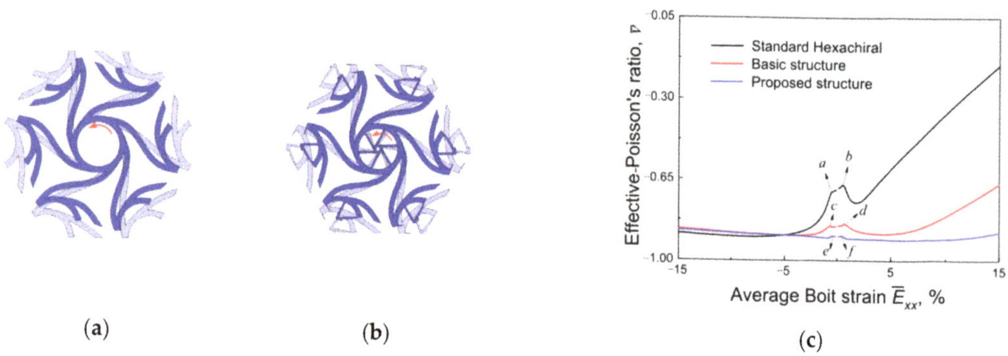

Figure 15. The hexachiral structure proposed by Zhu [26], and a comparison of the effects on the Poisson's ratio. (**a**) Basic structure. (**b**) Proposed structure. (**c**) Effect on Poisson's ratio under average strain for different hexachiral structures.

In this paper, the overall negative Poisson's ratio characteristics of a thickened-node honeycomb and a filled-node honeycomb were compared. It was found that both node-strengthening methods improve the overall negative Poisson's ratio characteristics. The reason for this improvement is that a greater stiffness of node configurations reduces the distortion of the nodes in the ligament-winding stage and promotes the rotational mechanism of chiral honeycombs (see Figure 16). Moreover, when compared with the thickened-node honeycomb, the internal cytosol of the filled-node honeycomb is more connected to the six ligaments on the node, and the inhibition of the node's distortion is better, leading to a greater improvement in the negative Poisson's ratio. This suggests that the negative Poisson's ratio performance of chiral honeycombs can also be improved via nodal thickening in addition to the node filling.

(a) (b) (c)

Figure 16. Comparison of cytosolic deformation of the three kinds of honeycomb structures. (**a**) Standard. (**b**) Node-thickened. (**c**) Node-filled.

6. Conclusions

To investigate the effect of node configuration on the six-ligament honeycomb, an in-plane compression test and its numerical simulation were conducted to study the macroscopic deformation, mechanical properties, and energy absorption characteristics of a standard honeycomb, a thickened-node honeycomb, and a filled-node honeycomb. The main conclusions drawn are as follows:

(1) Nodal thickening can improve the negative Poisson's ratio characteristics of honeycombs. The thickened-node honeycomb presented a greater wall thickness than that of the standard honeycomb, and this property suppressed the distortion of the node in the ligament-winding stage. Meanwhile, the cell elements inside the node played a supporting role for the nodes, preventing their distortion, of the filled-node honeycomb.

(2) Mechanical behaviors of the three kinds of honeycombs included an initial linear elasticity stage, a plateau stage, and a stress rise stage. The effect of nodal differences on deformation was reflected in both the plateau stage and the stress increase stage. The effect of node differences on the stress state was mainly reflected in the stress rise stage. The stress enhancement of the filled-node honeycomb was more obvious than that of the thickened-node honeycomb.

(3) The energy absorption of both the thickened-node honeycomb and the nodal-filled honeycombs was weaker than that of the standard honeycomb in the inelastic phase and platform phase and exceeded that of the standard honeycomb in the stress rise phase. This is because the strengthening of the nodes began to play a role in the stress rise stage and had no significant effect before that, while the mass rose due to the strengthening of the nodes leading to a decrease in the specific energy absorption.

(4) Through experimental and simulation studies, it was determined that the filled-node honeycomb had better energy-absorbing properties and more prominent negative Poisson's ratio characteristics at high strains, granting it better application prospects in relation to its lightweight and impact-resistant structure and greater significance for the design of the new protective structures for ships and marine structures.

Author Contributions: Z.W.: Investigation, Validation, Writing—original draft, Formal analysis. G.C. and W.C.: Supervision, Conceptualization, Methodology, Software, Validation, Formal analysis, Writing—original draft. X.C. and C.B.L.: Writing—review and editing, Validation. X.L.: Writing—review & editing, Supervision. All authors have read and agreed to the published version of the manuscript.

Funding: This research was funded by National Natural Science Foundation of China (52201334; 51979213).

Institutional Review Board Statement: Not applicable.

Informed Consent Statement: Not applicable.

Data Availability Statement: The authors declare that the data presented in this study are available on request.

Conflicts of Interest: The authors declare no conflict of interest.

References

1. Yang, D.-Q.; Ma, T.; Zhang, G.-L. A Novel Auxetic Broadside Defensive Structure for Naval Ships. *Explos. Shock Wave* **2015**, *35*, 243–248.
2. Lakes, R. Foam Structures with a Negative Poisson's Ratio. *Science* **1987**, *235*, 1038–1040. [CrossRef]
3. Zhang, J.; Lu, G.; You, Z. Large Deformation and Energy Absorption of Additively Manufactured Auxetic Materials and Structures: A Review. *Compos. Part B Eng.* **2020**, *201*, 108340. [CrossRef]
4. Novak, N.; Hokamoto, K.; Vesenjak, M.; Ren, Z. Mechanical Behaviour of Auxetic Cellular Structures Built from Inverted Tetrapods at High Strain Rates. *Int. J. Impact Eng.* **2018**, *122*, 83–90. [CrossRef]
5. Wang, Q.; Li, Z.; Zhang, Y. Ultra-Low Density Architectured Metamaterial with Superior Mechanical Properties and Energy Absorption Capability. *Compos. Part B Eng.* **2020**, *202*, 108379. [CrossRef]
6. Mousanezhad, D.; Haghpanah, B.; Ghosh, R.; Hamouda, A.M.; Nayeb-Hashemi, H.; Vaziri, A. Elastic Properties of Chiral, Anti-Chiral, and Hierarchical Honeycombs: A Simple Energy-Based Approach. *Theor. Appl. Mech. Lett.* **2016**, *6*, 81–96. [CrossRef]
7. Allen, T.; Hewage, T.; Newton-Mann, C.; Wang, W.; Duncan, O.; Alderson, A. Fabrication of Auxetic Foam Sheets for Sports Applications. *Phys. Status Solidi B* **2017**, 1700596. [CrossRef]
8. Miller, W.; Smith, C.W.; Evans, K.E. Honeycomb Cores with Enhanced Buckling Strength. *Compos. Struct.* **2011**, *93*, 1072–1077. [CrossRef]
9. Qiao, J.X.; Chen, C.Q. Impact Resistance of Uniform and Functionally Graded Auxetic Double Arrowhead Honeycombs. *Int. J. Impact Eng.* **2015**, *83*, 47–58. [CrossRef]
10. Bettini, P.; Airoldi, A.; Sala, G.; Di Landro, L.; Ruzzene, M.; Spadoni, A. Composite Chiral Structures for Morphing Airfoils: Numerical Analyses and Development of a Manufacturing Process. *Compos. Part B Eng.* **2010**, *41*, 133–147. [CrossRef]
11. Evans, K.E.; Alderson, K.L. Auxetic Materials: The Positive Side of Being Negative. *Eng. Sci. Amp Educ. J.* **2000**, *9*, 148–154.
12. Qiu, X.M.; Zhang, J.; Yu, T.X. Collapse of Periodic Planar Lattices under Uniaxial Compression, Part I: Quasi-Static Strength Predicted by Limit Analysis. *Int. J. Impact Eng.* **2009**, *36*, 1223–1230.
13. Qiu, X.M.; Zhang, J.; Yu, T.X. Collapse of Periodic Planar Lattices under Uniaxial Compression, Part II: Dynamic Crushing Based on Finite Element Simulation. *Int. J. Impact Eng.* **2009**, *36*, 1231–1241.
14. Ruan, D.; Lu, G.; Wang, B.; Yu, T.X. In-Plane Dynamic Crushing of Honeycombs—A Finite Element Study. *Int. J. Impact Eng.* **2003**, *28*, 161–182.
15. Wang, A.J.; Mcdowell, D.L. In-Plane Stiffness and Yield Strength of Periodic Metal Honeycombs. *J. Eng. Mater. Technol.* **2004**, *126*, 137–156.
16. Wang, Z.; Liu, J.; Yao, S. On Folding Mechanics of Multi-Cell Thin-Walled Square Tubes. *Compos. Part B Eng.* **2018**, *132*, 17–27.
17. Sun, D.; Zhang, W.; Zhao, Y.; Li, G.; Xing, Y.; Gong, G. In-Plane Crushing and Energy Absorption Performance of Multi-Layer Regularly Arranged Circular Honeycombs. *Compos. Struct.* **2013**, *96*, 726–735.
18. Hu, L.L.; Zhou, M.Z.; Deng, H. Dynamic Crushing Response of Auxetic Honeycombs under Large Deformation: Theoretical Analysis and Numerical Simulation. *Thin-Walled Struct.* **2018**, *131*, 373–384.
19. Hu, L.L.; Luo, Z.R.; Yin, Q.Y. Negative Poisson's Ratio Effect of Re-Entrant Anti-Trichiral Honeycombs under Large Deformation. *Thin-Walled Struct.* **2019**, *141*, 283–292.
20. Wang, H.; Lu, Z.; Yang, Z.; Li, X. A Novel Re-Entrant Auxetic Honeycomb with Enhanced in-Plane Impact Resistance. *Compos. Struct.* **2019**, *208*, 758–770.
21. Qi, C.; Jiang, F.; Yang, S. Advanced Honeycomb Designs for Improving Mechanical Properties: A Review. *Compos. Part B Eng.* **2021**, *227*, 109393.
22. Prawoto, Y. Seeing Auxetic Materials from the Mechanics Point of View: A Structural Review on the Negative Poisson's Ratio. *Comput. Mater. Ence* **2012**, *58*, 140–153.
23. Prall, D.; Lakes, R.S. Properties of a Chiral Honeycomb with a Poisson's Ratio of—1. *Int. J. Mech. Sci.* **1997**, *39*, 305–307.
24. Qi, C.; Jiang, F.; Yu, C.; Yang, S. In-Plane Crushing Response of Tetra-Chiral Honeycombs. *Int. J. Impact Eng.* **2019**, *130*, 247–265.
25. Dong, Z.; Li, Y.; Zhao, T.; Wu, W.; Liang, J. Experimental and Numerical Studies on the Compressive Mechanical Properties of the Metallic Auxetic Reentrant Honeycomb. *Mater. Des.* **2019**, *182*, 108036.
26. Zhu, Y.; Zeng, Z.; Wang, Z.P.; Poh, L.H.; Shao, Y.B. Hierarchical Hexachiral Auxetics for Large Elasto-Plastic Deformation. *Mater. Res. Express* **2019**, *6*, 085701.

27. Lu, Z.; Li, K. In-Plane Dynamic Crushing of Chiral and Anti-Chiral Honeycombs. *J. Vib. Shock* **2017**, *36*, 16–22.
28. Lu, H.; Wang, X.; Chen, T. In-Plane Dynamics Crushing of a Combined Auxetic Honeycomb with Negative Poisson's Ratio and Enhanced Energy Absorption. *Thin-Walled Struct.* **2021**, *160*, 107366.

Journal of
Marine Science and Engineering

MDPI

Article

Investigation of the Energy Absorption Characteristics and Negative Poisson's Ratio Effect of an Improved Star-Shaped Honeycomb

Qianning Li [1,2], Xiaofei Cao [3], Xingxing Wu [4], Wei Chen [1,2,*], Chunbao Li [1,2] and Xiaobin Li [1,2]

1 Key Laboratory of High Performance Ship Technology, Ministry of Education, Wuhan University of Technology, Wuhan 430063, China; lxbmark@163.com (X.L.)
2 School of Naval Architecture, Ocean and Energy Power Engineering, Wuhan University of Technology, Wuhan 430063, China
3 School of Science, Wuhan University of Technology, Wuhan 430063, China
4 China Ship Scientific Research Center, Wuxi 214082, China
* Correspondence: whutcw01@126.com

Abstract: An improved star-shaped honeycomb (ISSH) is a kind of honeycomb structure with excellent performance. The main objective of this study was to provide some ideas for the optimization of the ISSH structure in ships. As a result, 2D-ISSH specimens were fabricated using 3D printing technology, and a quasistatic compression test was carried out to investigate the deformation mode and mechanical properties. The experimental results showed that the 2D-ISSH structure exhibited "V"-shaped and "-"-shaped deformation patterns with a double-platform stress stage. To further utilize the excellent performance of the structure and obtain a better negative Poisson's ratio effect and broader application, based on the properties of the 2D-ISSH specimen, a 3D-ISSH structure was proposed and a finite element simulation was carried out. The simulation results of the 3D-ISSH structure showed different deformation patterns, including "X"- shaped and "-"-shaped patterns. According to the deformation mechanism of typical cells, the stress formula for the 3D-ISSH double platform was derived, and the theoretical results agreed well with the numerical results. The effects of the structural design, materials, and dimensions on the mechanical properties, such as the energy absorption and negative Poisson's ratio, of the ISSH and similar structures were explored. The combined performance of various honeycombs was evaluated from multiple perspectives.

Keywords: 3D star-shaped honeycomb; platform stress; stress–strain curve; specific energy absorption; negative Poisson's ratio

Citation: Li, Q.; Cao, X.; Wu, X.; Chen, W.; Li, C.; Li, X. Investigation of the Energy Absorption Characteristics and Negative Poisson's Ratio Effect of an Improved Star-Shaped Honeycomb. *J. Mar. Sci. Eng.* **2023**, *11*, 1799. https://doi.org/10.3390/jmse11091799

Academic Editor: Joško Parunov

Received: 24 August 2023
Revised: 10 September 2023
Accepted: 12 September 2023
Published: 15 September 2023

1. Introduction

Ships are inevitably exposed to the threat of various shocks, explosions, vibrations, and other intense loads in their normal operating environment. Therefore, the excellent performance of anti-explosion and anti-impact structures is the developmental direction of naval research. The existing design idea of the traditional protection structure is to alternate the arrangement of empty and liquid compartments [1]. Traditional protection design optimization ideas often rely on increasing the weight of the structure to improve the protection performance of the structure. However, a limited number of studies have focused on the design of lightweight protective structures. How to design lightweight protective structures is a popular research topic nowadays.

Lightweight porous structures have attracted much attention owing to their excellent mechanical properties, such as their light weight and high strength, and multifunctional properties, such as their vibration damping and energy absorption. A variety of porous structures have been developed, including dot-matrix, honeycomb, grid, and corrugated structures. Ullah et al. [2] investigated the energy absorption properties of face-centered cubic (FCC), body-centered cubic (BCC), face–body-centered cubic (FCC + BCC), and 3D-Kagome structures under compression and shear loading conditions. Zhang et al. [3] proposed an energy-absorbing enhancement design method for cardio-cubic dot structures. Kooistra et al. [4] proposed a new, multilevel corrugated sandwich structure with ten times the strength of a single-stage corrugated plate of the same mass. Pasqualino et al. [5] conducted simulation calculations and tests on the uniaxial compression of honeycomb sandwich panels in order to promote the integration of aluminum honeycomb sandwich structures into the structural design of ships. Due to their excellent performance, porous structures are used as advanced protective structures in marine fields.

Poisson's ratio is a typical structural characteristic. Generally, the macroscopic Poisson's ratio of a porous structure is positive. However, negative Poisson's ratio honeycombs have a negative Poisson's ratio effect that is not found in common porous structures, which gives them many excellent mechanical properties, such as relatively high load-carrying capacities [6,7], high shear modulus [8], high fracture toughness [9], excellent impact resistance [10,11] and superior specific energy absorption performance [12,13]. Under impact loading, the negative Poisson's ratio honeycomb allows for more material to accumulate near the point of force to absorb more impact energy than other structures, which gives it great application prospects in the field of impact protection. In recent years, many scholars have performed studies on negative Poisson's ratio honeycombs. Many distinctive negative Poisson's ratio structures have been designed, such as concave structures, rotating polygonal structures, chiral structures, lamellar pleated structures, perforated plate structures, and interlocking polygonal structures [14]. Among these structures, the concave structure has been highly developed over the years, and various structural forms with excellent performance have been designed. Gibson et al. [15] proposed a 2D concave hexagonal honeycomb structure in 1982, which was the first negative Poisson's ratio honeycomb structure that was designed. In 1985, Almgren [16] designed a concave hexagonal honeycomb structure and derived the analytical equations for its Young's modulus and Poisson's ratio, which amounted to -1. Larsen et al. [17] pioneered the design of negative Poisson's ratio structures using topology optimization methods and proposed an arrow-shaped honeycomb negative Poisson's ratio material, which was found to have a Poisson's ratio reaching -0.8, according to the theoretical analysis. Theocaris et al. [18] proposed a star-shaped negative Poisson's ratio honeycomb structure, which was widely recognized as a deformation of the traditional concave structure.

Although the traditional honeycomb has excellent performance, it is still incapable of facing certain complex and intense work scenarios. Thus, many scholars have designed various improved honeycombs by introducing other configurations based on traditional honeycombs. Wang et al. [19] combined a double-arrow structure with a conventional star structure to propose a novel tensile-expansive honeycomb, and investigated its in-plane elastic properties and energy absorption characteristics. Wei et al. [20–22] introduced triangles into the traditional star-shaped honeycomb and proposed a novel star–triangle honeycomb, and its deformation behavior at different velocities was investigated. The negative Poisson's ratio effect of the star–triangle honeycomb under pressure in two directions was analyzed using a simulation, and a theoretical model was established to predict its Young's modulus and Poisson's ratio more accurately. The star–triangle honeycomb was also extended to three dimensions, and its deformation mode was investigated by combining experiment and simulation, and the effect of the gradient design on the deformation and energy absorption of the structure was analyzed. Hu et al. [23] constructed an improved star-shaped honeycomb structure by introducing arrow-shaped and square-shaped structures, and by performing finite element simulation calculations. The testing results showed that the specific energy absorption of the improved star-shaped honeycomb is better than that of the star-shaped honeycomb, and the deformation is more stable under low-speed loading. Yang et al. [24] designed a new 3D star structure with a negative Poisson's ratio based on the traditional 2D star structure with a negative Poisson's ratio, and explored the effects of impact velocity and design angle on the dynamic deformation mode and mechanical response of the star model. Li et al. [25] designed a crossed star honeycomb by evolving the horizontal and vertical walls of a conventional star honeycomb into crossed sloping walls and explored the effects of the inclination angle on Poisson's ratio, Young's modulus, and the energy absorption. Hu et al. [26] printed models using 316 L steel and performed material tensile testing and honeycomb compression testing. They investigated the compressive mechanical properties of anti-tetrachiral honeycomb materials with different thickness ratios of the ligament to the cylindrical. They found a variety of deformation modes and obtained critical values for the transformation of the anti-tetrachiral honeycomb deformation modes. Wang et al. [27] printed models using 316 L steel and performed honeycomb compression testing, and also investigated the effect of nodal configuration on the mechanical properties of hexachiral honeycombs.

However, most studies focus on 2D negative Poisson's ratio materials; only a limited number of studies have focused on the 3D negative Poisson's ratio structure. Moreover, these studies are mainly limited to numerical simulations and lack detailed experimental studies and validation. To obtain a better negative Poisson's ratio and energy absorption performance, further playing with the role of structural design and broadening the application scenarios for the structure are needed. Here, an improved star-shaped honeycomb structure is designed, and its deformation pattern and mechanical properties are analyzed using a combination of experiments and finite element simulations. Moreover, there is a discussion of the structure used in this paper and similar structures to evaluate the performance advantages and disadvantages of the honeycomb from multiple perspectives. The main research of this paper is as follows: quasistatic compression tests of the 2D-ISSH are conducted in Section 2 to reveal its deformation pattern and mechanical properties. In Section 3, the validity of the finite element simulation calculation method is verified and finite element simulation calculations are carried out on the 3D-ISSH to reveal its deformation patterns and mechanical properties. In Section 4, the theoretical derivation of the double-platform stresses of the 3D-ISSH under quasistatic compression is presented; the effects of the structural design, materials, and dimensions on the mechanical properties, such as the energy absorption and negative Poisson's ratio, of the ISSH and similar structures are explored. And the combined performance of the various honeycombs is evaluated from multiple perspectives. The main conclusions of this paper are presented in Section 5.

2. An Improved Star-Shaped Honeycomb

2.1. Model Design, Preparation, and Testing

Based on the traditional star-shaped honeycomb, an improved star-shaped honeycomb structure (ISSH) was constructed by introducing arrow-shaped and rectangular structures, using arrow-shaped structures instead of the horizontal walls of the star-shaped honeycomb, and bringing the thin-walled rectangles into contact with the four concave angles. The configuration and geometric parameters of the cell are shown in Figure 1. The main geometric parameters are shown in Table 1. The honeycomb had seven and eight cells along the x and y directions, with lengths of 99.61 mm and 80 mm, respectively. To prevent instability during compression, the length of the honeycomb was extended along the outwards direction of the face by 80 mm. A quasistatic compression test was carried out using a material testing system (MTS), as shown in Figure 2.

Table 1. Geometric and material parameters of the 2D-ISSH.

	H (mm)		α (°)		t (mm)	
2D-ISSH	10		30		0.25	
Materials	Density (kg/m^3)	Young's modulus (GPa)	Poisson's ratio	Yield strength (MPa)	Tensile strength (MPa)	Failure strain
316 L steel	7980	205	0.3	742	1042	0.33

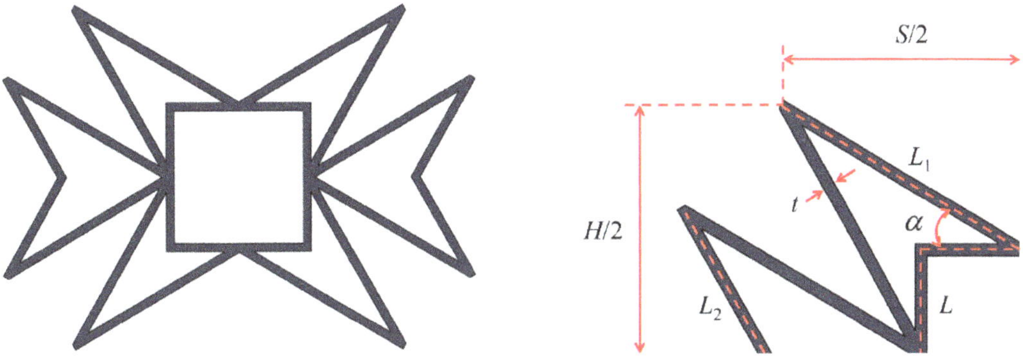

Figure 1. Design of the 2D-ISSH.

(a) (b)

Figure 2. Quasistatic compression test set. (**a**) 2D-ISSH test model. (**b**) Quasistatic compression test.

The tested honeycomb sample was fabricated using a selective laser melting molding process, with monolithic printing and no welding splicing; 316 L steel was selected as the base material. The 316 L steel material parameters are shown in Table 1. During the fabrication process, a high-powered laser was used to completely melt the metal powder layer-by-layer until the samples were completed. In addition, to eliminate some defects in the material structure and to fully utilize the performance potential of the material, the honeycomb was heat-treated.

2.2. Deformation Mode

As shown in Figure 3, the main deformation mode of the 2D-ISSH specimen under quasistatic compression is the rotational deformation of the tilted wall of the cell around the plastic hinge. The rotation of the tilted wall leads to the contraction of the cells, which induces a negative Poisson's ratio effect. Below a compression strain of 0.02, the honeycomb mainly undergoes elastic deformation, and an overall contraction is observed. As compression proceeds, the honeycomb enters the plastic phase, and the lower part of the honeycomb is first deformed. Due to the presence of the intermediate rectangle, the cells cannot be compressed directly, and the walls of the cells fit diagonally, resulting in an inverted "V"-shaped region of localized deformation. Subsequently, the deformed cells gradually spread upwards, and a "V"-shaped localized deformation region appears on the impact side. At this stage, the rectangular structure in the middle of the cells is compressed into a hollow bamboo shape. Moreover, part of the inclined wall becomes horizontal, and the honeycomb produces a significant contraction. As the specimen is further compressed, the arrow-shaped structures on both sides of the cells deform, and the horizontal walls of the honeycomb overlap, showing a "-"-shaped dense region. Subsequently, the middle region of the cytosol collapses and comes into close contact with the upper and lower cytosolic walls, at which point the honeycomb densifies.

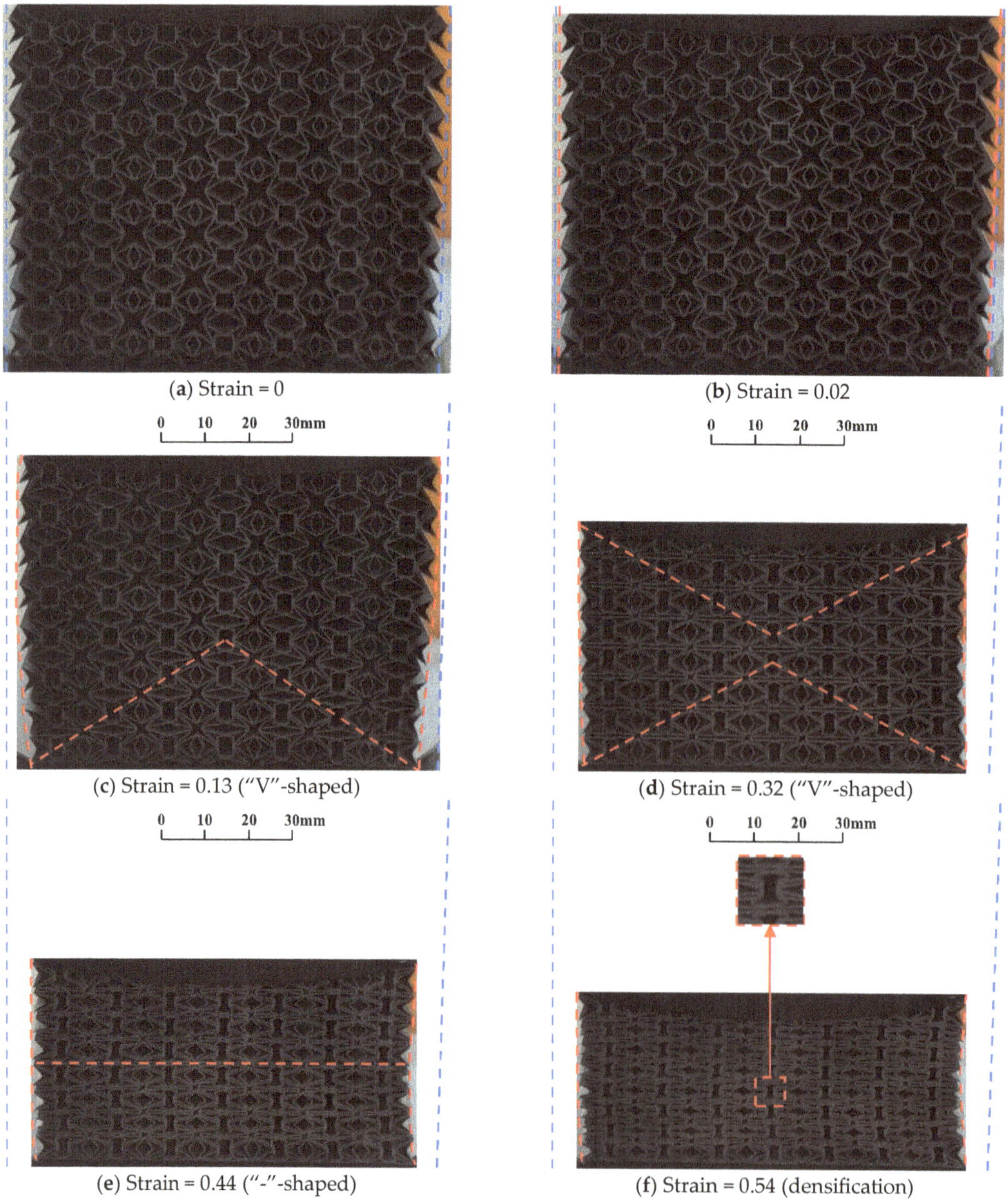

Figure 3. Deformation patterns for quasistatic compression tests of the 2D-ISSH specimen.

2.3. Mechanical Properties

The stress–strain curve of the 2D-ISSH specimen under quasistatic compression is shown in Figure 4, which can be divided into elastic, first plateau, transition, second plateau, and densification stages. After the elastic phase, due to the yielding of the walls of the star-shaped cells, the first plateau stage appears with a "V"-shaped deformation pattern. This deformation pattern flattens the stress curve, avoids peak stresses, and improves the honeycomb's impact resistance performance. Subsequently, the star-shaped cell wall deforms to a horizontal degree, and the arrow-shaped structure yields and deforms. With the gradual increase in the yielded cell wall, the stress–strain curve reaches the transition stage, and the stress gradually increases. In this state, the honeycomb exhibits a "-"-shaped deformation pattern. This deformation pattern prevents the honeycomb from experiencing a sudden increase in stress prior to densification. As the number of yielding cell walls stabilizes, the second plateau stage appears. Subsequently, the cells are gradually compacted, the "-"-shaped regions gradually overlap, and the whole honeycomb is pressed into a square shape to achieve densification.

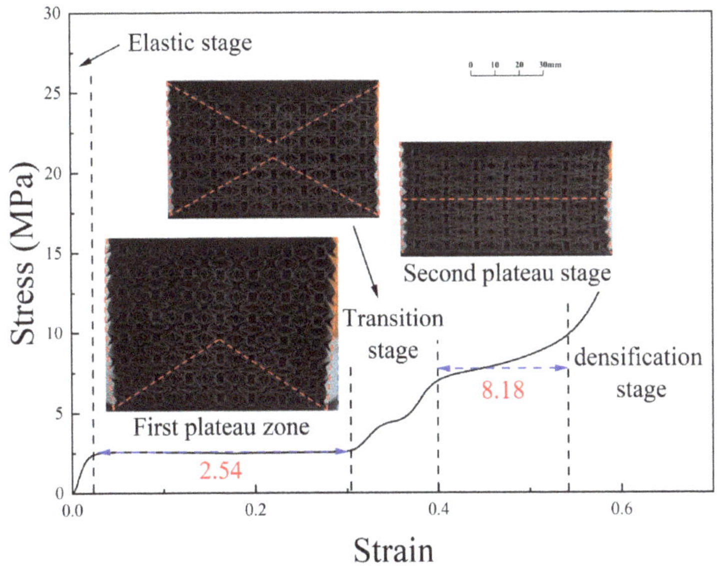

Figure 4. Stress–strain curve of the 2D-ISSH specimen.

3. An Improved Star-Shaped Honeycomb

3.1. Method Validation and Structural Design

Based on the improved star-shaped honeycomb structure, a 3D improved star-shaped honeycomb structure was obtained by cross-combining two 2D-ISSHs. The width of the cross-cells was b. As a result of the structural crossover, the tilted wall above the cell became shorter (from L1 to L2), and the angle between the tilted wall and the horizontal became larger (from α to θ). The other parameters remained the same as in the 2D-ISSH. The geometric configuration and the geometric parameters of the cells are shown in Figure 5a, and the main geometric parameters are shown in Table 2. The finite element model was built based on Abaqus (2022) software, as shown in Figure 5b. Two plates were arranged at the top and bottom of the honeycomb, which were set as discrete rigid bodies. Fixed constraints were added to the lower plate and the upper plate was impacted downwards at a constant velocity. General contact with a friction coefficient of 0.2 occurred between the honeycomb itself and the upper and lower plates. The honeycomb took 7, 8, and 7 cells along the x, y, and z directions, with lengths of 99.61 mm, 80 mm, and 99.61mm, respectively. The whole structure was modeled with S4R shell elements. When the cell size is reduced, equivalent stresses tend to converge and, considering accuracy and computational efficiency, this paper chose 0.5mm mesh for the finite element calculation, and a total of 310,464 meshes were generated. The base material was 316L steel. The material properties are shown in Table 1.

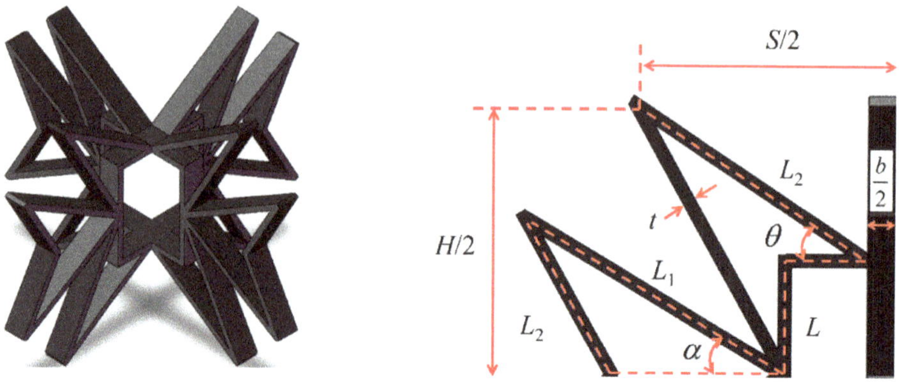

(**a**) Design of the 3D-ISSH

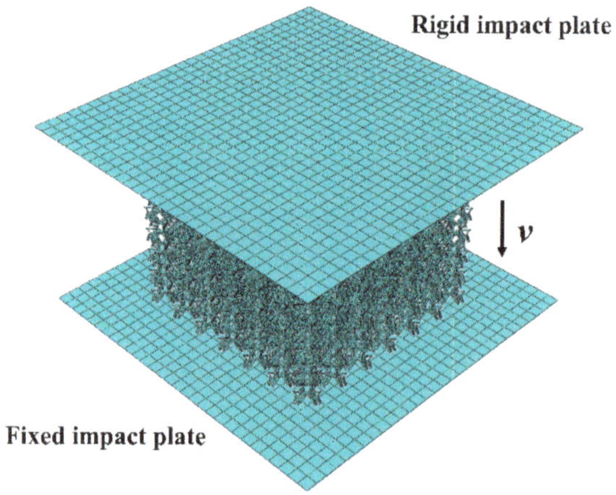

(**b**) Finite element model of the honeycomb structure

Figure 5. Geometric and finite element models for the 3D-ISSH. (**a**) Design of the 3D-ISSH. (**b**) Finite element model of the honeycomb structure.

Table 2. Geometric parameters of the 3D-ISSH.

	H (mm)	α (°)	t (mm)	b (mm)
3D-ISSH	10	30	0.25	1

To verify the accuracy and reliability of the simulation method, finite element simulation calculations were carried out on the two-dimensional honeycomb test, and the test and simulation results were compared. As shown in Figure 6, the deformation patterns of the two structures were consistent, and the stress value under the same strain was increasingly consistent, indicating the accuracy of the finite element simulation method.

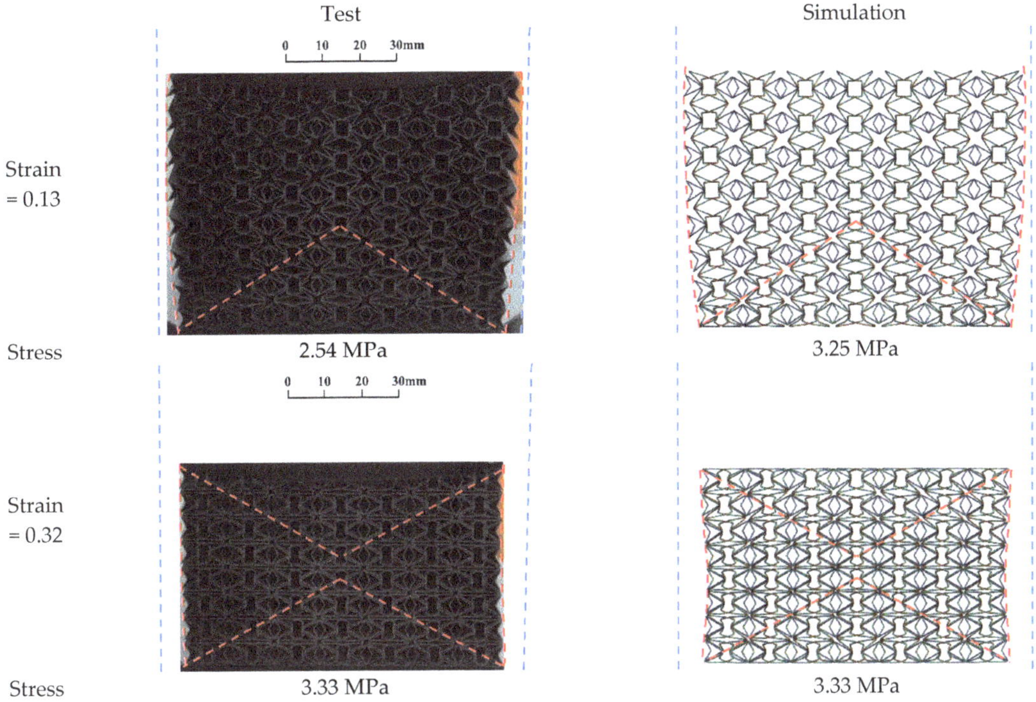

Figure 6. Comparison of the deformation mode between the numerical simulation and experimental test.

3.2. Deformation Mode

The deformation pattern of the 3D-ISSH specimen is shown in Figure 7. The 3D-ISSH shrinks simultaneously in the x and z directions. Before a compression strain of 0.02, the honeycomb undergoes mainly elastic deformation and the whole structure shrinks. After the elastic stage, the middle region of the 3D-ISSH shrinks first. Due to the presence of the middle rectangle, the cells cannot be compressed directly and the cell walls rotate with compression. When the strain reaches 0.13, an "X"-shaped region of localized deformation appears in the honeycomb. At this time, the degrees of deformation of the cells in the left and right regions of the "X"-shape are large, and the degrees of deformation of the cells in the upper and lower regions are small. When the specimen is compressed further, the upper and lower portions of the honeycomb shrink. When the strain reaches 0.4, the upper and lower regions of the "X"-shaped cytosol undergo large degrees of deformation, the middle rectangle of the cell is compressed into a hollow bamboo shape, and part of the inclined wall is pressed to the horizontal. When the strain is 0.5, the arrow-shaped structure around the cells deforms, and the horizontal walls of the honeycomb overlap, showing a "-"-shaped dense region. Subsequently, the middle rectangle of the cells collapses and is in close contact with the upper and lower cell walls, and the "-"-shaped regions gradually overlap; at this time, the honeycomb is densified. The 3D-ISSH top-view observation resembles the cross-assembly of plate parts, and a phenomenon similar to the yielding of plates occurs under pressure, resulting in the bending and rotation of some of the surrounding cells.

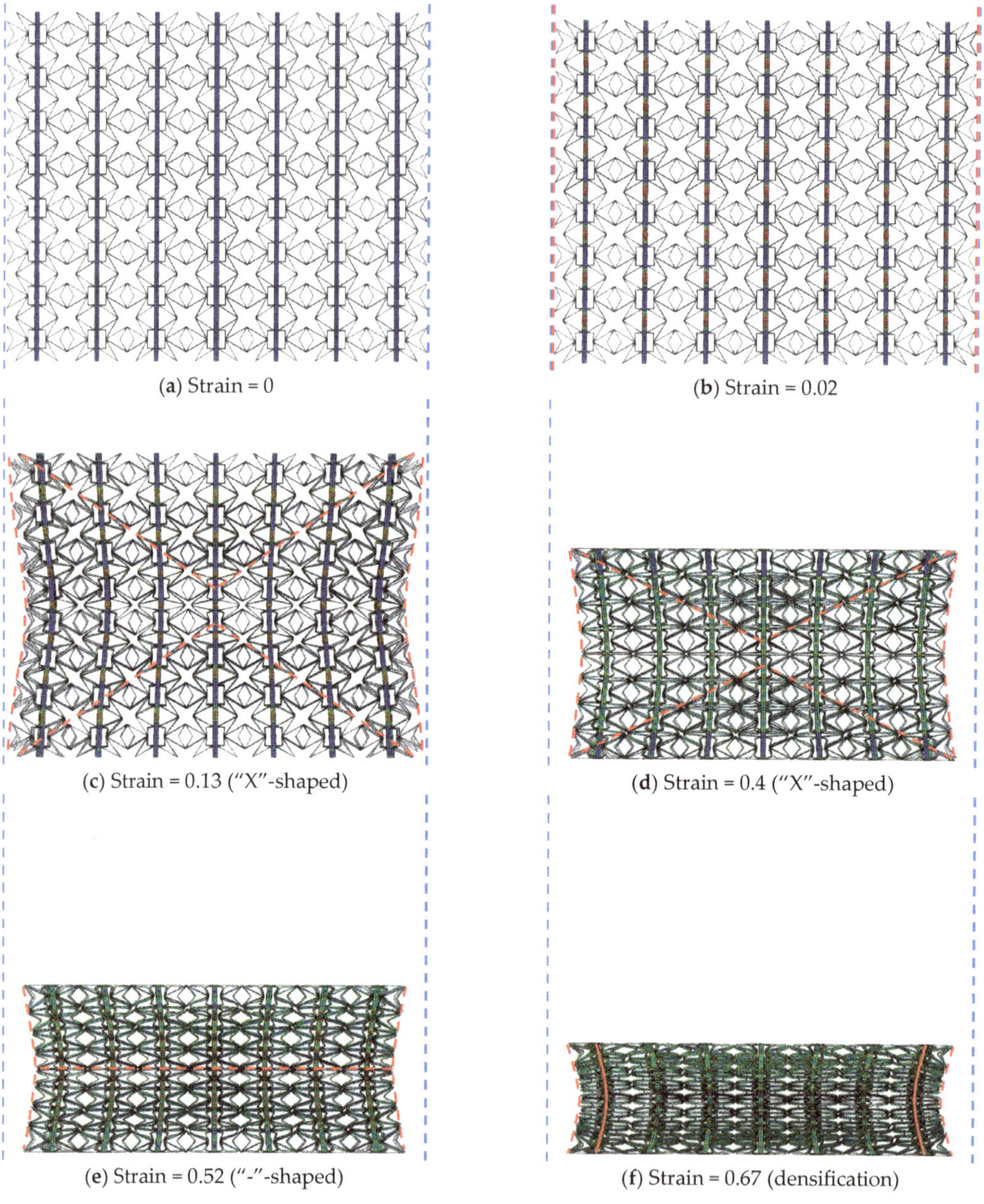

Figure 7. Deformation pattern of the 3D-ISSH specimen.

3.3. Mechanical Properties

As shown in Figure 8, the stress–strain curve of the 3D-ISSH specimen under quasistatic compression is similar to that of the 2D-ISSH specimen, and both can be divided into elastic, first plateau, transition, second plateau, and densification stages. After the elastic stage, due to the yielding of the walls of the star-shaped cells, the first plateau stage appears, where the main deformation pattern of the honeycomb is "X" shaped. This deformation pattern flattens the stress curve, avoids peak stresses, and improves the honeycomb's impact resistance performance. Subsequently, the wall of the star-shaped cell is deformed to a horizontal degree, and the arrow-shaped structure around the cell yields and deforms. With the gradual increase in the yielding cell wall, the stress–strain curve reaches the transition stage. At this moment the stress gradually increases, and the honeycomb shows a "-"-shaped deformation pattern. This deformation pattern prevents the honeycomb from experiencing a sudden increase in stress prior to densification. As the number of yielding cytosolic walls stabilizes, the second plateau region appears. Subsequently, the cells gradually collapse, and the "-"-shaped regions gradually overlap. In this stage, the whole honeycomb is densified.

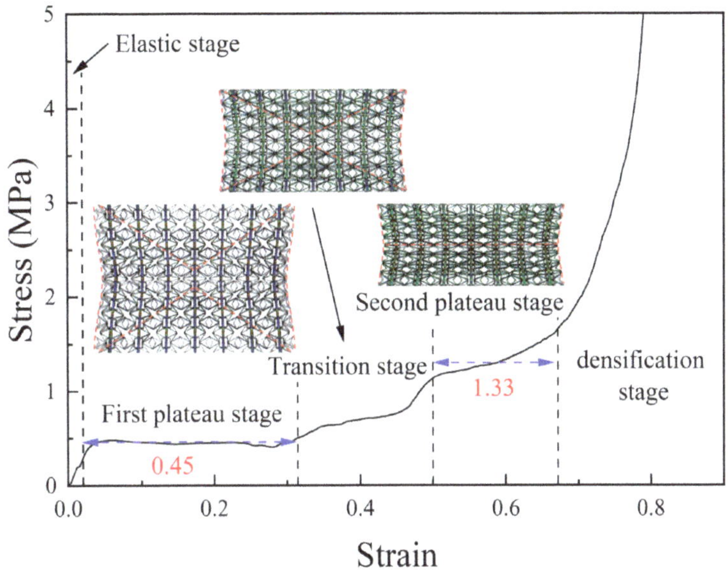

Figure 8. Stress–strain curve of the 3D-ISSH specimen.

4. Discussion

4.1. Theoretical Deduction

Platform stress is an important indicator of structural energy absorption. Based on the law of conservation of energy and plastic hinge theory, the stress equations for the quasistatic platform of the 3D-ISSH specimen are derived, and the predicted results are utilized for comparison with the simulation results.

The typical cell morphologies at the start and end stages of the first plateau are shown in Figure 9a,b. Due to the symmetry of the model, the inclined wall AB is rotated around points A and B until it reaches the horizontal direction. Similarly, the inclined wall BC is rotated around B and C until it comes into contact with CD. The thin square wall in the center also rotates around points A and C due to contraction. The 24 plastic hinges appearing in this plane of the cell at this stage are highlighted by red circles in Figure 9a, where the angle of rotation of the inclined wall AB is $\Delta\varphi$, the angle of rotation of the inclined wall BC is $\Delta\phi$, and the angle of rotation of the square thin-walled plastic hinge is $\Delta\delta$. It is assumed that the angle of rotation of each plastic hinge in a thin square wall is the same, and that the length of each inclined wall is constant, with due consideration for the wall thickness of the cell. The plastic hinge angle is calculated as follows:

$$\Delta\varphi = \theta \tag{1}$$

$$\Delta\phi = \alpha \tag{2}$$

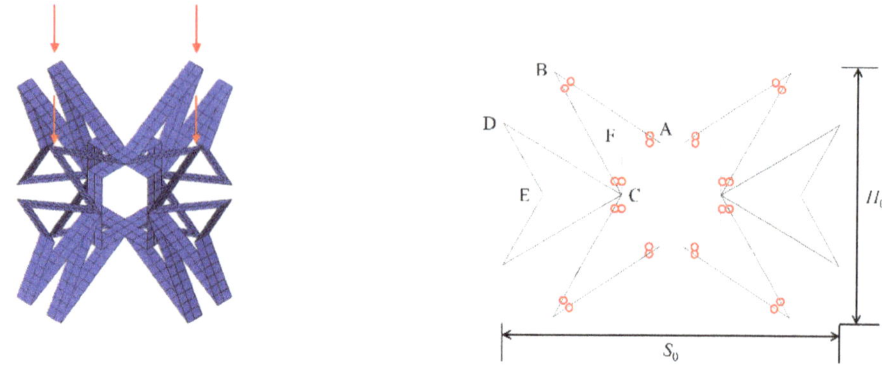

(**a**)

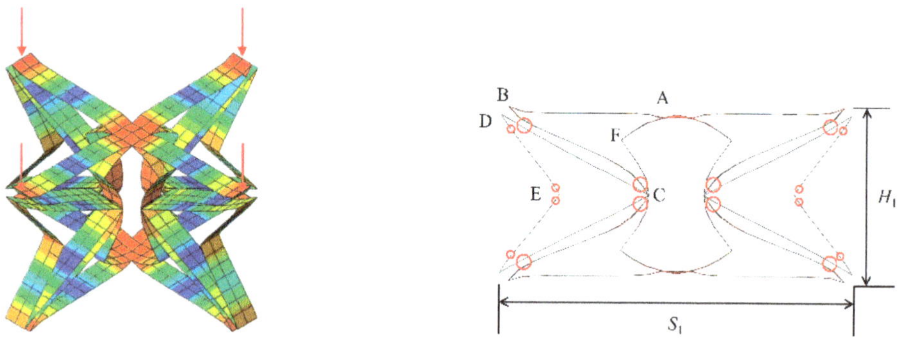

(**b**)

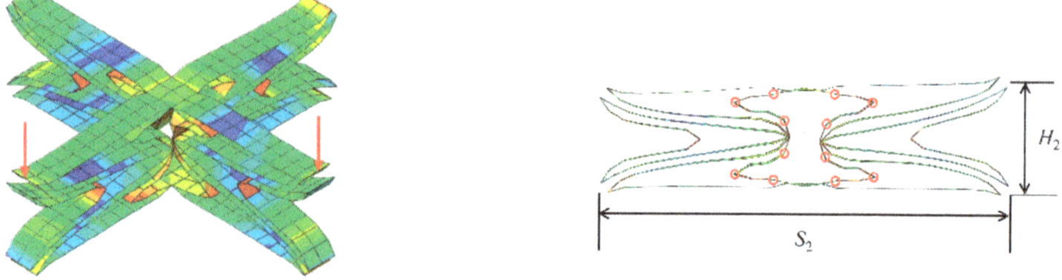

(**c**)

Figure 9. Deformation stages of the typical cells of 3D-ISSH under quasistatic compression. (**a**) Initial form. (**b**) Middle form. (**c**) Densified form.

The work conducted by the external force on the honeycomb is equal to the energy dissipated by the plastic hinges on the sloping and thin square walls of the honeycomb. Based on the conservation of energy, the following equations can be obtained:

$$W_p = \sigma_{P1} {s_0}^2 \Delta y \tag{3}$$

$$\Delta y = H_0 - H_1 = H - \frac{H \sin \alpha}{\cos \alpha} - 2t \tag{4}$$

where s_0 is the initial transverse length of the cell, σ_{P1} is the stress at the first platform of the honeycomb, Δy is the relative displacement in the compression direction, H_0 is the initial height of the cell, and H_1 is the height of the cell at the end of the first platform stage.

The total energy dissipation from the initial phase to the intermediate phase consists of two parts: the inclined-wall plastic hinges, E_1, and the square thin-wall plastic hinges, E_2, and can be expressed as follows:

$$E_1 = 16M_1 \Delta \varphi + 16M_1 \Delta \phi \tag{5}$$

$$E_2 = 16M_1 \Delta \delta = 16M_1 \arcsin\left(\frac{\Delta x}{L - 2t}\right) \tag{6}$$

The plastic ultimate bending moment, M_1, of a rectangular section beam is as follows:

$$M_1 = \sigma_0 b t^2 / 4 \tag{7}$$

σ_0 is the flow stress of the material, and is described as follows:

$$\sigma_0 = \sqrt{\frac{\sigma_y \sigma_u}{1+n}} \tag{8}$$

where n is the strain intensification index of the material. Δx is the transverse shrinkage displacement:

$$\Delta x = S_0 - S_1 = H\left(2 - \tan \alpha - \frac{1}{\cos \alpha}\right) \tag{9}$$

According to the conservation of energy, we obtain the following equation:

$$W_p = E_1 + E_2 \tag{10}$$

In summary, the first plateau stress for the 3D-ISSH specimen under quasistatic compression can be expressed as follows:

$$\sigma_{p1} = \frac{16M_1 \Delta \varphi + 16M_1 \Delta \phi + 16M_1 \Delta \delta}{{S_0}^2 \Delta y} = \frac{4\sigma_0 b t^2 \left[\theta + \alpha + \arcsin\left(\frac{\Delta x}{H(1-\tan \alpha) - 2t}\right)\right]}{[H(1 - \tan \alpha) - 2t][H(2 - \tan \alpha)]^2} \tag{11}$$

When the inclined wall is adhered to the arrow-shaped long edge, as shown in the middle form of Figure 9b, the stresses start to increase and the second plateau stage is initiated. In the second platform stage, the inclined wall BC is rotated around points B and C, and the arrow-shaped long side CD is rotated around points C and D. The two fit together and rotate at the same angle. The arrow-shaped short side DE is rotated around points D and E. The number of plastic hinges in a thin square wall increases from 8 to 12 in a single plane. The 28 plastic hinges appearing in this plane of the cytosol at this stage are highlighted by the red circles in Figure 9b,c.

The starting height of the cytosol at the second platform stage and the height at the time of densification are as follows:

$$H_1 = H \tan \alpha + 2t \tag{12}$$

$$H_2 = 11t \tag{13}$$

The initial cytosolic transverse length is as follows:

$$S_1 = \frac{H}{\cos\alpha} \tag{14}$$

The plastic hinge rotation angle of the tilted wall BC of the cell and the arrow-shaped long edge CD is as follows:

$$\Delta\varphi' = \alpha - \arcsin(H_2/2L_1) \tag{15}$$

The angle of rotation of the plastic hinge of the short edge of the arrowhead shape of the cell is as follows:

$$\Delta\phi' = \frac{\pi}{2} - \alpha - \arcsin(H_2/2L_1) \tag{16}$$

The thin wall AF in the square is rotated around points A and F, and CF is rotated around points C and F. It is easy to know that the angle of rotation of the plastic hinge at point F is $\pi/2$, according to the characteristics of squares, and the sum of the angles of rotation of the plastic hinges in the $1/4$ square structure is π.

The second platform stress is still obtained from the law of the conservation of energy, which is expressed as follows:

$$\sigma_{P2} = \frac{16M_2\Delta\varphi\prime + 16M_1\Delta\phi' + 16M_1\Delta\delta'}{{S_1}^2(H_1 - H_2)} = \frac{\sigma_y bt^2\left[\frac{14\pi}{3} - 12\arcsin\left(\frac{11t\cos\alpha}{H}\right)\right]}{\left(\frac{H}{\cos\alpha}\right)^2(H\tan\alpha - 9t)} \tag{17}$$

where M_2 is the plastic ultimate bending moment of the overlapping inclined wall, and $M_2 = 2M_1$.

A comparison of the theoretical and simulation results for the platform stress is shown in Table 3. The theoretical analysis results of the platform stress in the two segments are in good agreement with the numerical simulation results. During the second platform stage, the honeycomb is compressed to the late stage, where the cells have a certain degree of rotation. Affected by the boundary effect, the degree of rotation from the center to the surroundings is gradually intensified. Due to the rotation from the absorption of part of the energy, the stress produces a tendency to enhance the stress, and the stress begins to exceed the theoretical value.

Table 3. Comparison of theoretical and simulation results for platform stress.

	First Platform Stress, MPa (Average Value)	Second Platform Stress, MPa (Average Value)	Second Platform Stress, MPa (Starting Value)
Simulation	0.45	1.33	1.18
Theory	0.46	1.19	1.19

4.2. *Specific Energy Absorption*

The specific energy absorption values of the 2D-ISSH and 3D-ISSH specimens under quasistatic compression are shown in Figure 10. When the compression strain is less than 0.6, the deformation mode is similar for the 2D-ISSH and 3D-ISSH structures. Thus, the trend of stress is basically the same, inducing an equivalent specific energy absorption value. At a later stage, as the 3D-ISSH cytosol rotates, additional energy absorption occurs and the specific energy absorption values of the 3D-ISSH structure are larger than those of the 2D-ISSH structure. Relative to the 2D-ISSH, the specific energy absorption value of the 3D-ISSH is increased by 5.02%.

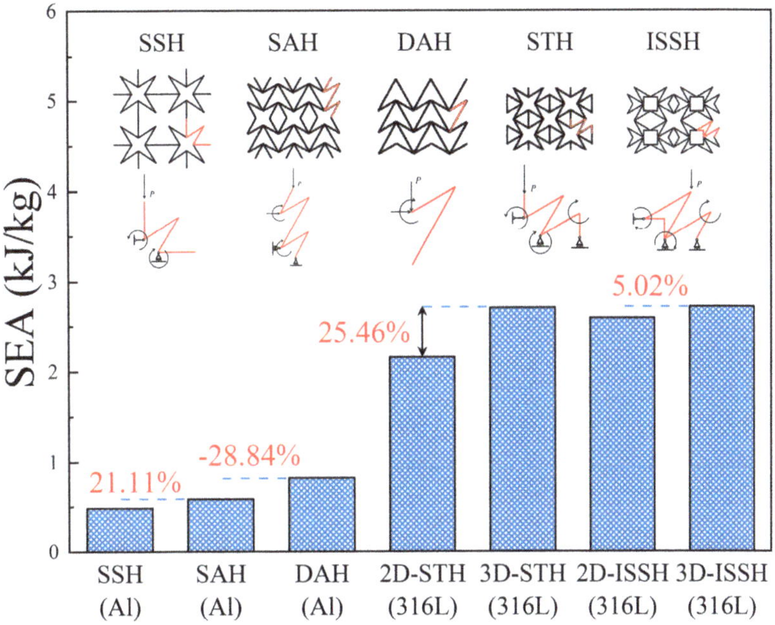

Figure 10. Comparison of the SEA of the existing configurations at densification [19,22].

To explore the relationship between the structural design and energy absorption in depth, and to analyze the effects of the structural materials on the specific energy absorption, other honeycombs are introduced. Figure 10 shows the specific absorption energy of the 2D-ISSH, 3D-ISSH, star-shaped honeycomb (SSH), star–arrowhead honeycomb (SAH), double arrowhead honeycomb (DAH), 2D star–triangular honeycomb (2D-STH), and 3D star–triangular honeycomb (3D-STH) at densification. It can be observed that the SAH adds an arrow-shaped structure to the SSH and replaces the straight wall in it, and the deformation of the arrow-shaped structure increases the number of plastic hinges that absorb additional energy; therefore, the specific energy absorption is improved by 21.11% compared to that of the SSH. The specific absorption energy of the SAH is still lower than that of the DAH because the number of plastic hinges in the star-shaped structure is less than that in the arrow-shaped structure. The SSH and SAH are made of aluminum, which is weaker. However, for the 2D-STH and 2D-ISSH, the material is 316 L steel, which is relatively strong; thus, each plastic hinge can absorb more energy. The specific energy absorption levels of the SSH and SAH are significantly weaker than those of the 2D-STH and 2D-ISSH. When both the 2D-STH and 2D-ISSH are expanded to three dimensions, the specific energy absorption is increased, which is mainly attributed to the fact that some of the cells rotate after the 3D honeycomb is compressed, increasing the energy absorption level. The 3D-ISSH has a 5.02% improvement in specific energy absorption over the 2D-ISSH.

4.3. Elasticity and Deformation

The transverse contraction displacement of the honeycomb during compression can be obtained by arranging the measurement points uniformly on the left and right sides of the honeycomb model, and the transverse strain of the honeycomb can be obtained by dividing the average value by the original length of the honeycomb. Figure 11 illustrates Poisson's ratio–strain curves of the 2D-ISSH, 3D-ISSH, SSH, SAH, DAH, 2D-STH, and 3D-STH structures. The negative Poisson's ratio effect is most pronounced in the early stage of DAH due to the special deformation trajectory of the arrow-shaped structure. But the advantages and disadvantages of a structure cannot be evaluated from one angle of negative Poisson's ratio alone; they have to be analyzed from multiple factors. According to the use scenarios of the honeycomb structures, in addition to the negative Poisson's ratio effect and the specific energy absorption mentioned above, the load-carrying capacity and regularity during deformation should be considered as important evaluation criteria.

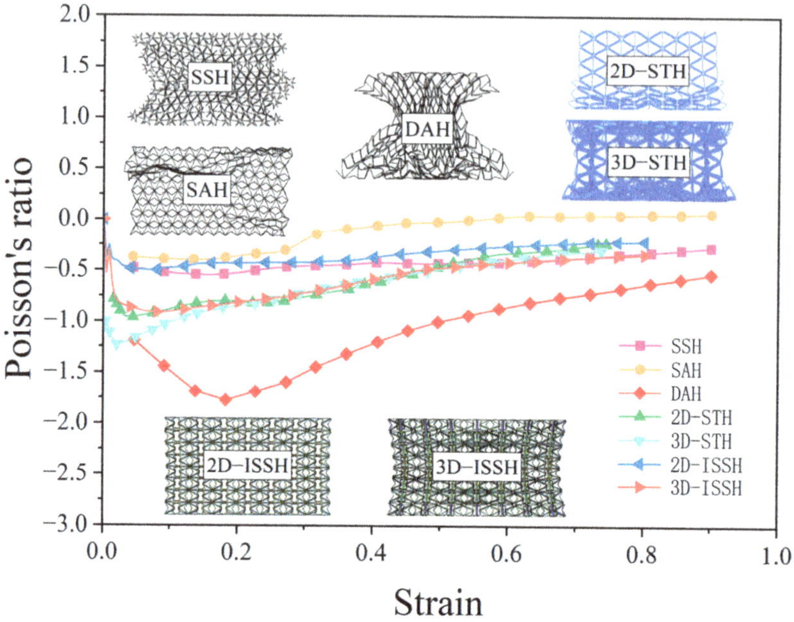

Figure 11. Comparison of the Poisson's ratio–strain curve between existing configurations [19,22].

The carrying capacity of a honeycomb structure is evaluated by E^*. The greater the value of E^* is, the greater the carrying capacity of the honeycomb.

$$E^* = \frac{E_1}{E_s\overline{\rho}} \tag{18}$$

where E_1 is Young's modulus of the honeycomb structure, E_s is Young's modulus of the substrate, and $\overline{\rho}$ is the relative density of the honeycomb structure.

Deformation regularity is evaluated using the maximum deviation of unidirectional horizontal strains (*MDUHS*) [25]. The smaller the value of *MDUHS* is, the more regular the overall deformation of the honeycomb. *MDUHS* can be expressed as follows:

$$MDUHS = \max\left\{\frac{\max|X_{Li}| - \min|X_{Li}|}{L}, \frac{\max|X_{Ri}| - \min|X_{Ri}|}{L}\right\} \tag{19}$$

where $|X_{Li}|$ and $|X_{Ri}|$ are the displacements in the x-direction on the left and right sides of the honeycomb, and L is the original horizontal width of the honeycomb. The overall deformation regularity of the honeycomb is evaluated by taking the *MDUHS* of each honeycomb strain as 0.4–0.5 when the cells are fully deformed. To visualize the results of the representation, $1/MDUHS$ is used as the rubric. In addition, to visualize the combined performance of each honeycomb, the maximum Poisson's ratio, E^*, and $1/MDUHS$ are used to plot the radargrams in Figure 12.

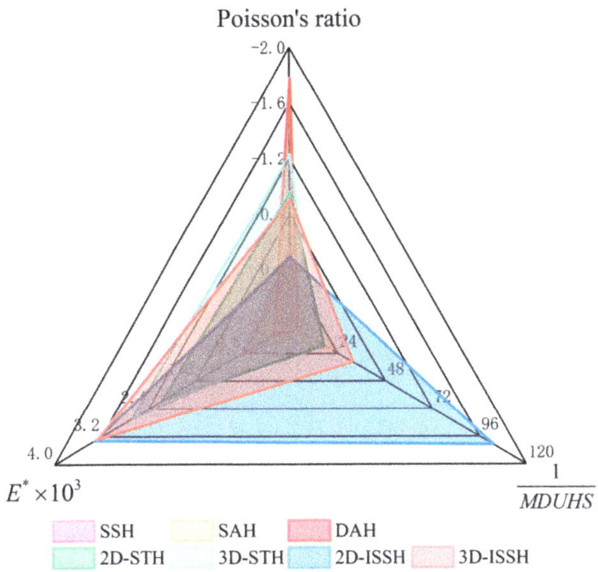

Figure 12. Comprehensive evaluation mapping the existing honeycomb structures [19,22].

As shown in Figure 12, due to the arrow-shaped structure being more stable compared to that of a straight wall, an increase in regularity during deformation, but a decrease in the negative Poisson's ratio performance and load-carrying capacity are observed in the SAH structure relative to the SSH structure. The negative Poisson's ratio effect is most pronounced for DAH due to the special deformation trajectory of the arrow-shaped structure. However, because of this, DAH's load-carrying capacity and deformation regularity are poor. Compared to the SSH and SAH structures, the 2D-STH structure designs the triangular structure on the left and right sides of the cell, with improved negative Poisson's ratio performance, load-carrying capacity, and deformation regularity. Moreover, the 2D-ISSH in this paper uses an arrow-shaped structure to replace the horizontal wall and adds a rectangular structure in the center of the star, which significantly improves the overall performance of the honeycomb, especially in terms of load-carrying capacity and deformation regularity. The most obvious change in the expansion of the 2D honeycomb to a 3D honeycomb is the improvement in the negative Poisson's ratio performance, which is attributed to the biaxial contraction property. The Poisson's ratio value of the 3D-ISSH is close to twice that of the 2D-ISSH. However, due to the rotation of some of the 3D honeycomb cells, the regularity of the 3D honeycomb deformation decreases.

5. Conclusions

In this paper, a 3D-ISSH was first designed based on a 2D-ISSH, and a comprehensive study of the 2D-ISSH and 3D-ISSH structures was carried out through a combination of experiments and numerical simulations under quasistatic compression loading conditions. There was also a discussion of similar structures, which explored the effects of structural materials on the energy absorption and negative Poisson's ratio properties. For the deformation process of a typical cell in a 3D-ISSH structure, the platform stress formula was derived. The following conclusions can be drawn from this study:

(1) In the quasistatic compression process, the 2D-ISSH and 3D-ISSH presented different deformation patterns. The lower part of the 2D-ISSH was first deformed, showing a "V" shape, and this deformation was gradually transferred to the upper part of the honeycomb during the compression process. The 3D-ISSH deformation first occurred in the middle position and then developed simultaneously up and down in an "X"-

shape. At the end of the first plateau region, both honeycombs showed a "-"-shape deformation pattern. As the specimen was further compressed, the "-" shapes gradually overlapped and the honeycombs were crushed. The two honeycombs exhibited similar stress stages during deformation, and both had double platforms (the 2D-ISSH double platform stress averages were, respectively, 2.54MPa and 8.18MPa; the 3D-ISSH double platform stress averages were, respectively, 0.45MPa and 1.33MPa).

(2) According to the deformation mechanisms of typical cells, the theoretical formula for the 3D-ISSH dual-platform stress was derived based on the law of energy conservation and the plastic hinge theory, validating the accuracy of the numerical model.

(3) After comparing multiple types of honeycomb structures, it was found that the ordinary star-shaped honeycomb could increase the plastic hinge during deformation by introducing certain structures, such as an arrow-shaped structure, which could increase the specific energy absorption of the structure. The expansion of the 2D honeycomb to a 3D honeycomb not only lightened the structure but also improved the specific energy absorption, which was mainly attributed to the fact that some of the cells rotated during the deformation of the three-dimensional honeycomb, absorbing additional energy. The 3D-ISSH has a 5.02% improvement in specific energy absorption over the 2D-ISSH.

(4) After considering the above honeycombs in multiple dimensions, it was found that, for star-shaped honeycombs, the replacement of the original straight wall with certain structures, such as arrows and triangles, could reduce the occurrence of instability and improve the regularity of the honeycomb during deformation. In addition, the adding of an arrow-shaped structure and a center rectangle in the 2D-ISSH greatly improved the loading capacity and the regularity of deformation. When the 2D honeycomb was expanded to 3D, the negative Poisson's ratio effect was greatly enhanced due to the biaxial shrinkage property. The Poisson's ratio value of the 3D-ISSH is close to twice that of the 2D-ISSH.

Author Contributions: Q.L.: investigation, validation, writing—original draft, and formal analysis. W.C.: supervision, conceptualization, methodology, software, validation, formal analysis, and writing—original draft. X.C., X.W. and C.L.: writing—review and editing and validation. X.L.: writing—review and editing and supervision. All authors have read and agreed to the published version of the manuscript.

Funding: This research was funded by the National Natural Science Foundation of China (52201334; 51979213; 12302482).

Institutional Review Board Statement: Not applicable.

Informed Consent Statement: Not applicable.

Data Availability Statement: The authors declare that the data presented in this study are available on request.

Conflicts of Interest: The authors declare no conflict of interest.

References

1. Wu, Y.B. Design of Protective Structure in Ship under Impact Loading and Its Numerical Technology. Harbin Engineering University. 2010. Available online: https://kns.cnki.net/KCMS/detail/detail.aspx?dbname=CMFD2011&filename=1011019864.nh (accessed on 24 July 2023). (In Chinese).
2. Ullah, I.; Brandt, M.; Feih, S. Failure and energy absorption characteristics of advanced 3D truss core structures. *Mater. Des.* **2016**, *92*, 937–948. [CrossRef]
3. Zhang, H.R.; Zhou, H.; Zhou, Z.X.; Zeng, H.Z.; Zhang, X.Y.; Yang, J.Z.; Lei, H.S.; Han, F.S. Energy absorption diagram characteristic of metallic self-supporting 3D lattices fabricated by additive manufacturing and design method of energy absorption structure. *Int. J. Solids Struct.* **2021**, *226–227*, 111082. [CrossRef]
4. Kooistra, G.W.; Deshpande, V.; Wadley, H.N. Hierarchical corrugated core sandwich panel concepts. *J. Appl. Mech.* **2007**, *74*, 259–268. [CrossRef]
5. Corigliano, P.; Palomba, G.; Crupi, V.; Garbatov, Y. Stress–Strain Assessment of Honeycomb Sandwich Panel Subjected to Uniaxial Compressive Load. *J. Mar. Sci. Eng.* **2023**, *11*, 365. [CrossRef]

6. Qiao, J.X.; Chen, C.Q. Analyses on the In-Plane Impact Resistance of Auxetic Double Arrowhead Honeycombs. *J. Appl. Mech.* **2015**, *82*, 051007. [CrossRef]
7. Qiao, J.X.; Chen, C.Q. Impact resistance of uniform and functionally graded auxetic double arrowhead honeycombs. *Int. J. Impact Eng.* **2015**, *83*, 47–58. [CrossRef]
8. Mousanezhad, D.; Haghpanah, B.; Ghosh, R.; Hamouda, A.M.; Nayeb-Hashemi, H.; Vaziri, A. Elastic properties of chiral, anti-chiral, and hierarchical honeycombs: A simple energy-based approach. *Theor. Appl. Mech. Lett.* **2016**, *6*, 81–96. [CrossRef]
9. Novak, N.; Hokamoto, K.; Vesenjak, M.; Ren, Z. Mechanical behaviour of auxetic cellular structures built from inverted tetrapods at high strain rates. *Int. J. Impact Eng.* **2018**, *122*, 83–90. [CrossRef]
10. Qi, D.; Lu, Q.; He, C.; Li, Y.; Wu, W.; Xiao, D. Impact energy absorption of functionally graded chiral honeycomb structures. *Extrem. Mech. Lett.* **2019**, *32*, 100568. [CrossRef]
11. Wang, T.; An, J.; He, H.; Wen, X.; Xi, X. A novel 3D impact energy absorption structure with negative Poisson's ratio and its application in aircraft crashworthiness. *Compos. Struct.* **2021**, *262*, 113663. [CrossRef]
12. Zhang, J.; Lu, G.; You, Z. Large deformation and energy absorption of additively manufactured auxetic materials and structures: A review. *Compos. Part B Eng.* **2020**, *201*, 108340. [CrossRef]
13. Xiang, J.; Du, J. Energy absorption characteristics of bio-inspired honeycomb structure under axial impact loading. *Mater. Sci. Eng. A* **2017**, *696*, 283–289. [CrossRef]
14. Ren, X.; Zhang, X.Y.; Xie, Y.M. Research progress in auxetic materials and structures. *Chin. J. Theor. Appl. Mech.* **2019**, *51*, 656–687. Available online: https://kns.cnki.net/kcms/detail/11.2062.O3.20190121.1555.002.html (accessed on 24 July 2023). (In Chinese)
15. Gibson, L.J.; Ashby, M.F.; Schajer, G.S.; Robertson, C.I. The mechanics of two-dimensional cellular materials. Proceedings of the Royal Society of London. A. *Math. Phys. Sci.* **1982**, *382*, 25–42. [CrossRef]
16. Almgren, R.F. An isotropic three-dimensional structure with Poisson's ratio = -1. *J. Elast.* **1985**, *15*, 427–430. [CrossRef]
17. Larsen, U.D.; Signund, O.; Bouwsta, S. Design and fabrication of compliant micromechanisms and structures with negative Poisson's ratio. *J. Microelectromech. Syst.* **1997**, *6*, 99–106. [CrossRef]
18. Theocaris, P.S.; Stavroulakis, G.E.; Panagiotopoulos, P.D. Negative Poisson's ratios in composites with star-shaped inclusions: A numerical homogenization approach. *Arch. Appl. Mech.* **1997**, *67*, 274–286. [CrossRef]
19. Wang, H.; Lu, Z.X.; Yang, Z.Y.; Li, X. In-plane dynamic crushing behaviors of a novel auxetic honeycomb with two plateau stress regions. *Int. J. Mech. Sci.* **2018**, *151*, 746–759. [CrossRef]
20. Wei, L.L.; Zhao, X.; Yu, Q.; Zhu, G.H. A novel star auxetic honeycomb with enhanced in-plane crushing strength. *Thin-Walled Struct.* **2020**, *149*, 106623. [CrossRef]
21. Wei, L.L.; Zhao, X.; Yu, Q.; Zhu, G.H. Quasi-static axial compressive properties and energy absorption of star-triangular auxetic honeycomb. *Compos. Struct.* **2021**, *267*, 113850. [CrossRef]
22. Wei, L.L.; Xu, S.W.; Zhu, G.H.; Zhao, X.; Shi, P.L. In-plane compression behavior of a novel 3D auxetic honeycomb. *Mater. Today Commun.* **2023**, *35*, 105729. [CrossRef]
23. Hu, J.S.; Lin, Y.S.; Chen, W.; Li, X.B.; Wu, W.G. In-plane dynamic response and energy absorption characteristics of improved star-shape honeycomb. *J. Vib. Shock.* **2022**, *41*, 119–128. (In Chinese) [CrossRef]
24. Yang, Z.S.; Xue, Y.X.; Liu, A.R. Study on the impact dynamics of three-dimensional star-shaped structure with negative Poisson's ratio. *Eng. Mech.* **2022**, *39*, 356–363. Available online: https://kns.cnki.net/kcms/detail/detail.aspx?FileName=GCLX2022S1045&DbName=CJFQ2022 (accessed on 24 July 2023). (In Chinese)
25. Li, X.L.; Li, Z.Z.; Guo, Z.Y.; Mo, Z.L.; Li, J. A novel star-shaped honeycomb with enhanced energy absorption. *Compos. Struct.* **2023**, *309*, 116716. [CrossRef]
26. Hu, J.S.; Lin, Y.S.; Chen, W. Studies on the Compressive Mechanical Properties of Antitetrachiral Honeycombs with Different Thickness Ratios of Ligament to Cylinder. *Acta Mech. Solida Sin.* **2022**, *35*, 470–480. [CrossRef]
27. Wang, Z.; Chen, G.; Cao, X.; Chen, W.; Li, C.B.; Li, X. Study on the Effect of Nodal Configuration on the Mechanical Properties of Hexa-Ligamentous Chiral Honeycombs. *J. Mar. Sci. Eng.* **2023**, *11*, 1692. [CrossRef]

Journal of
Marine Science and Engineering

MDPI

Article

Shear Force and Bending Moment Tuning Algorithm of Shuttle Tanker Model for Global Structural Analysis

Chaeog Lim [1], Ik-seung Han [2], Ju-Young Kang [3], Im-jun Ban [1], Byungkeun Lee [1], Jun Soo Park [4] and Sung-chul Shin [1,*]

[1] Department of Naval Architecture and Ocean Engineering, Pusan National University, Busan 46241, Republic of Korea; orc@pusan.ac.kr (C.L.); june3373@pusan.ac.kr (I.-j.B.); optilee@pusan.ac.kr (B.L.)

[2] Marine Renewable Energy Research Division, Korea Research Institute of Ships and Ocean Engineering, Daejeon 34103, Republic of Korea; ihan77@kriso.re.kr

[3] Design Team, Bridge International Patent & Law Firm, Busan 48058, Republic of Korea; jyoung98@hanmail.net

[4] Department of Naval Architecture and Ocean System Engineering, Kyungnam University, Changwon 51767, Republic of Korea; junsoopark@kyungnam.ac.kr

* Correspondence: scshin@pusan.ac.kr

Abstract: Global ship analysis is conducted using a finite element model (FE model) for ship design and construction, which involves structural, motion, and vibration analyses. It is crucial to examine the structural safety of the hull and motion response. In the ship FE model used in global ship analysis, weight distribution is employed to adjust the light weight and center of gravity (COG), which are required to perform the analysis. Further, the FE model needs to satisfy the required longitudinal shear force (SF) and bending moment (BM) under the loading conditions of the ship. Moreover, the SF and BM in the ship Trim and Stability data are utilized to perform shear force tuning (SFT) and bending moment tuning (BMT) for the ship FE model. This ensures the ship model exhibits curves of the SF and BM that coincide with those of the ship. The SFT and BMT for the ship FE model are time-consuming and costly. Thus, to address these limitations, we propose an effective and accurate algorithm and program for SFT and BMT. Accordingly, we developed a C#-based algorithm to tune the weight, SF, BM, and COG of the ship FE model to the required target value. Finally, the accuracy of the newly developed algorithm was analyzed and compared by applying it to the shuttle tanker FE model under the ballast and full load conditions. Accuracy was within tolerance in both loading conditions. The average errors of SF and BM were smaller in the ballast condition than in the full load condition, and the errors were smaller at the bow than at the stern.

Keywords: longitudinal strength; ship FE model; shear force; bending moment; global ship analysis; algorithm

Citation: Lim, C.; Han, I.-s.; Kang, J.-Y.; Ban, I.-j.; Lee, B.; Park, J.S.; Shin, S.-c. Shear Force and Bending Moment Tuning Algorithm of Shuttle Tanker Model for Global Structural Analysis. *J. Mar. Sci. Eng.* **2023**, *11*, 1900. https://doi.org/10.3390/jmse11101900

Academic Editors: Bin Liu, Kun Liu and Chenfeng Li

Received: 13 August 2023
Revised: 22 September 2023
Accepted: 27 September 2023
Published: 29 September 2023

1. Introduction

With the increase in ship size and structural complexity, global ship analysis is conducted for shipbuilding upon the request of classification agencies and ship owners to evaluate structural safety and motion performance (Cho et al., 2008; Valtonen et al., 2020) [1–3]. High precision is required in global ship analysis to satisfy the needs of ship owners and achieve classification approval. Ship owners require global ship analysis results that can apply more specific and diverse loading conditions than before. Because the analysis results are used for classification approval and represent the competitiveness of the shipyard's analysis technology, global ship analysis has gained increasing importance for use in shipyards. The analysis is conducted using a finite element (FE)-based global ship model to consider the complexity of the analysis and various loading conditions (Kim et al., 2005; Park et al., 2007) [4,5]. A global ship FE model (referred to below as ship

FE model) implies an FE model of full size, and it consists of the ship structural members and main cargo tank. As the model excludes small secondary tanks, liquids, the outfit, and electric systems, its weight is different from that of the ship. Moreover, the ship FE model must have the same light weight and center of gravity (COG) as the actual ship to perform an accurate analysis, and the respective required target values are ensured through weight distribution (WD) (Lim et al., 2022) [6]. The target COG and weight are the value of the actual ship, and these values must be consistent with Trim and Stability (TnS) data of the ship. However, when the cargo state with the loading conditions for analysis is applied to the ship FE model that completed the WD, the longitudinal shear force (SF) curve and bending moment (BM) curve of the model do not match with those of the ship. For accurate analysis, the ship FE model must satisfy the longitudinal SF and BM curves of the ship under the loading conditions.

Owing to alternating hogging and sagging during the voyage, the BM of the ship is closely related to its structural safety. Thus, the SF and BM are significant factors in determining the structural safety of the ship (Kim et al., 2018; Liu and Guedes Soares, 2020; Wang et al., 2019) [7–9]. However, as the distribution of the SF and BM of the ship also affects the structural analysis results, a reliable method of tuning the SF and BM is needed. Therefore, shear force tuning (SFT) and bending moment tuning (BMT) are required for the ship FE model to tune the longitudinal SF and BM curves of the model to those of the ship TnS data. Here, the model COG and weight after SFT and BMT must satisfy the target values required for analysis, which are identical to those of the model that completed the WD.

Several studies on SF and BM have been conducted. Pesterev et al. (2001) [10] proposed a new method for calculating the BM and SF of proportionally damped beams owing to the movement of concentrated loads. This method was explained through its application to the moving oscillator problem. Wang et al. (2019) [9] designed a 10,000 TEU (Twenty-Feet Equivalent Unit) similar-scale model of a container ship and performed a numerical analysis on the structural strength and similar-scale models under the conditions of combined torsion and bending, pure torsion, and pure hogging bending. Based on the results and similarity theory, a similarity in ultimate strength between container ships and similar-scale models was proposed. Yun et al. (2020) [11] conducted a study to improve the existing analytical model used to forecast the nonlinear bending behavior of flexible risers. In contrast to previous analytical models, in the present study, a series of equilibrium equations were derived considering the radial and shear deformations of multiple layers. Further, to verify the proposed model, bending analysis was performed on a 2.5 in flexible riser, and the results were compared with the FE results. Kim et al. (2020) [12] proposed a multi-hysteresis modeling method to analyze the BM behavior of the riser extended to the global area. The proposed method used a realistic and accurate bending hysteresis curve considering shear deformation and axisymmetric load. Waskito et al. (2020) [13] measured the wave load on the surface of a ship model using a fiber Bragg grating pressure sensor attached to the ship. Comparisons and an analysis of the measured values and computational fluid dynamics results revealed them as being consistent with the nonlinear vertical BM. Kövesdi et al. (2016) [14] studied the effect of additional normal stress on the bending stress of trapezoidally corrugated webs. In particular, a study was performed on the determination of the transverse BM and its effect on the load transfer of trapezoidally corrugated webs. In addition, the structural behavior of a trapezoidally corrugated web girder under BM and SF was studied. Recupero et al. (2005) [15] conducted a generalization study of prestressed and reinforced concrete models subjected to the interaction of SF, BM, and axial load. Biondi et al. (2004) [16] proposed two methods to capture discontinuities due to moving mass in the BM and SF laws along a continuous structure with varying accuracy. Kim et al. (2018) [7] calculated the secondary BM induced by sea waves using the secondary strip theory. It was found that the secondary BM changed significantly according to the change in the ship side angle. Liu and Guedes Soares (2020) [8] reported that the hull girder under the condition of the cyclic BM exhibited weaker ultimate strength than the monotonically increasing BM. Further, the collapse modes of a hull girder under monotonic and cyclic loading were compared, revealing the cyclic ultimate load of the hull structure as being approximately 10% lower than the existing ultimate strength. Rahman and Chowdhury (1996) [17] developed the calculating method of the ultimate value of the longitudinal BM on the cross-section in the girder of the ship or box. The calculating method was developed using a FORTRAN program and tested on several box girder models and actual ships. Xu and Haddara (2001) [18] developed an artificial neural network (ANN) model to predict the vertical BM of the ship due to waves. The ANN learned the heave and pitch motions to estimate the vertical BM. Xu et al. (2019) [19] experimented on the ultimate strength of an inland catamaran under the condition of vertical BM. Further, an FE analysis of a scale model indicated that the ultimate BM obtained through analysis coincided with the experiment results. Mansour and d'Oliveira (1975) [20] developed a computerized procedure to predict the hull vertical BM owing to slamming. The developed equation was validated through its application to a mariner ship. Moreira and Soares (2020) [21] proposed a time-domain technique based on the ANN to estimate the regular wave-induced vertical BM and SF. The ANN was used to model the time-domain relationship between BM, SF, and ship motion. Lee et al. (2010) [22] suggested a three-dimensional source distribution method to estimate the motion and vertical BM of a ship advancing in a regular wave. In addition, the validity of the method was confirmed through comparisons of results with experimental values. Kwon (1996) [23] studied the effect of the hull WD on still-water bending and wave-induced BM. Kim and Paik (2003) [24] calculated and analyzed the wave-induced BM due to long- and short-term responses subjected to various sailing conditions and sea states for a very large crude oil carrier and capsize bulk carrier. Thus, most studies have emphasized the development of a calculation method for the SF and BM of a ship, numerical calculation of the SF and BM responses for the structure using FE analysis, and measurement of the SF and BM through experiments. Therefore, owing to the scarcity of studies related to practical methods for the SFT and BMT of a ship FE model, further research and development are necessary.

In shipyards, NASTRAN software has been primarily used for global ship analysis (MSC, 2018) [25]. Although Det Norske Veritas (DNV) classification increases due to the demands of ship owners, requests for structural research in shipyards, and classification approval, global ship analysis must be conducted using the DNV's Sesam tool instead of NASTRAN in certain cases. In the case of global ship analysis using NASTRAN, general-purpose software based on NASTRAN is available for SFT and BMT. In contrast, for global ship analysis using DNV Sesam, general-purpose software has not been developed; thus, several calculation sheets arbitrarily created in the preliminary design stage are used in each work process and project for SFT and BMT. This increases the time consumed and cost because the designer performs tuning through trial and error using calculation sheets. In addition, as calculation sheets are not created using a systematic method for SFT and BMT, human errors and bottlenecks can occur in DNV Sesam analysis using calculation sheets. To address this limitation, an effective algorithm and software that can quickly and accurately perform SFT and BMT must be developed. Therefore, in this study, an algorithm and software to tune the SF and BM of a ship FE model were developed for global ship analysis using DNV Sesam. The newly developed algorithm and software are used such that the target SF and BM, as well as the target weight and COG, are satisfied. C# language was used to build the program, and a full-scale ship FE model was used to validate the performance and accuracy of the algorithm.

2. Research Method

2.1. Algorithm Overview

The five-step algorithm for SFT and BMT is shown in Figure 1. The weight and COG were tuned to satisfy the required target values during SFT and BMT. In the first step, the data of a ship FE model and TnS information of the ship were saved. The TnS data comprise the actual ship information (weight, trim, and stability) and are calculated using only the weight distribution information. Next, the ship FE model was apportioned according to the required locations of the SF and BM checks. Thereafter, the weight of the ship FE model was calculated, while the SF and BM were calculated at each division position (DP). Finally, the SF and BM of the ship FE model were changed and tuned, and the ship FE model was saved. The dimensions of all elements and the COG for the ship model are represented using the coordinate system in Figure 2.

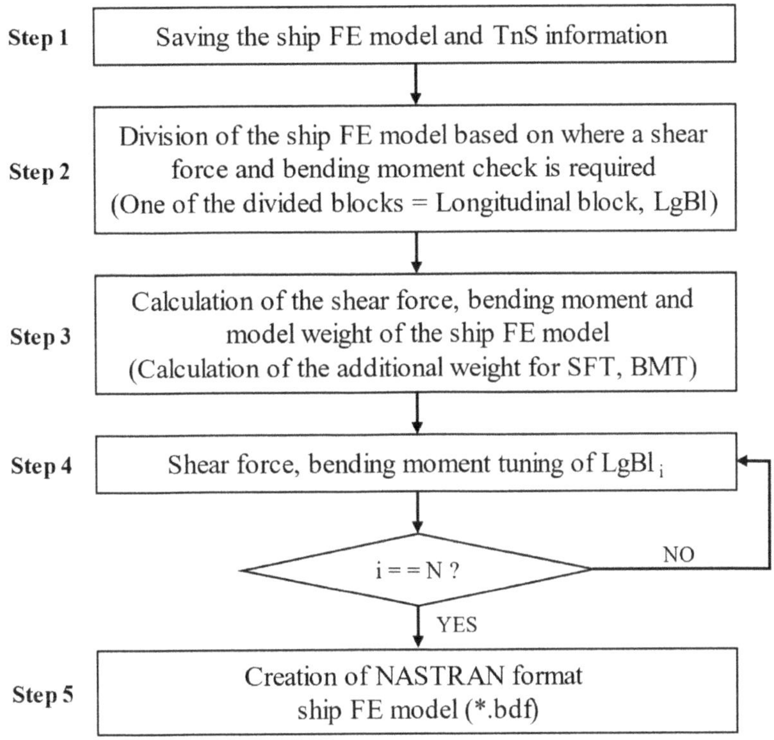

Figure 1. SFT and BMT algorithm.

Step 1: First, the TnS data and ship FE model were saved. The ship FE model comprises the coordinates, material properties, geometry, and dimensions of the elements. The weight of each structural member and the SF and BM of the ship FE model were calculated using these data and information in each step of the algorithm. The MSC NASTRAN 2018 Quick Reference Guide was used for calculating and handling the element data of the ship FE model (MSC, 2018). The TnS information contains the longitudinal WD information for each system, tank, liquid, and cargo of the ship and the information on longitudinal SF and BM. Step 2: Next, the ship FE model was divided in the longitudinal direction. The DPs were selected as the locations to check the SF and BM. The data on each divided model were then saved. An SF and BM check was required at each DP. Based on these positions, the ship FE model was split into N blocks. Each divided block is called a longitudinal block (LgBl), with a different length and size depending on the DP. Step 3: Using the data stored in each LgBl, the weight and COG of each LgBl and the ship FE model with the completed WD were calculated. The ship FE model weight was equal to the sum of the structural elements' weight and the additional weight ($AddWt$) added to the model because of the WD. Further, for model weight calculation, after calculating the element weight of each structural member, the weight was equally distributed to the grid of each element model that represented the coordinate data. The method of calculating the weight of the grid for a rectangular-shaped shell and a T cross-section beam is shown in Figure 3 and

Equations (1)–(1.b). The shell weight M_{shell} was divided into 4, and M_{shell} was allocated to *Grid*1, *Grid*2, *Grid*3, and *Grid*4 of the shell element as $\frac{M_{shell}}{4}$ (Equation (1)). Similarly, the weight of a T cross-section beam is M_{beam}, and it consists of two grids. A weight of $\frac{M_{beam}}{2}$ was assigned to *Grid*4 and *Grid*5 of the beam (Equation (1.a)). *Grid*4, which is included in both the beam and shell, has an $\frac{M_{beam}}{2} + \frac{M_{shell}}{4}$ weight, as it takes the beam and shell. Further, the weight of $\frac{M_{beam}}{2}$ is distributed to *Grid*5, and the weight of $\frac{M_{shell}}{4}$ is distributed to *Grid*1, *Grid*2, and *Grid*3. Thus, the final weight was $\frac{M_{shell}}{4} + AddWt_{g1}$ for *Grid*1, $\frac{M_{shell}}{4} + AddWt_{g2}$ for *Grid*2, $\frac{M_{shell}}{4} + AddWt_{g3}$ for *Grid*3, $\frac{M_{beam}}{2} + \frac{M_{shell}}{4} + AddWt_{g4}$ for *Grid*4, and $\frac{M_{beam}}{2} + \frac{M_{beam}}{2} + AddWt_{g5}$ for *Grid*5 (Equation (1.b)). Following the WD to all grids, the weight and COG of each LgBl model and the ship FE model weight and COG were calculated using the grid weight and coordinates of the model. In addition, the weight of the ship FE model was equal to the sum of the weights of all LgBls.

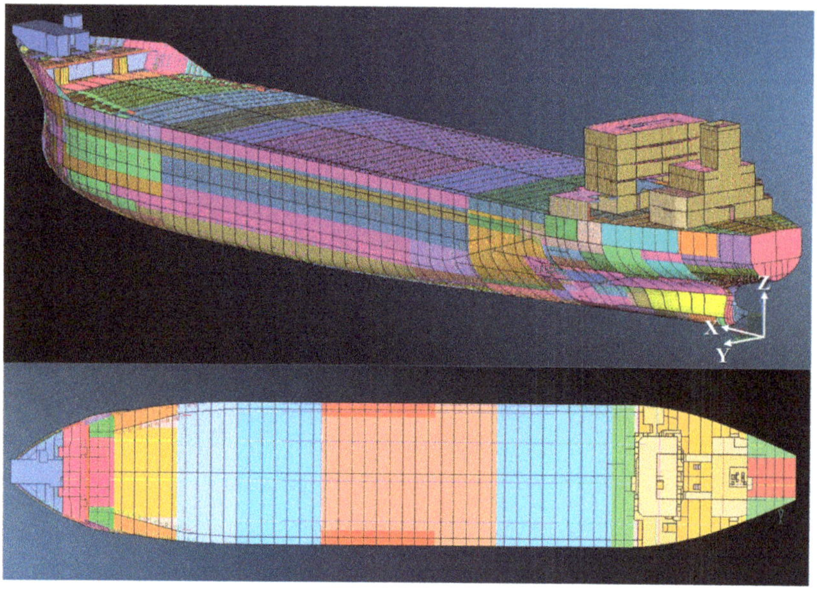

Figure 2. Coordinate system for all dimensions of the ship FE model.

$$m_{g1\ of\ shell}, m_{g2\ of\ shell}, m_{g3\ of\ shell}, m_{g4\ of\ shell} = \frac{M_{shell}}{4} \tag{1}$$

$$m_{g4\ of\ beam}, m_{g5\ of\ beam} = \frac{M_{beam}}{2} \tag{1.a}$$

$$\begin{gathered} m_{g1} = \frac{M_{shell}}{4} + AddWt_{g1},\ m_{g2} = \frac{M_{shell}}{4} + AddWt_{g2},\ m_{g3} = \frac{M_{shell}}{4} + AddWt_{g3} \\ m_{g4} = \frac{M_{beam}}{2} + \frac{M_{shell}}{4} + AddWt_{g4},\ m_{g5} = \frac{M_{beam}}{2} + AddWt_{g5} \end{gathered} \tag{1.b}$$

where

$$M_{shell} = Shell\ weight$$

$$M_{beam} = T - beam\ weight$$

$$m_{gj} = m_{gj\ of\ shell} + m_{gj\ of\ beam} + AddWt_{gj} : Grid_j\ weight$$

$$AddWt_{gj} = Additional\ weight\ of\ Grid_j\ by\ WD$$

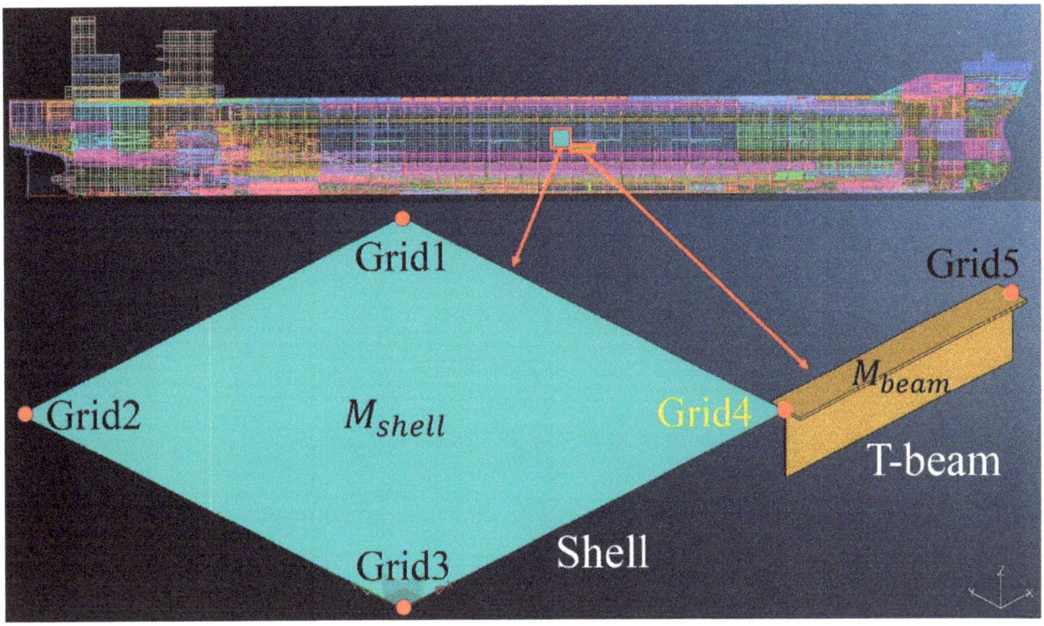

Figure 3. Calculation of grid weight of shell and T-beam elements with Equation (1).

The values of the SF and BM among the analysis results of the ship FE model with the WD completed were saved. Consequently, based on these values and the longitudinal SF and BM values of the ship obtained from the TnS data, the SFT and BMT values at each DP of the ship FE model were calculated. Here, the ship's weights, SF, BM, and COG in the TnS data are target values of the ship FE model. The ship FE model satisfies the target value through SFT and BMT. Subsequently, considering these values and the grid weight information calculated above, the α additional weight (αWt) to be added or subtracted from each LgBl of the ship FE model for SFT and BMT was calculated. Steps 4 and 5: SFT and BMT were performed by distributing αWt to each LgBl in the direction from stern to bow. The SF and BM of the ship FE model that has completed tuning coincided with the SF and BM in TnS; that is, the SF and BM of the ship were the target SF and BM to be satisfied by the ship FE model. Finally, the algorithm was iterated until SFT and BMT were performed for all LgBls through the distribution of αWt. After tuning, the αWt information was saved as CONM2 cards, that is, the point mass of the grids, and consequently, a NASTRAN file format ship FE model containing the weight information was created (Lim et al., 2022) [6].

2.2. Detailed Process of the SFT and BMT

In Steps 3, 4, and 5 of Figure 1, the detailed process of SFT and BMT for the ship FE model that has completed the WD is as follows. Instead of performing SFT and BMT simultaneously, BMT was performed only when required after SFT. This is because a ship FE model that has completed the WD and SFT often satisfies the target values of the weight, COG, and SF at each DP, and thus has also satisfied the value of the BM under these conditions. However, BMT is required when the BM deviates from the target tolerance because of a slight difference in structural geometry between the model and ship, precision of the model data, or accuracy of the SFT values. Figure 4 shows the detailed process of SFT and BMT. If the target values of the SF and BM are obtained after SFT, tuning is completed.

Otherwise, BMT is performed, and tuning is completed if the target value is obtained after BMT. In the ship FE model coordinate system, "Xcog," "Ycog," and "Zcog" are the X—(longitudinal), Y—(transverse), and Z—(height) components of the COG location.

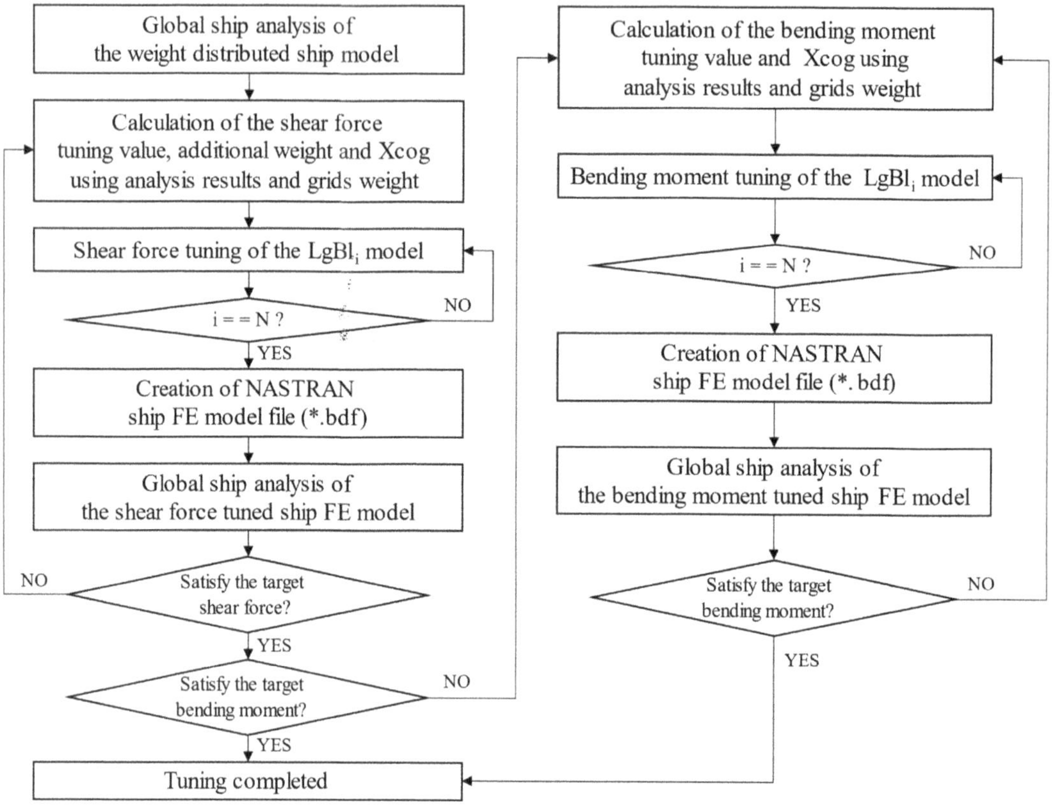

Figure 4. Detailed process for SFT and BMT in Steps 3, 4, and 5 of Figure 1. "Xcog" is the X-component of the ship FE model COG location.

2.2.1. Detailed Process of SFT

First, only SFT was performed for the ship FE model. Before the SF and BM were tuned for the ship FE model that completed the WD, they were significantly different from the target values. Thus, the SFT values at each DP of the ship FE model were calculated using the SF values of the ship FE model with WD analysis results, the longitudinal SF curve values of the TnS data, and the grid weight information. This is required to tune the SF of the current ship FE model to that of the actual ship. Based on the calculated value, αWt was added or subtracted from each LgBl of the ship FE model for SFT. Figure 5 and Equations (2)–(2.e) show the SFT value and the αWt calculation method when the number of DPs is three for the ship FE model, and the SF is calculated in the direction from the stern to the bow. Regarding $LgBl_1$ and the DP of p_1, the SF increases by ΔSF_1, and as a result, it changes from SF_1 to the target value SF_1' (Equation (2)). Further, the weight of $LgBl_1$ changes from m_1 to m_1' (Equation (2.a)). Moreover, when the SFs of $LgBl_2$ and p_2 and that of $LgBl_3$ and p_3 are adjusted in a similar manner, the SF of the ship FE model coincides with the TnS SF, which is the target value, and the total weight becomes equal to the target value, that is, the weight after WD Equations (2.b)–(2.e).

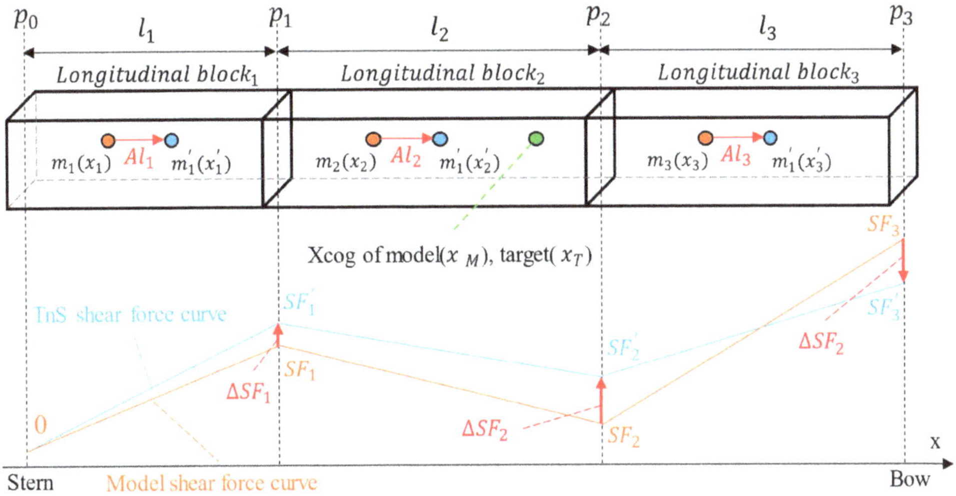

Figure 5. Calculation of the SFT value (ΔSF_i) using the TnS and model SF curve by global ship analysis. Calculation of the $LgBl_i$ target Xcog (x_i') using a ratio A, to correspond to the ship FE model Xcog (x_M) to the target Xcog (x_T) through the SFT.

Prior to SFT, the ship FE model satisfies the target weight and COG because the WD has been completed. However, during SFT, the ship FE model COG is affected by the change in the COG of each LgBl. Therefore, when the SF is tuned by distributing the previously calculated αWt to each LgBl, the target COG that must be satisfied by each LgBl is calculated to tune the ship FE model COG to the target COG. When the target SF, weight, and Xcog are satisfied through SFT for all LgBls, those values associated with the completed SFT of the ship FE model will also satisfy the target values. Subsequently, the Ycog and Zcog of each LgBl were also tuned to the Ycog and Zcog of the ship FE model. The calculating method for the target Xcog of all LgBls when the ship FE model is separated into three LgBl parts is detailed in Figure 5 and Equations (3)–(3.c). The direction of the positive x-axis is the bow direction in Figure 5. Equation (3.c) shows the calculation method of A, which means the coefficient of Xcog moving the length of the LgBl. Further, regarding $LgBl_1$, the Xcog prior to SFT was x_1 with a weight of m_1 and a length of l_1. After SFT, the Xcog moved by Al_1 from x_1 to x_1' and the weight was m_1'. Thus, when the Xcog of all LgBls is changed by Al_i in the same method, the Xcog of the ship FE model, x_M, coincides with the target value of x_T. Moreover, the total weight of the ship FE model, M, is maintained and coincides with the target value of M'. In addition, as the Ycog and Zcog of each LgBl are identical to the target values, the transverse (Ycog) and height (Zcog) components of the ship FE model COG location also coincide with the target values.

$$\Delta SF_1 = SF_1' - SF_1 \tag{2}$$

$$m_1' = m_1 + \Delta SF_1 \tag{2.a}$$

$$\Delta SF_2 = SF_2' - SF_2 \tag{2.b}$$

$$m_2' = m_2 + \Delta SF_2 - \Delta SF_1 \tag{2.c}$$

$$\Delta SF_3 = SF_3' - SF_3 \tag{2.d}$$

$$m'_3 = m_3 + \Delta SF_3 - \Delta SF_2 \tag{2.e}$$

$$x_M = \frac{m_1x_1 + m_2x_2 + m_3x_3}{M} = x_T \tag{3}$$

$$x_T = \frac{m'_1x'_1 + m'_2x'_2 + m'_3x'_3}{M'} = x_M \tag{3.a}$$

$$\begin{aligned} x_TM' &= m'_1(x_1 + Al_1) + m'_2(x_2 + Al_2) + m'_3(x_3 + Al_3)) \\ &= A\left(m'_1l_1 + m'_2l_2 + m'_3l_3\right) + m'_1x_1 + m'_2x_2 + m'_3x_3 \end{aligned} \tag{3.b}$$

$$A = \frac{x_TM' - \left(m'_1x_1 + m'_2x_2 + m'_3x_3\right)}{m'_1l_1 + m'_2l_2 + m'_3l_3} \tag{3.c}$$

where

$$m_i = Weight\ of\ LgBl_i\ model\ before\ SFT$$

$$m'_i = m_i + \alpha Wt_i\ (Weight\ of\ LgBl_i\ model\ after\ SFT)$$

$$\alpha Wt_i = \alpha\ additional\ weight\ of\ LgBl_i\ for\ SFT\ (\alpha Wt)$$

$$p_i = i\ DP$$

$$x_i = Xcog\ of\ LgBl_i\ model\ before\ SFT$$

$$x'_i = x_i + Al_i\ (Xcog\ of\ LgBl_i\ model\ after\ SFT)$$

$$SF_i = Model\ shear\ force\ at\ p_i$$

$$SF'_i = TnS\ shear\ force\ at\ p_i$$

$$\Delta SF_i = STF\ value$$

$$x_M = Xcog\ of\ the\ global\ ship\ model$$

$$x_T = Target\ Xcog\ of\ the\ global\ ship\ model\ (x_M = x_T,\ in\ SFT)$$

$$M = m_1 + m_2 + m_3\ (Weight\ of\ the\ global\ ship\ model\ before\ SFT)$$

$$M' = m'_1 + m'_2 + m'_3$$

$$(Weight\ of\ the\ global\ ship\ model\ after\ SFT)(M = M',\ in\ SFT)$$

$$l_i = Length\ of\ LgBl_i\ model$$

$$A = Length\ of\ a\ ratio\ used\ to\ set\ the\ Xcog\ of\ LgBl\ model\ after\ SFT$$

Figures 6 and 7 show the detailed process of weight addition to each $LgBl_i$ for SFT, and the description is as follows. Figure 7a–c show one $LgBl_i$, where the current Xcog of the LgBl is x_i, and the target Xcog is x_i'. The LgBl was divided into two based on x_i, as shown in Figure 7a, and the right side of the LgBl is referred to as partial block 1 ($PB1$), while the left side is partial block 2 ($PB2$). The weight and Xcog of $PB1$ are m_{x1} and x_{p1}, and those of $PB2$ are m_{x2} and x_{p2}, respectively. The weight added to $LgBl_i$ for SFT was αWt_i, calculated using Equations (2)–(3.c). At this time, the weights added to $PB1$ and $PB2$ are Δm_{x1} and Δm_{x2}, respectively. As shown in Figure 7, for x_i to coincide with x_i', x_i must be moved to the right by Al_i. Furthermore, Δm_{x1} and Δm_{x2} were calculated using Equation (4), and only Δm_{x2} was added to x_{p2} of $PB2$. In this case, Δm_{x1} was not added to $PB1$ because its value is zero. By contrast, if x_i' is on the left side of x_i, only Δm_{x1} is added to x_{p1} of $PB1$ because x_i must be moved to the left by Al_i. Similarly, Δm_{x2} is not added to $PB2$ because its value is zero. To satisfy the Xcog, Ycog, and Zcog target values, only 10% of Δm_{x1} and Δm_{x2} were added to the partial block. Further, 10% weight was added to each grid included in the partial block, as calculated using Equations (5)–(5.a). The weight added to the grid was calculated based on the ratio of the current grid weight to the current LgBl weight. Thus, the above process is a method to achieve the Xcog target value during SFT. The SFT, Ycog, and Zcog adjustments are similar to that of the above Xcog method and are shown in Figure 7b,c, respectively. As evident, when adjusting the Ycog, the LgBl is divided into left and right partial blocks based on the current Ycog (Figure 7b), whereas, when adjusting the Zcog, the LgBl is divided into top and bottom blocks based on the current Zcog (Figure 7c). Furthermore, when SFT and adjustments are conducted in the order Xcog → Zcog → Ycog (1 iteration), and if COG $(x_i,\ y_i,\ z_i)$ does not meet the target value $(x_i',\ y_i',\ z_i')$, the process is repeated.

$$\Delta m_{x1} = 0,\ \Delta m_{x2} = -\frac{(m_{x1} + \Delta m_{x1})(l_{x1} + Al_i) + m_{x2}(l_{x2} + Al_i)}{(l_{x2} + Al_i)} \tag{4}$$

where

$$l_{x1} = x_i - x_{p1}$$

$$l_{x2} = x_i - x_{p2}$$

$$Al_i = x_i' - x_i$$

$$x_i = Current\ Xcog\ of\ LgBl_i\ model$$

$$x_i' = Target\ Xcog\ of\ LgBl_i\ model\ for\ SFT$$

$$\Delta m_{x1} = \alpha\ additional\ weight\ of\ PB1\ (Green\ area)\ for\ SFT$$

$$\Delta m_{x2} = \alpha\ additional\ weight\ of\ PB2\ (Blue\ area)\ for\ SFT$$

$$m_{x1} = Current\ weight\ of\ PB1$$

$$m_{x2} = Current\ weight\ of\ PB2$$

$$\begin{aligned} \alpha Wt_{gjx,k} &= \frac{Current\ weight\ of\ Grid_j}{Current\ weight\ of\ LgBl_i} \times 0.1\ \Delta m_{x2}\ (When:\ \Delta m_{x1} = 0) \\ &= \frac{Current\ weight\ of\ Grid_j}{Current\ weight\ of\ LgBl_i} \times 0.1\ \Delta m_{x1}\ (When:\ \Delta m_{x2} = 0) \end{aligned} \tag{5}$$

$$\alpha Wt_{gj,k} = \alpha Wt_{gjx,k} + \alpha Wt_{gjy,k} + \alpha Wt_{gjz,k} \tag{5.a}$$

where

$$\alpha Wt_{gj,k} = k^{th}\ \alpha\ additional\ weight\ of\ Grid_j\ for\ SFT$$

$$\alpha Wt_{gjx,k} = k^{th}\ \alpha\ additional\ weight\ of\ Grid_j\ for\ SFT, Xcog$$

$$\alpha Wt_{gjy,k} = k^{th}\ \alpha\ additional\ weight\ of\ Grid_j\ for\ SFT, Ycog$$

$$\alpha Wt_{gjz,k} = k^{th}\ \alpha\ additional\ weight\ of\ Grid_j\ for\ SFT, Zcog$$

$$k = Iteration\ number$$

$$\Delta m_{x2} = \alpha\ additional\ weight\ of\ PB2\ for\ SFT, Xcog$$

$$\Delta m_{x1} = \alpha\ additional\ weight\ of\ PB1\ for\ SFT, Xcog$$

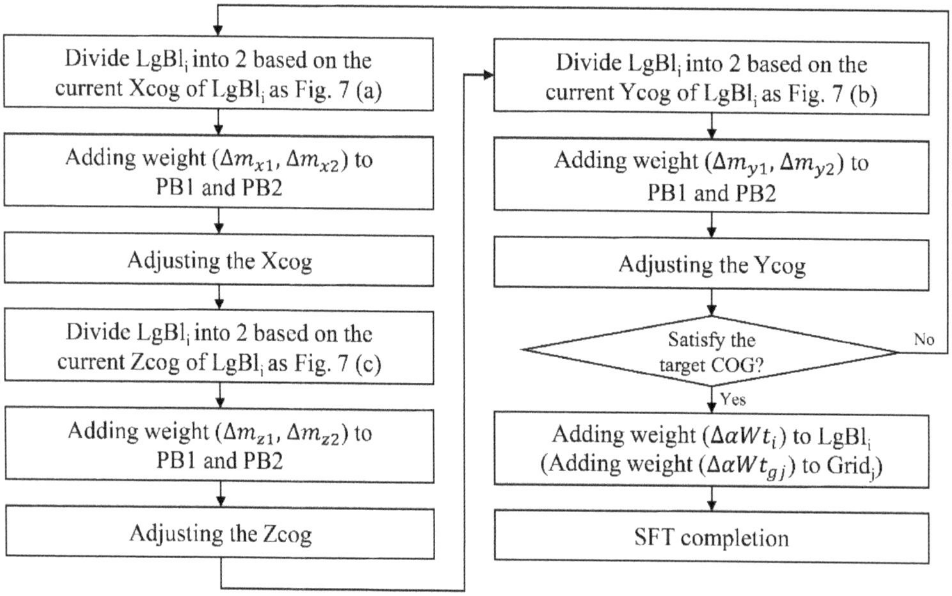

Figure 6. Detailed process of adding weight to $LgBl_i$ and adjusting the COG for the SFT.

In Equations (5)–(5.a), $\alpha Wt_{gjx,k}$ is the weight added to $Grid_j$ for the SFT and adjustment of Xcog in the kth iteration, and $\alpha Wt_{gjy,k}$ and $\alpha Wt_{gjz,k}$ are the weights added for the SFT and adjustment of Ycog and Zcog, respectively. Figure 8 shows the weight added to $Grid_j$ in the kth iteration, prior to the addition of the weight; the grid weight is m_{gj}, whereas it is $m_{gj} + \alpha Wt_{gj,k}$ after weight addition.

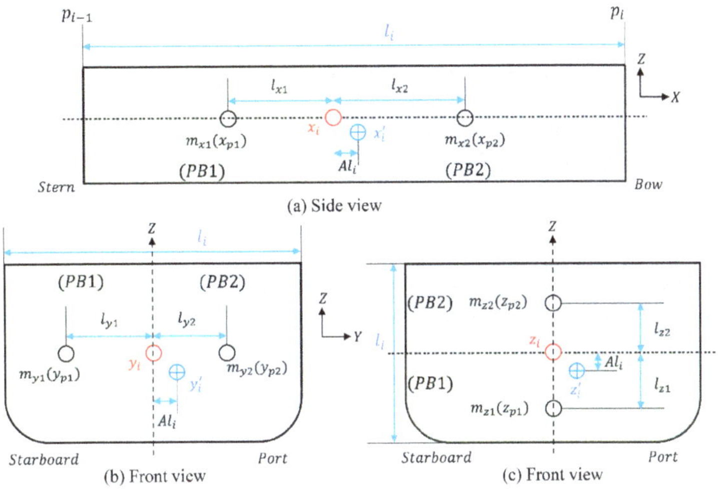

Figure 7. Dividing $LgBl_i$ into two partial blocks based on the current $LgBl_i$ COG location for the SFT. The LgBl is divided according to the current (**a**) Xcog (longitudinal), (**b**) Ycog (transverse), and (**c**) Zcog (height) locations within the block.

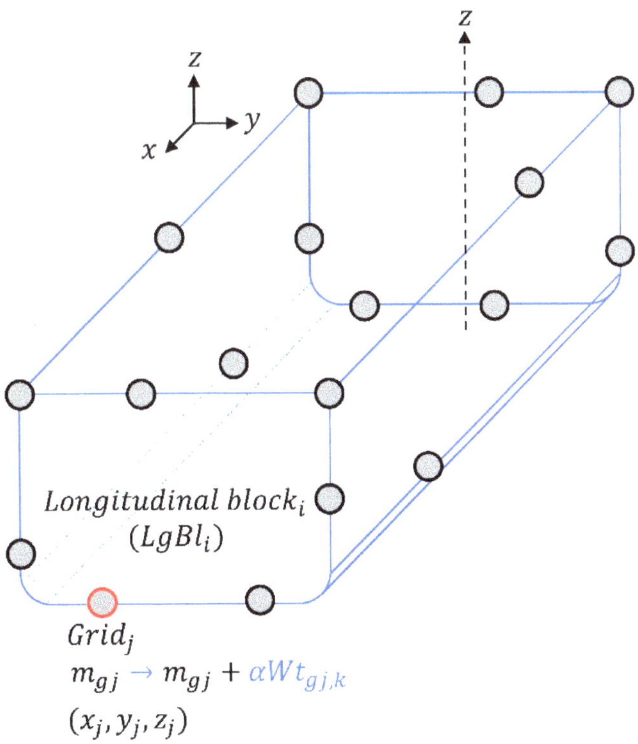

Figure 8. Change in the weight of the grid in $LgBl_i$ for SFT in the kth iteration.

The weight added to $LgBl_i$ in the kth iteration in Equations (5)–(5.a) and (6)–(6.a) is $\alpha Wt_{i,k}$, which is equal to the sum of the weight added $\left(\alpha Wt_{gj,k}\right)$ to $Grid_j$ in the kth iteration. Further, the total weight added to the grid during the entire iteration is the same as the weight added $(\sum_{k=1}^{m} \alpha Wt_{i,k})$ to $LgBl_i$ during the entire iteration. When completing the iteration, the COG satisfies the target value; however, the SF does not satisfy the target value. Therefore, the weight ($\Delta\alpha Wt_i$) is added to $LgBl_i$ using Equations (6)–(6.b) such that the total weight added to the $LgBl_i$ becomes αWt_i, and consequently, the SF is tuned to the target value. At this moment, to keep the COG of the LgBl to the target value, the weight ($\Delta\alpha Wt_i$) is distributed and assigned to the grid included in the LgBl using Equations (6)–(6.b), and the added weight to the grid is $\Delta\alpha Wt_{gj}$.

$$\begin{aligned}\Delta\alpha Wt_i &= \alpha Wt_i - \sum_{k=1}^{m} \alpha Wt_{i,k}\ (m = Total\ iteration\ nember\ by\ SFT\ at\ LgBl_i) \\ &= \sum_{j=1}^{n} \Delta\alpha Wt_{gj}\ (n = Number\ of\ grids\ in\ LgBl_i)\end{aligned} \tag{6}$$

$$\alpha Wt_{i,k} = \sum_{j=1}^{n} \alpha Wt_{gj,k}\ (n = Number\ of\ grids\ in\ LgBl_i) \tag{6.a}$$

$$\Delta\alpha Wt_{gj} = \frac{Current\ weight\ of\ Grid_j}{Current\ weight\ of\ LgBl_i} \times \Delta\alpha Wt_i \tag{6.b}$$

where

$$\Delta\alpha Wt_i = \alpha\ additional\ weight\ of\ LgBl_i\ after\ completion\ of\ iteration\ for\ SFT$$

$$\alpha Wt_i = (Total)\ \alpha\ additional\ weight\ of\ LgBl_i\ for\ SFT$$

$$\alpha Wt_{i,k} = k^{th}\ \alpha\ additional\ weight\ of\ PB\ in\ LgBl_i\ for\ SFT$$

$$k = Iteration\ number$$

$$\Delta\alpha Wt_{gj} = \alpha\ additional\ weight\ of\ Grid_j\ after\ completion\ of\ iteration\ for\ SFT$$

If the SF and COG of all LgBls are adjusted in this way, the total weight, SF, and COG of the ship FE model can meet the target values. Upon completion of SFT, the SFT results and the BM value are examined through global ship analysis. If the total weight, COG, and SF curve of the ship FE model satisfy the target values, the BM value will also satisfy the target value. However, there exist cases where the BM cannot satisfy the target tolerance. In such cases, BMT is performed.

2.2.2. Detailed Process of the BMT

The detailed process of BMT for the ship FE model that has completed SFT is as follows. Because the ship FE model that has completed SFT satisfies the target values for the weight, COG, and SF, the COG of each LgBl must be adjusted for BMT while maintaining the current status of these values. Thus, the BMT value at each DP of the ship FE model is calculated using the BM values obtained from the global ship model analysis, longitudinal BM curve of the ship obtained from the TnS data, and grid weight information. This is required to tune the current BM of the ship FE model to the TnS ship BM. To perform BMT using the calculated values, the movement distance required for the COG of each LgBl is calculated. Figure 9 and Equations (7)–(7.g) show the method of calculating the BMT value and COG when the ship FE model is separated into three sections. The BM is calculated in the direction from the stern to the bow, as in SFT. In contrast to SFT, wherein αWt is added or subtracted, the COG is moved by adjusting the position and ratio of αWt distributed in each LgBl without the addition of αWt for BMT. Moreover, regarding $LgBl_1$ and the DP of p_1, when Xcog changes from x_1' to x_1'' during BMT, the BM changes by ΔBM_1 and meets the target value of BM_1' at BM_1, while the weight of $LgBl_1$ is maintained at m_1'. Here, as the

Ycog and Zcog of each LgBl were tuned to the target Ycog and Zcog values, the transverse (Ycog) and height (Zcog) components of the ship FE model COG location also coincided with the target values. Further, if the BM of $LgBl_2$ and p_2 and that of $LgBl_3$ and p_3 are tuned in the same method, the BM of the ship FE model coincides with the target TnS BM. Moreover, the weight and COG are maintained at the target values.

$$\begin{aligned} \Delta BM_1 &= BM_1' - BM_1 \\ &= m_1'\left(p_1 - x_1''\right) - m_1'\left(p_1 - x_1'\right) \\ &= m_1'\left(x_1' - x_1''\right) \end{aligned} \tag{7}$$

$$\begin{aligned} x_1'' &= \frac{m_1' x_1' - \Delta BM_1}{m_1'} \\ &= x_1' - \frac{m_1'}{\Delta BM_1} \end{aligned} \tag{7.a}$$

$$\begin{aligned} \Delta BM_2 &= BM_2' - BM_2 \\ &= m_1'\left(p_2 - x_1''\right) + m_2'\left(p_2 - x_2''\right) - \left\{m_1'\left(p_2 - x_1'\right) + m_2'\left(p_2 - x_2'\right)\right\} \\ &= m_1'\left(x_1' - x_1''\right) + m_2'\left(x_2' - x_2''\right) \end{aligned} \tag{7.b}$$

$$\begin{aligned} x_2'' &= \frac{m_2' x_2' - \Delta BM_2}{m_2'} + \frac{m_1'\left(x_1' - x_1''\right)}{m_2'} \\ &= x_2' - \frac{\Delta BM_2}{m_2'} + \frac{m_1'\left(x_1' - x_1''\right)}{m_2'} \end{aligned} \tag{7.c}$$

$$\begin{aligned} \Delta BM_3 &= BM_3' - BM_3 \\ &= m_1'\left(p_3 - x_1''\right) + m_2'\left(p_3 - x_2''\right) + m_3'\left(p_3 - x_3''\right) \\ &\quad - \left\{m_1'\left(p_3 - x_1'\right) + m_2'\left(p_3 - x_2'\right) + m_1'\left(p_3 - x_3'\right)\right\} \\ &= m_1'\left(x_1' - x_1''\right) + m_2'\left(x_2' - x_2''\right) + m_3'\left(x_3' - x_3''\right) \end{aligned} \tag{7.d}$$

$$x_3'' = x_3' - \frac{\Delta BM_3}{m_3'} + \frac{m_1'\left(x_1' - x_1''\right) + m_2'\left(x_2' - x_2''\right)}{m_3'} \tag{7.e}$$

$$x_M = \frac{m_1' x_1' + m_2' x_2' + m_3' x_3'}{M'} = x_T \tag{7.f}$$

$$x_T = \frac{m_1' x_1'' + m_2' x_2'' + m_3' x_3''}{M'} = x_M \tag{7.g}$$

where

$$m_i' = Weight\ of\ LgBl_i\ model\ before\ BMT$$

$$M' = m_1' + m_2' + m_3'\ (Weight\ of\ the\ global\ ship\ model)$$

$$x_i' = Xcog\ of\ LgBl_i\ model\ before\ BMT$$

$$x_i'' = Xcog\ of\ LgBl_i\ model\ after\ BMT$$

$$BM_i = Model\ bending\ moment\ at\ p_i$$

$$BM_i' = TnS\ bending\ moment\ at\ p_i$$

$$\Delta BM_i = BMT\ value\ at\ p_i$$

$$x_M = Xcog\ of\ the\ global\ ship\ model$$

$$x_T = Target\ Xcog\ of\ the\ global\ ship\ model$$

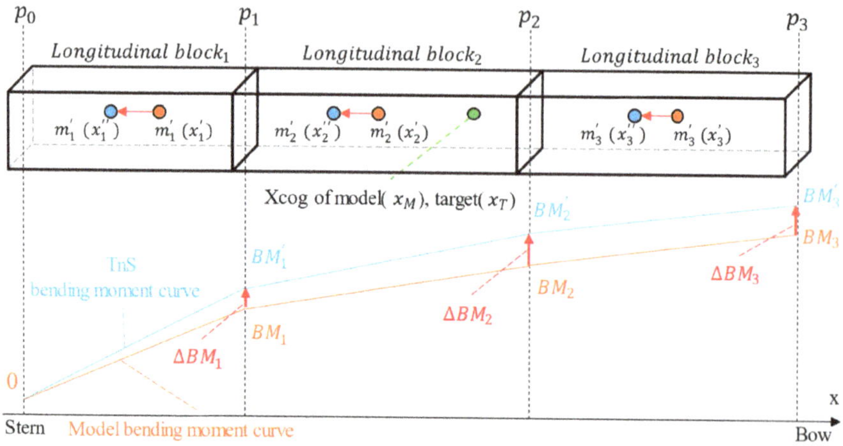

Figure 9. Calculation of the BMT value (ΔBM_i) using the TnS and model BM by global ship analysis. Calculation of the $LgBl_i$ target Xcog x_i'' using ΔBM_i to correspond to the ship FE model Xcog (x_M) to the target Xcog (x_T) through the BMT.

The detailed method for adjusting the COG of each LgBl for BMT is the same as that of the SFT method explained in Figure 6. Similar to 10% of Δm_{x1} or Δm_{x2} being added to $PB1$ or $PB2$ for SFT, 10% of the current LgBl weight (m_i') was added to $PB1$ or $PB2$ for BMT. The weight added to the grid included in each PB is calculated using Equations (8) and (8.a), and that added to $Grid_j$ in the hth iteration is $\alpha Wt_{gj,h}$. This process is repeated until the COG (x_i', y_i', z_i') of the LgBl satisfies the target value (x_i'', y_i'', z_i''). Upon completion of the iterations, although the COG satisfies the target value, the BM cannot satisfy the target value.

$$\alpha Wt_{gjx,h} = \frac{Current\ weight\ of\ Grid_j}{m_i'} \times 0.1 m_i' \quad (8)$$

$$\alpha Wt_{gj,h} = \alpha Wt_{gjx,h} + \alpha Wt_{gjy,h} + \alpha Wt_{gjz,h} \quad (8.a)$$

where

$$\alpha Wt_{gj,h} = h^{th}\ \alpha\ additional\ weight\ of\ Grid_j\ for\ BMT$$

$$\alpha Wt_{gjx,h} = h^{th}\ \alpha\ additional\ weight\ of\ Grid_j\ for\ BMT, Xcog$$

$$\alpha Wt_{gjy,h} = h^{th}\ \alpha\ additional\ weight\ of\ Grid_j\ for\ BMT, Ycog$$

$$\alpha Wt_{gjz,h} = h^{th}\ \alpha\ additional\ weight\ of\ Grid_j\ for\ BMT, Zcog$$

$$h = Iteration\ number$$

$$m_i' = Current\ weight\ of\ LgBl_i$$

Therefore, the excess weight $(\Delta\gamma Wt_i)$ is removed from the LgBl such that the weight of the LgBl is equal to the weight prior to BMT (weight after SFT). At this moment, excess weight $(\Delta\gamma Wt_{gj})$ is removed from the grid in the LgBl using Equations (9)–(9.b) to maintain the COG of the LgBl at the target value.

$$\begin{aligned}\Delta\gamma Wt_i &= \sum_{h=1}^{o} \alpha Wt_{i,h} \ (o = Total\ iterations\ by\ BMT\ at\ LgBl_i) \\ &= \sum_{j=1}^{n} \Delta\gamma Wt_{gj} \ (n = Number\ of\ grids\ in\ LgBl_i)\end{aligned} \quad (9)$$

$$\alpha Wt_{i,h} = \sum_{j=1}^{n} \alpha Wt_{gj,h} \ (n = Number\ of\ grids\ in\ LgBl_i) \quad (9.a)$$

$$\Delta\gamma Wt_{gj} = \frac{Current\ weight\ of\ Grid_j}{Current\ weight\ of\ LgBl_i} \times \Delta\gamma Wt_i \quad (9.b)$$

where

$$\Delta\gamma Wt_i = \gamma\ removal\ weight\ of\ LgBl_i\ after\ completion\ of\ iteration\ for\ BMT$$

$$\alpha Wt_{i,h} = h^{th}\ \alpha\ additional\ weight\ of\ PB\ in\ LgBl_i\ by\ BMT$$

$$h = Iteration\ number$$

$$\Delta\gamma Wt_{gj} = \gamma\ removal\ weight\ of\ Grid_j\ after\ completion\ of\ iteration\ for\ BMT$$

Equations (10) and (10.a) represent the final αWt that is added to $LgBl_i$ and $Grid_j$ in $LgBl_i$ following the application of SFT and BMT. If the SF, BM, and COG satisfy the target values by SFT, the weight $\sum_{h=1}^{o} \alpha Wt_{gj,h} - \Delta\gamma Wt_{gj}$ that is added to the grid based on the BMT in Equation (10.a) becomes zero.

$$\begin{aligned}\alpha Wt_i &= \sum_{k=1}^{m} \alpha Wt_{i,k} + \Delta\alpha Wt_i \quad \left(\sum_{h=1}^{o} \alpha Wt_{i,h} - \Delta\gamma Wt_i = 0\right) \\ &= \sum_{j=1}^{n} \alpha Wt_{gj} (n = Number\ of\ grids\ in\ LgBl_i)\end{aligned} \quad (10)$$

$$\begin{aligned}&\alpha Wt_{gj} = \sum_{k=1}^{m} \alpha Wt_{gj,k} + \Delta\alpha Wt_{gj} + \sum_{h=1}^{o} \alpha Wt_{gj,h} - \Delta\gamma Wt_{gj} \\ &(m = Total\ iterations\ by\ SFT) \\ &(o = Total\ iterations\ by\ BMT)\end{aligned} \quad (10.a)$$

where

$$\alpha Wt_{ij} = (Total)\ \alpha\ additional\ weight\ of\ LgBl_i\ by\ SFT\ and\ BMT$$

$$\alpha Wt_{gj} = (Total)\ \alpha\ additional\ weight\ of\ Grid_j\ by\ SFT\ and\ BMT$$

Thus, by adjusting the BM and COG of all LgBls, the total weight, COG, SF, and BM of the ship FE model can coincide with the required target values.

3. Results

3.1. Global Ship FE Model for Tuning the SF and BM

The accuracy of the SF, BM, weight, and COG were examined before and after performing SFT and BMT for the ship FE model using the developed algorithm. An oil shuttle tanker model, shown in Figure 10, was used to test and verify the accuracy of the SFT and BMT algorithms. The full-size FE model geometry and main dimensions of the shuttle tanker are in Table 1 (Lim et al., 2022) [6]. The ship FE model had an overall length (L_{OA}), breadth (B), and depth of 277.8, 49, and 24.5 m, respectively, divided into 168,120 grids with 357,673 elements.

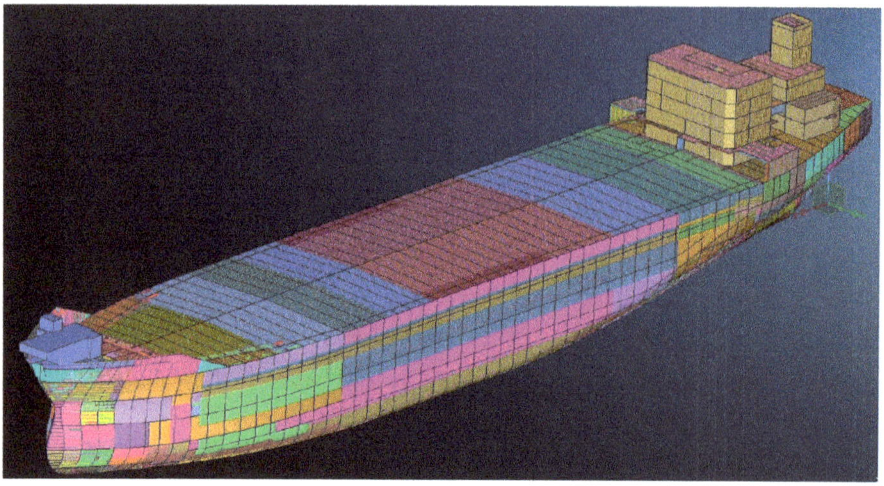

Figure 10. Global ship FE model of oil shuttle tanker used for algorithm validation.

Table 1. Main dimensions and information of ship FE model.

Ship Type	L_{OA}	Breadth	Depth
Oil shuttle tanker	277.8 m	49 m	24.5 m

Table 2 lists the target weight and COG component values for the ship FE model along with the weight and COG before SFT (after the WD). The weight of the ship FE model relative to the target weight is expressed using the unit PU, calculated using Equation (11). The distances between each coordinate (Xcog, Ycog, and Zcog) of the target COG and the COG of the ship FE model before SFT were 0.2707, 0.0171, and 0.0766 m, respectively, with a model weight of 1.0063 PU of the target weight. Thus, the condition of the ship FE model before SFT sufficiently satisfied the target values for the global ship analysis.

Table 2. Weight and COG components for the target and ship FE model before SFT.

Value	Weight (PU)	Xcog (m)	Ycog (m)	Zcog (m)
Target	1.0000	119.9770	−0.1880	14.4790
Before SFT	1.0063	120.2477	−0.1709	14.4024

Table 3 presents the model weight and COG results before and after the application of SFT and BMT. Figures 11–16 show the SF and BM results for the ship FE model before and after the application of SFT and BMT. The associated graph data and errors are also presented in Tables 4–9. In each graph, the x-axis represents the longitudinal direction position of the ship; the x-coordinate of zero corresponds to the after perpendicular of the stern. The SF and BM values are expressed in PU using the weight notation method of the ship FE model. In addition, of all the TnS values, which are the target values for each tuning case, the largest absolute value was used as the reference value. The errors in each table were calculated using the differences between the model and TnS values. The PU for the SF and BM and the error were calculated using Equations (11)–(11.b).

$$Weight\ (PU) = \frac{Weight_{Model}}{Weight_{Target}} \tag{11}$$

$$Shear\ force,\ Bending\ moment\ (PU) = \frac{Model\ value_i - TnS\ value_i}{TnS\ value_{ABS\ max}} \tag{11.a}$$

$$Error = \left| \frac{Model\ value_i - TnS\ value_i}{TnS\ value_i} \right| \times 100 \tag{11.b}$$

where

$$Weight_{Model} = Weight\ of\ global\ ship\ model$$

$$Weight_{Target} = Target\ weight$$

$$Model\ value_i = Model\ shear\ force\ or\ bending\ moment\ at\ ith\ division\ position$$

$$TnS\ value_i = TnS\ shear\ force\ or\ bending\ moment\ at\ ith\ division\ position$$

$$TnS\ value_{ABS\ max} = Maximum\ absolute\ value\ of\ TnS\ shear\ force\ or\ bending\ moment$$

Table 3. Weight, COG target value, and results of SFT and BMT.

Status	Ballast Condition				Full Load Condition			
	Weight (PU)	Xcog (m)	Ycog (m)	Zcog (m)	Weight (PU)	Xcog (m)	Ycog (m)	Zcog (m)
Target value	1.0000	119.9770	−0.1880	14.4790	1.0000	119.9770	−0.1880	14.4790
Before SFT	1.0063	120.2477	−0.1709	14.4024	1.0063	120.2477	−0.1709	14.4024
After SFT	1.0063	120.2477	−0.1710	14.4023	1.0063	120.2475	−0.1707	14.4028
After BMT	1.0063	120.2477	−0.1723	14.4190	1.0063	120.2475	−0.1708	14.4030

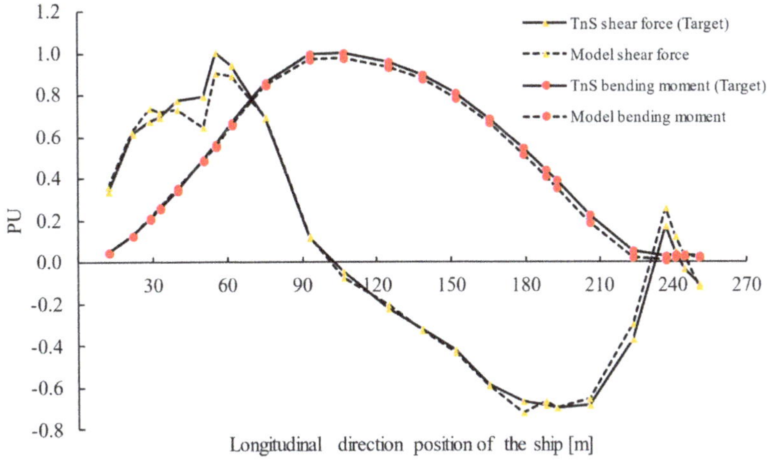

Figure 11. SF and BM curves of the TnS data and ship FE model at each DP before SFT under the ballast condition.

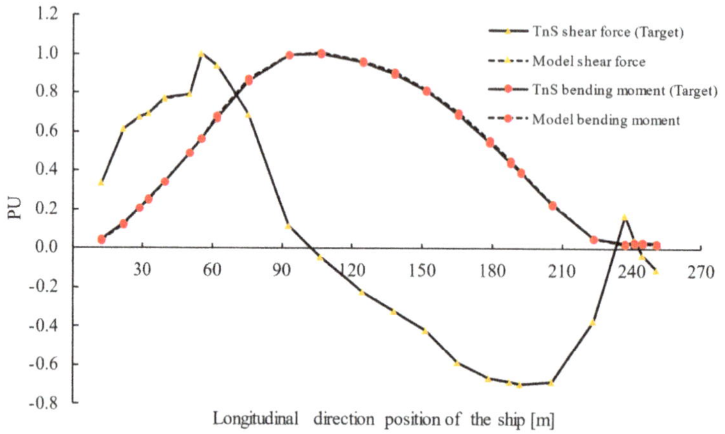

Figure 12. SF and BM curves of the TnS data and ship FE model at each DP after SFT (before BMT) under the ballast condition.

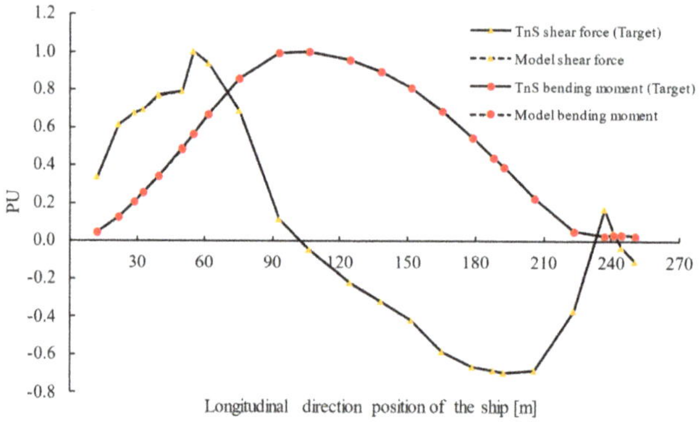

Figure 13. SF and BM curves of the TnS data and ship FE model at each DP after BMT under the ballast condition.

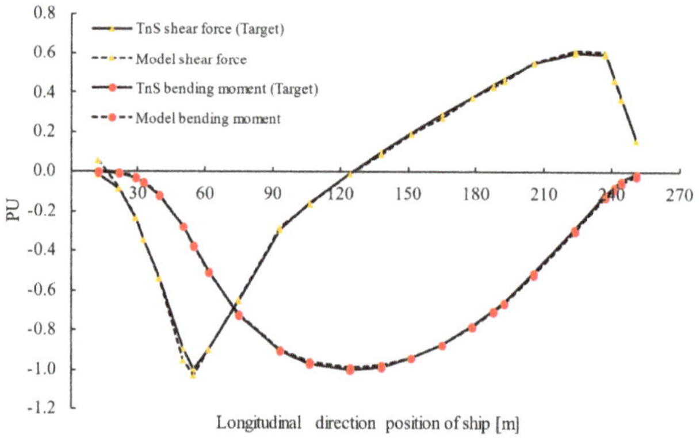

Figure 14. SF and BM curves of the TnS data and ship FE model at each DP before SFT under the full load condition.

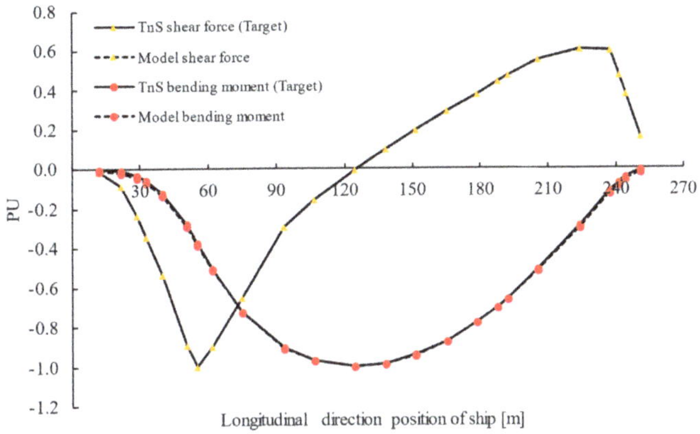

Figure 15. SF and BM curves of the TnS data and ship FE model at each DP after SFT (before BMT) under the full load condition.

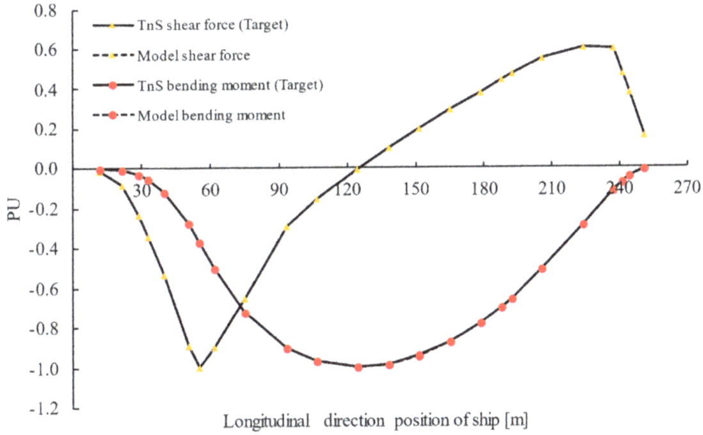

Figure 16. SF and BM curves of the TnS data and ship FE model at each DP after BMT under the full load condition.

Table 4. SF, BM, and error of the TnS data and ship FE model before SFT under the ballast condition.

DP (m)	SF			BM		
	TnS (PU)	Model (PU)	Error (%)	TnS (PU)	Model (PU)	Error (%)
12	0.3385	0.3641	7.5500	0.0449	0.0463	3.2606
21.6	0.6123	0.6283	2.6156	0.1264	0.1281	1.3824
28.8	0.6739	0.7348	9.0418	0.2070	0.2157	4.2228
32.8	0.6937	0.7155	3.1438	0.2541	0.2649	4.2613
40	0.7730	0.7311	5.4138	0.3438	0.3539	2.9276
50.4	0.7915	0.6438	18.6606	0.4873	0.4852	0.4350
55.2	1.0000	0.9011	9.8899	0.5613	0.5529	1.4917

Table 4. *Cont.*

DP (m)	SF			BM		
	TnS (PU)	Model (PU)	Error (%)	TnS (PU)	Model (PU)	Error (%)
61.7	0.9368	0.8874	5.2787	0.6696	0.6574	1.8324
75.2	0.6874	0.6914	0.5933	0.8613	0.8453	1.8641
93.2	0.1160	0.1203	3.6578	0.9914	0.9692	2.2401
106.7	−0.0455	−0.0732	60.9939	1.0000	0.9788	2.1188
124.7	−0.2225	−0.2066	7.1364	0.9587	0.9330	2.6803
138.2	−0.3200	−0.3242	1.2960	0.8961	0.8786	1.9554
151.7	−0.4197	−0.4345	3.5415	0.8111	0.7831	3.4499
165.2	−0.5851	−0.5916	1.1095	0.6901	0.6661	3.4754
178.7	−0.6654	−0.7217	8.4626	0.5457	0.5135	5.9153
187.7	−0.6818	−0.6676	2.0897	0.4425	0.4074	7.9450
192.2	−0.6964	−0.6972	0.1185	0.3898	0.3548	8.9901
205.7	−0.6863	−0.6538	4.7360	0.2280	0.1881	17.4778
223.7	−0.3692	−0.2943	20.2785	0.0526	0.0237	54.9932
237.2	0.1702	0.2540	49.2646	0.0270	0.0119	55.7607
241.2	0.0497	0.1205	142.4491	0.0346	0.0250	27.8002
244.4	−0.0328	0.0378	215.3379	0.0352	0.0288	18.2176
250.8	−0.1057	−0.1189	12.4643	0.0276	0.0250	9.5497

Table 5. SF, BM, and error of the TnS data and ship FE model after SFT (before BMT) under the ballast condition.

DP (m)	SF			BM		
	TnS (PU)	Model (PU)	Error (%)	TnS (PU)	Model (PU)	Error (%)
12	0.3385	0.3389	0.1182	0.0449	0.0435	3.0940
21.6	0.6123	0.6123	0.0017	0.1264	0.1219	3.5271
28.8	0.6739	0.6739	0.0017	0.2070	0.2045	1.2152
32.8	0.6937	0.6937	0.0017	0.2541	0.2507	1.3419
40	0.7730	0.7660	0.9023	0.3438	0.3404	0.9862
50.4	0.7915	0.7915	0.0012	0.4873	0.4896	0.4791
55.2	1.0000	1.0000	0.0007	0.5613	0.5666	0.9377
61.7	0.9368	0.9368	0.0009	0.6696	0.6782	1.2845
75.2	0.6874	0.6874	0.0007	0.8613	0.8715	1.1839
93.2	0.1160	0.1160	0.0032	0.9914	0.9947	0.3340
106.7	−0.0455	−0.0455	0.0044	1.0000	1.0071	0.7140
124.7	−0.2225	−0.2225	0.0002	0.9587	0.9645	0.6016
138.2	−0.3200	−0.3200	0.0008	0.8961	0.9091	1.4430
151.7	−0.4197	−0.4197	0.0005	0.8111	0.8157	0.5658
165.2	−0.5851	−0.5851	0.0005	0.6901	0.7014	1.6489
178.7	−0.6654	−0.6654	0.0005	0.5457	0.5555	1.7876

Table 5. *Cont.*

DP (m)	SF			BM		
	TnS (PU)	Model (PU)	Error (%)	TnS (PU)	Model (PU)	Error (%)
187.7	−0.6818	−0.6818	0.0005	0.4425	0.4534	2.4551
192.2	−0.6964	−0.6964	0.0004	0.3898	0.4002	2.6488
205.7	−0.6863	−0.6863	0.0006	0.2280	0.2313	1.4447
223.7	−0.3692	−0.3692	0.0009	0.0526	0.0517	1.6756
237.2	0.1702	0.1702	0.0006	0.0270	0.0222	17.9298
241.2	0.0497	0.0497	0.0016	0.0346	0.0298	13.8203
244.4	−0.0328	−0.0328	0.0015	0.0352	0.0298	15.3904
250.8	−0.1057	−0.1057	0.0005	0.0276	0.0229	16.9229

Table 6. SF, BM, and error of the TnS data and ship FE model after BMT under the ballast condition.

DP (m)	SF			BM		
	TnS (PU)	Model (PU)	Error (%)	TnS (PU)	Model (PU)	Error (%)
12	0.3385	0.3391	0.1691	0.0449	0.0448	0.1534
21.6	0.6123	0.6123	0.0040	0.1264	0.1262	0.0982
28.8	0.6739	0.6739	0.0035	0.2070	0.2069	0.0660
32.8	0.6937	0.6937	0.0038	0.2541	0.2539	0.0537
40	0.7730	0.7659	0.9165	0.3438	0.3437	0.0412
50.4	0.7915	0.7915	0.0042	0.4873	0.4872	0.0280
55.2	1.0000	1.0000	0.0036	0.5613	0.5612	0.0251
61.7	0.9368	0.9369	0.0040	0.6696	0.6695	0.0198
75.2	0.6874	0.6874	0.0064	0.8613	0.8612	0.0158
93.2	0.1160	0.1161	0.0426	0.9914	0.9913	0.0102
106.7	−0.0455	−0.0454	0.1159	1.0000	0.9999	0.0086
124.7	−0.2225	−0.2225	0.0246	0.9587	0.9586	0.0073
138.2	−0.3200	−0.3200	0.0176	0.8961	0.8961	0.0067
151.7	−0.4197	−0.4196	0.0127	0.8111	0.8110	0.0068
165.2	−0.5851	−0.5850	0.0083	0.6901	0.6900	0.0058
178.7	−0.6654	−0.6654	0.0070	0.5457	0.5457	0.0056
187.7	−0.6818	−0.6818	0.0060	0.4425	0.4425	0.0064
192.2	−0.6964	−0.6963	0.0058	0.3898	0.3898	0.0057
205.7	−0.6863	−0.6863	0.0049	0.2280	0.2280	0.0071
223.7	−0.3692	−0.3692	0.0060	0.0526	0.0526	0.0163
237.2	0.1702	0.1702	0.0070	0.0270	0.0270	0.0120
241.2	0.0497	0.0497	0.0184	0.0346	0.0346	0.0070
244.4	−0.0328	−0.0328	0.0221	0.0352	0.0352	0.0057
250.8	−0.1057	−0.1057	0.0038	0.0276	0.0276	0.0062
12	−0.0127	0.0583	559.8927	0.0005	0.0031	587.5897
21.6	−0.0828	−0.0800	3.3737	−0.0067	0.0013	120.1511

Table 7. SF, BM, and error of the TnS data and ship FE model before SFT under the full load condition.

DP (m)	SF			BM		
	TnS (PU)	Model (PU)	Error (%)	TnS (PU)	Model (PU)	Error (%)
28.8	−0.2356	−0.2258	4.1622	−0.0301	−0.0223	25.8677
32.8	−0.3400	−0.3418	0.5557	−0.0544	−0.0472	13.1915
40	−0.5331	−0.5417	1.6211	−0.1212	−0.1153	4.8087
50.4	−0.8946	−0.9513	6.3369	−0.2758	−0.2770	0.4554
55.2	−1.0000	−1.0324	3.2370	−0.3720	−0.3759	1.0509
61.7	−0.8962	−0.8990	0.3171	−0.5027	−0.5067	0.7942
75.2	−0.6528	−0.6460	1.0284	−0.7249	−0.7199	0.6880
93.2	−0.2957	−0.2818	4.7007	−0.9063	−0.8983	0.8822
106.7	−0.1561	−0.1614	3.3437	−0.9701	−0.9582	1.2260
124.7	−0.0048	−0.0081	70.6826	−1.0000	−0.9898	1.0193
138.2	0.1024	0.0870	15.0214	−0.9861	−0.9788	0.7398
151.7	0.1981	0.1887	4.7091	−0.9431	−0.9410	0.2232
165.2	0.2888	0.2753	4.6792	−0.8732	−0.8751	0.2241
178.7	0.3730	0.3738	0.2053	−0.7786	−0.7840	0.6950
187.7	0.4402	0.4272	2.9478	−0.7010	−0.7073	0.8961
192.2	0.4714	0.4590	2.6262	−0.6575	−0.6661	1.2939
205.7	0.5460	0.5547	1.6107	−0.5117	−0.5234	2.2689
223.7	0.6046	0.6152	1.7516	−0.2881	−0.3006	4.3185
237.2	0.5963	0.6075	1.8625	−0.1168	−0.1282	9.7491
241.2	0.4684	0.4663	0.4456	−0.0718	−0.0831	15.7269
244.4	0.3749	0.3703	1.2333	−0.0432	−0.0554	28.0790
250.8	0.1598	0.1583	0.9361	−0.0078	−0.0199	156.3072

Table 8. SF, BM, and error of the TnS data and ship FE model after SFT (before BMT) under the full load condition.

DP (m)	SF			BM		
	TnS (PU)	Model (PU)	Error (%)	TnS (PU)	Model (PU)	Error (%)
12	−0.0127	−0.0120	5.1498	0.0005	−0.0072	1687.3114
21.6	−0.0828	−0.0828	0.0245	−0.0067	−0.0163	142.9273
28.8	−0.2356	−0.2355	0.0138	−0.0301	−0.0408	35.8468
32.8	−0.3400	−0.3399	0.0115	−0.0544	−0.0662	21.7294
40	−0.5331	−0.5335	0.0806	−0.1212	−0.1335	10.2235
50.4	−0.8946	−0.8945	0.0077	−0.2758	−0.2879	4.3821
55.2	−1.0000	−0.9999	0.0077	−0.3720	−0.3828	2.9009
61.7	−0.8962	−0.8961	0.0095	−0.5027	−0.5118	1.8116
75.2	−0.6528	−0.6527	0.0152	−0.7249	−0.7259	0.1307
93.2	−0.2957	−0.2956	0.0407	−0.9063	−0.9078	0.1639
106.7	−0.1561	−0.1560	0.0825	−0.9701	−0.9690	0.1044

Table 8. *Cont.*

DP (m)	SF			BM		
	TnS (PU)	Model (PU)	Error (%)	TnS (PU)	Model (PU)	Error (%)
124.7	−0.0048	−0.0046	2.8505	−1.0000	−0.9988	0.1156
138.2	0.1024	0.1025	0.1311	−0.9861	−0.9853	0.0826
151.7	0.1981	0.1982	0.0665	−0.9431	−0.9437	0.0642
165.2	0.2888	0.2889	0.0427	−0.8732	−0.8745	0.1586
178.7	0.3730	0.3731	0.0303	−0.7786	−0.7812	0.3344
187.7	0.4402	0.4403	0.0235	−0.7010	−0.7034	0.3406
192.2	0.4714	0.4715	0.0207	−0.6575	−0.6609	0.5140
205.7	0.5460	0.5460	0.0146	−0.5117	−0.5171	1.0545
223.7	0.6046	0.6046	0.0084	−0.2881	−0.2979	3.3859
237.2	0.5963	0.5964	0.0050	−0.1168	−0.1285	10.0436
241.2	0.4684	0.4684	0.0049	−0.0718	−0.0839	16.9518
244.4	0.3749	0.3749	0.0048	−0.0432	−0.0561	29.7041
250.8	0.1598	0.1598	0.0062	−0.0078	−0.0202	159.8586

Table 9. SF, BM, and error of the TnS data and ship FE model after BMT under the full load condition.

DP (m)	SF			BM		
	TnS (PU)	Model (PU)	Error (%)	TnS (PU)	Model (PU)	Error (%)
12	−0.0127	−0.0113	10.5398	0.0005	0.0013	188.5672
21.6	−0.0828	−0.0828	0.0058	−0.0067	−0.0062	6.7113
28.8	−0.2356	−0.2356	0.0022	−0.0301	−0.0297	1.2199
32.8	−0.3400	−0.3400	0.0015	−0.0544	−0.0540	0.6274
40	−0.5331	−0.5319	0.2211	−0.1212	−0.1209	0.2312
50.4	−0.8946	−0.8946	0.0006	−0.2758	−0.2756	0.0794
55.2	−1.0000	−1.0000	0.0006	−0.3720	−0.3718	0.0487
61.7	−0.8962	−0.8962	0.0006	−0.5027	−0.5026	0.0308
75.2	−0.6528	−0.6528	0.0009	−0.7249	−0.7249	0.0072
93.2	−0.2957	−0.2957	0.0019	−0.9063	−0.9062	0.0022
106.7	−0.1561	−0.1561	0.0031	−0.9701	−0.9701	0.0030
124.7	−0.0048	−0.0048	0.1033	−1.0000	−1.0001	0.0090
138.2	0.1024	0.1024	0.0044	−0.9861	−0.9862	0.0134
151.7	0.1981	0.1981	0.0020	−0.9431	−0.9433	0.0181
165.2	0.2888	0.2888	0.0013	−0.8732	−0.8734	0.0236
178.7	0.3730	0.3730	0.0009	−0.7786	−0.7788	0.0310
187.7	0.4402	0.4402	0.0005	−0.7010	−0.7013	0.0371
192.2	0.4714	0.4714	0.0006	−0.6575	−0.6578	0.0407
205.7	0.5460	0.5460	0.0004	−0.5117	−0.5120	0.0552
223.7	0.6046	0.6046	0.0001	−0.2881	−0.2884	0.0930
237.2	0.5963	0.5963	0.0001	−0.1168	−0.1170	0.1839

Table 9. *Cont.*

DP (m)	SF			BM		
	TnS (PU)	Model (PU)	Error (%)	TnS (PU)	Model (PU)	Error (%)
241.2	0.4684	0.4684	0.0001	−0.0718	−0.0720	0.2694
244.4	0.3749	0.3749	0.0001	−0.0432	−0.0434	0.4017
250.8	0.1598	0.1598	0.0001	−0.0078	−0.0079	1.6419

3.2. Results of SFT and BMT under the Ballast Condition

The results of the analysis of the ship FE model under the ballast condition, before and after the application of SFT, are shown in Figures 11 and 12, respectively, while the results after the BMT are shown in Figure 13. As shown in Figure 11, the SF and BM curves of the ship FE model subjected only to the WD exhibit similar trends as those of the TnS data; however, the target values were not satisfied in certain areas before SFT. In particular, the SF curve of the accommodation area located on the stern side within the range of 12–75 m was significantly different from that of the TnS data. Moreover, the SF curve on the bow side at positions > 170 m exhibited a significant difference from that of the TnS data. Table 4 shows that the SF error was the largest (215.3379%) at 244.4 m. Here, the SF was approximately 3.3% of the maximum absolute value of the TnS SF. Further, the BM curve on the bow side at positions > 90 m was less consistent with the TnS data compared to that on the stern side at positions $\leq$ 90 m. In particular, Table 4 shows that the BM error was the largest (55.7607%) at 237.2 m. In this instance, the BM was approximately 1.2% of the maximum absolute value of the TnS BM. Thus, the SF and BM curves were found to have different error occurrence locations.

As shown in Figure 12, the SF and BM curves of the ship FE model were more consistent with those of the TnS data after SFT than before SFT. In particular, the SF exhibited very high accuracy. The dotted line expressing the SF curve of the ship FE model is highly consistent with the solid line curve of the TnS data, resulting in the two curves appearing as one curve along the entire length. Table 5 shows that the SF error was the largest (0.9023%) at 40 m. Here, the model SF was relatively high, approximately 76.60% of the maximum absolute value of the TnS SF. In addition, the error did not exceed 0.0044% along the entire length, except for the 0.9023 and 0.1182% errors at the 40 and 12 m positions, respectively. The accuracy of the BM increased following the SFT application. In particular, the error between the BM curve of the ship FE model and that of the TnS data decreased on the bow side at positions > 90 m, and the number of sections where the dotted line overlapped the solid line increased. As shown in Table 5, the maximum error of the BM in the midship part of the hull did not exceed 2.7%, except for the stern side positions < 21.6 m and those on the bow side at >237.2 m. Further, the bow side positions > 237.2 m exhibited larger errors than those on the stern side at <21.6 m, and the maximum error (17.9298%) was observed at 237.2 m. Thus, the accuracy of the SF was found to increase significantly following the application of SFT, while the accuracy of the BM also improved.

As shown in Figure 13, the SF and BM curves of the ship FE model were highly consistent with those of the TnS data following the BMT application. The respective SF and BM curves appear as one curve each, along their entire lengths, indicating that the BMT was successful and that the high accuracy of the SF curve was maintained. Table 6 shows that the largest SF error (0.9165%) occurred at 40 m, which is the same position that exhibited a high error in the results associated with performing only SFT, as given in Table 5. Further, for the ship FE model, the maximum SF and BM errors did not exceed 0.9165 and 0.1534%, respectively. Moreover, the average SF and BM errors were 0.0591 and 0.0258%, respectively, indicating that very high accuracy was obtained using the algorithm to perform SFT and BMT for the ship FE model under a ballast condition. In addition, Table 3 shows that not only the weight of the ship FE model under the ballast condition but also the COG remained constant before and after SFT and after BMT.

3.3. Results of SFT and BMT under the Full Load Condition

The analysis results of the full load condition of the ship FE model before and after the application of SFT are shown in Figures 14 and 15, respectively, while the results after BMT are shown in Figure 16. As shown in Figure 14, the SF and BM curves of the ship FE model before SFT were analogous to those of the TnS data. Similar to the ballast condition, the BM curve was more consistent than the SF curve, which exhibited low accuracy on the stern side at positions <60 m. Table 7 shows that the SF and BM errors exceeded 500% at 12 m. This is because the model SF was 5.83% of the maximum absolute value of the TnS SF, with a sign opposite to that of the TnS SF. In addition, the BM error was relatively large because both the ship FE model and TnS data had very small BM values. Thus, when the SF and BM values are very small compared to the maximum absolute values of the TnS data, substantial errors may occur despite the small differences.

As shown in Figure 15, the SF and BM curves of the ship FE model were more consistent with those of the TnS data after SFT than before SFT. In particular, the SF curve was more consistent than the BM curve. The results of the BM curve were more consistent under the full load condition than the ballast condition. Consequently, the respective SF and BM curves for the ship FE model and TnS data each appear as one curve along the entire length. Comparing the BM curves of the ship FE model and TnS data, slight differences were observed on the stern side at positions < 60 m. Table 8 shows that the SF error decreased from 559.8927% before SFT to 5.1498% after SFT at the 12 m position; however, the BM error increased approximately 2.8 times from 587.5897 to 1687.3114% at this position. Moreover, because the TnS BM is extremely small (0.05% of the maximum absolute value of the TnS BM) at 12 m, a large error occurred despite fine tuning. In addition, the BM error was large (159.8586%) at the 250.8 m position on the bow side. The TnS BM at this position was also very small, 0.78% of the maximum absolute value of the TnS BM. Thus, the SFT results under the full load condition were found to be more accurate than or similar to the SFT results under the ballast condition, except for the values associated with the positions at both ends, the bow and stern.

Figure 16 shows the BMT results. The SF and BM curves of the ship FE model were highly consistent with those of the TnS data. Further, the results in Figure 16 were more consistent than those in Figure 15, and even the small errors observed on the bow and stern sides were not visible, rendering the respective SF and BM curves as one curve each, along their entire lengths. This indicates that the BMT was successful and that the high accuracy of the SF curve was maintained. Table 9 shows that the BM error decreased by approximately 8.9 times from 1687.3114 to 188.5672% at the 12 m position associated with the largest error. In contrast, at the same position, the SF error increased from 5.1498 to 10.5398%. Both the SF and BM exhibited maximum errors at the 12 m location. Further, all the errors on the bow side at locations > 237.2 m decreased, and the maximum values did not exceed 1.6419%. In addition, the maximum errors of the 32.8–241.2 m positions, excluding those positions where the SF and BM values were relatively small, did not exceed 0.66274%. The average SF and BM errors were 0.4538 and 8.3478%, respectively. Thus, except for the 12 m position where the SF and BM values were very small, the accuracy obtained using the algorithm to perform SFT and BMT for the ship FE model under the full load condition can be concluded to be very high. Table 3 shows that not only the weight of the ship FE model under the full load condition but also the COG remained constant before and after SFT and after BMT.

3.4. Limitations of the Research

The errors in the tuning results of the SF and BM under ballast and full load conditions are mainly due to the method of distributing the weight of elements divided at the LgBl DP to the grids.

Two methods of distributing and calculating the weight of structural beam elements to the grid can be used. One is the method used in this study, and the other is an improved method. The method of distributing the weight used in the study is shown in Figure 17 and Equations (12)–(12.c). The vertical dotted lines represent the DPs. The weight of the structural beam element (M_{beam}) across the LgBl DPs is equally allocated to *Grid*1 and *Grid*2, and αWt is calculated and distributed to both grids using Equations (12)–(12.c). Figure 18 and Equations (13)–(13.g) describe the improved method. After dividing and calculating the beam weight considering the DPs, the beam weight is distributed to A1 and A2, which are the COGs of both divided beams. Subsequently, αWt is calculated and distributed to both grids. In the first method (Figure 17), because the WD and COG tuning of the structural member considering the DPs are inaccurate, the αWt distribution is also inaccurate, causing errors in the SF and BM. Therefore, errors also occur in the ship FE model. In the improved method (Figure 18), if the beam weight can be distributed to A1 and A2, and αWt is calculated and distributed considering the DPs, the ship FE model weight, COG, SF, and BM errors will be reduced, and more accurate tuning is possible.

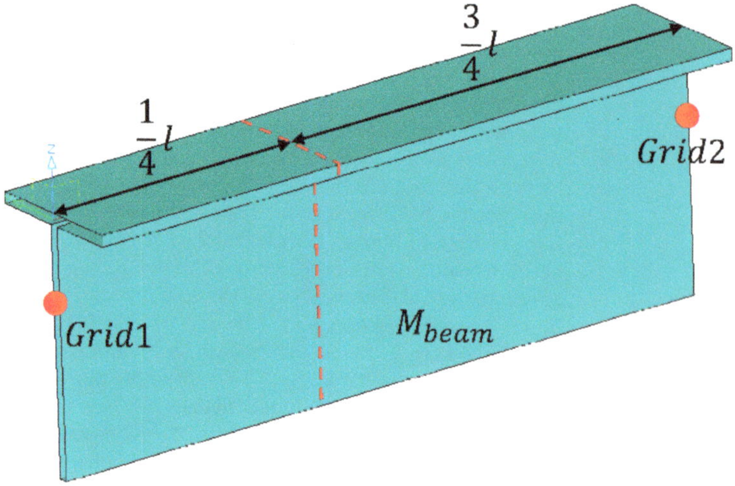

Figure 17. A beam weight distribution method using Equation (12) in this study.

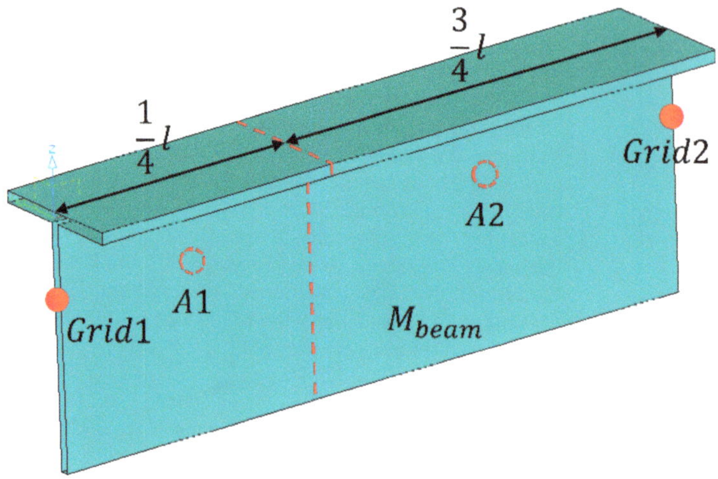

Figure 18. An improved beam weight distribution method using Equation (13).

As indicated in Table 6, which presents the SFT and BMT results for the ship FE model under a ballast condition, the SF and BM at each DP satisfied a tolerance of less than 5%.

However, Table 9, which presents the ship FE model results under a full load condition, shows that the SF and BM errors did not satisfy the tolerance at the 12 m position. In particular, the BM showed a considerable error of 188.5672%. In Figure 16, illustrating the results in Table 9, the BM curves of the ship FE model and TnS appear to coincide at 12 m. However, at this position, the BM values of the ship FE model and TnS were very small. The model BM was 0.13% of the maximum absolute value of the TnS BM, while the TnS BM was 0.05% of the maximum absolute value. Consequently, these very small values caused relatively large errors despite a small change. In addition, the BM at 10 m showed an error of 6.7113%, which was higher than the tolerance, and exhibited the same characteristics as described above.

For the assessment of structural safety using the ship FE model, areas with relatively small SF and BM values, such as the 12 m position, are not areas of interest. The main areas of interest are those with large absolute values of SF and BM. Therefore, even if the SF and BM values exceed the error tolerance, if the values themselves are small, they are not considered rigorously in the structural safety assessment and, thus, are regarded as exceptions. The SFT and BMT results under ballast and full load conditions show that the values at positions where the error exceeded 5% were mostly very small compared to the maximum absolute value. Further, the tuning results exhibited larger errors under a full load condition than under a ballast condition, while the stern showed larger errors than the bow under each condition.

$$Grid1\ weight_{(Before\ tuning)} = \frac{M_{beam}}{2} + AddWt_{g1} \tag{12}$$

$$Grid2\ weight_{(Before\ tuning)} = \frac{M_{beam}}{2} + AddWt_{g2} \tag{12.a}$$

$$Grid1\ weight_{(After\ tuning)} = \frac{M_{beam}}{2} + AddWt_{g1} + \alpha Wt_{g1} \tag{12.b}$$

$$Grid2\ weight_{(After\ tuning)} = \frac{M_{beam}}{2} + AddWt_{g2} + \alpha Wt_{g2} \tag{12.c}$$

$$A1\ weight_{(Before\ tuning)} = \frac{1}{4} M_{beam} \tag{13}$$

$$A2\ weight_{(Before\ tuning)} = \frac{3}{4} M_{beam} \tag{13.a}$$

$$Grid1\ weight_{(Before\ tuning)} = AddWt_{g1} \tag{13.b}$$

$$Grid2\ weight_{(Before\ tuning)} = AddWt_{g2} \tag{13.c}$$

$$A1\ weight_{(After\ tuning)} = \frac{1}{4} M_{beam} \tag{13.d}$$

$$A2\ weight_{(After\ tuning)} = \frac{3}{4} M_{beam} \tag{13.e}$$

$$Grid1\ weight_{(After\ tuning)} = AddWt_{g1} + \alpha Wt_{g1} \tag{13.f}$$

$$Grid2\ weight_{(After\ tuning)} = AddWt_{g2} + \alpha Wt_{g2} \tag{13.g}$$

where

$$M_{beam} = T - Beam\ weight$$

$$AddWt_{gj} = Additional\ weight\ of\ Grid_j\ by\ weight\ distribution$$

$$\alpha Wt_{gj} = \alpha\ additional\ weight\ of\ Grid_j\ by\ SFT\ or\ BMT$$

4. Conclusions

The ship FE model used in global ship analysis must satisfy the longitudinal SF and BM under the loading conditions of the ship. For this, SFT and BMT must be performed for the ship FE model using the TnS data of the ship to allow the model to exhibit SF and BM curves that coincide with those of the ship. In addition, the ship FE model weight and COG must be tuned to the required target values. Thus, in this study, we proposed and developed a C#-based algorithm of SFT and BMT. The accuracy of the new algorithm was analyzed and compared using the ship FE model of an oil shuttle tanker, and the accuracy under a ballast condition was compared with that under a full load condition. The longitudinal positions where an SF and BM check are required for the ship FE model were designated. Based on each position, the model was divided into LgBls, and SFT and BMT were performed. The results are as follows:

(1) Before SFT, the SF and BM curves of the ship FE model under a ballast condition exhibited tendencies similar to those of TnS; however, the SF and BM curves significantly deviated from the TnS values (target values) in the accommodation area on the stern side and the area on the bow side, respectively. However, following SFT, the SF and BM curves of the ship FE model more closely coincided with those of TnS. Although the BM curve exhibited slight differences, the SF curve of the ship FE model was highly consistent with that of the TnS. Further, the curves of the model and TnS appeared as one curve because they were highly consistent with each other. Following BMT, the SF and BM curves of the ship FE model became more consistent with the TnS curves, indicating that the BMT was successful and that it maintained the high accuracy of the SF curve.

Regarding the SF and BM curves of the ship FE model under a full load condition before SFT, the latter was more consistent than the former in a manner similar to the ballast condition. Following SFT, the SF and BM curves of the ship FE model were more consistent with those of TnS. Although slight differences were observed on the stern side similar to in the ballast condition, the curves of the model overlapped those of TnS, thus appearing as one curve. Following BMT, the SF and BM curves of the ship FE model were highly consistent with the TnS curves in the entire area, except for the errors at the 12 m position.

(2) Under a ballast condition, the ship FE model exhibited a maximum SF error of 215.3379% at 244.4 m and a maximum BM error of 55.7607% at 237.2 m before SFT. After SFT, the SF error at 244.4 m significantly decreased from 215.3379% to 0.0015%, and the maximum error was 0.1182% at 12 m. The BM error at 237.2 m significantly decreased from 55.7607% to 17.9298%, which was the maximum error. After BMT, the SF error at 12 m slightly increased from 0.1182% to 0.1691%, which was the maximum error. The BM error at 237.2 m significantly decreased from 17.9298% to 0.0120%, and a maximum error of 0.1534% occurred at 12 m, which exhibited the maximum SF error.

Under a full load condition, the ship FE model exhibited a maximum SF error of 559.8927% at 12 m and a maximum BM error of 587.5897% at the same position before SFT. After SFT, the SF error at 12 m significantly decreased from 559.8927% to 5.1498%, which was the maximum error. The BM error at the same position significantly increased from 587.5897% to 1687.3114%, which was the maximum error. After BMT, the SF error at 12 m increased from 5.1498% to 10.5398%, which was the maximum error. The BM error at the same position decreased from 1687.3114% to 188.5672%, which was the maximum error. Under a ballast condition and a full load condition, not only the weight of the ship FE model but also the COG remained constant at the target values before and after SFT and after BMT.

Under a ballast condition, the ship FE model exhibited very small errors and high accuracy at all positions following the application of SFT and BMT. In contrast, under a full load condition, the SF and BM of the model exceeded the tolerance at 12 m. However, the SF and BM at the position exhibited large errors despite a relatively small change because they were very small compared to the maximum absolute values. Such areas with small values are regarded as exceptions even if the values exceed the tolerance because they are outside the area of interest and thus are not considered carefully in the assessment of structural safety using global ship analysis.

In the above results, the errors of the SF and BM occurred due to WD errors that did not consider the LgBl DPs. Accuracy can be further improved if the WD, SF, and BM are adjusted in consideration of the LgBl DPs. The proposed algorithm is a generalized algorithm for a shuttle tanker (liquid cargo ship). A program including automatic functions was developed to easily use the proposed algorithm. The time required for SFT and BMT using the proposed algorithm takes from a few minutes to an hour, whereas not using the algorithm takes a few days. In addition, the advantage of using the algorithm is that it can achieve high accuracy. Further, the SFT and BMT results under ballast and full load conditions satisfy the weight, SF, BM, and COG conditions required for global ship analysis; therefore, the developed SFT and BMT algorithms are expected to be utilized in global ship analysis. In the future, an SFT and BMT algorithm for the global ship analysis of container ships and offshore plants other than oil carriers must be developed. Therefore, research on an SFT and BMT algorithm for container ships will be conducted.

Author Contributions: C.L.: conceptualization, data curation, investigation, methodology, software; I.-s.H.: data curation, funding acquisition, validation; J.-Y.K.: data curation, validation; I.-j.B. and B.L.: data curation, validation; J.S.P.: validation, writing—review and editing; S.-c.S.: conceptualization, funding acquisition, investigation, project administration, supervision. All authors have read and agreed to the published version of the manuscript.

Funding: This work was supported by the Korea Institute of Energy Technology Evaluation and Planning (KETEP), the Ministry of Trade, Industry, and Energy (MOTIE) of the Republic of Korea (no. 20224000000090), and the Korea Evaluation Institute Of Industrial Technology (KEIT) and the Ministry of Trade, Industry, and Energy (MOTIE) of the Republic of Korea (no. 20018667, Development of Integrated Ship Design System for Hull, Compartment, Basic Calculation, and Loading Guidance Based on Artificial Intelligent Technology). The funders had no role in the study design, data collection and analysis, decision to publish, or preparation of the manuscript.

Institutional Review Board Statement: Not applicable.

Informed Consent Statement: Not applicable.

Data Availability Statement: Not applicable.

Conflicts of Interest: The authors declare no conflict of interest.

References

1. Biondi, B.; Muscolino, G.; Sidoti, A. Methods for Calculating Bending Moment and Shear Force in the Moving Mass Problem. *J. Vib. Acoust.* **2004**, *126*, 542–552. [CrossRef]
2. Cho, H.; Choi, S.; Joo, K. A Study on the Global Vibration Analysis Method of the Commercial Vessels. *Trans. Korean Soc. Noise Vib. Eng.* **2008**, *18*, 509–515. [CrossRef]
3. Kim, D.-m. Effects of Operational Condition and Sea States on Wave-Induced Bending Moments of Large Merchant Vessels. *J. Soc. Nav. Arch. Korea* **2003**, *40*, 60–67. [CrossRef]
4. Kim, I.I.; Choi, J.H.; Jo, H.J.; Suh, H.W. A development of data structure and mesh generation algorithm for global ship analysis modeling system. *Korean J. Comp. Des. Eng.* **2005**, *10*, 61–69. [CrossRef]
5. Du Kim, J.; Jang, B.-S.; Yun, R.-H.; Jeong, H.-S. Improvement of the bending behavior of a flexible riser: Part I I–Hysteretic Modeling of Bending Stiffness in Global Dynamic Analysis. *Appl. Ocean Res.* **2020**, *101*, 102249. [CrossRef]
6. Kim, S.; Ryue, J.; Park, I.-K. Prediction of the wave induced second order vertical bending moment due to the variation of the ship side angle by using the quadratic strip theory. *Int. J. Nav. Arch. Ocean Eng.* **2018**, *10*, 259–269. [CrossRef]

7. Kim, Y.S.; Kim, J.H.; Kim, B.W. A study on Wave Loads acting on a Containership in Waves (II). In Proceedings of the Korean Institute of Navigation and Port Research Autumn Conference, Incheon, Republic of Korea, 16–17 December 2008; Volume 32, pp. 284–285.
8. Kövesdi, B.; Jáger, B.; Dunai, L. Bending and shear interaction behavior of girders with trapezoidally corrugated webs. *J. Constr. Steel Res.* **2016**, *121*, 383–397. [CrossRef]
9. Kwon, Y.S. On the characteristics of still-water and wave bending moments with the variations of ship weight distribution. *J. Ocean Eng. Technol.* **1996**, *10*, 3–13.
10. Lee, S.C.; Doh, D.H.; Goo, J.S. Analysis of wave loads of ships with advancing speed in regular waves. *J. Korea Soc. Power Syst. Eng.* **2010**, *14*, 53–58.
11. Lim, C.; Han, I.-S.; Park, B.-C.; Oh, S.-J.; Kim, G.-Y.; Shin, S.-C. Weight distribution algorithm for global ship analysis in preliminary design stage. *J. Comput. Des. Eng.* **2022**, *9*, 907–918. [CrossRef]
12. Liu, B.; Soares, C.G. Ultimate strength assessment of ship hull structures subjected to cyclic bending moments. *Ocean Eng.* **2020**, *215*, 107685. [CrossRef]
13. Mansour, A.; D'Oliveira, J.M. Hull Bending Moment Due to Ship Bottom Slamming in Regular Waves. *J. Ship Res.* **1975**, *19*, 80–92. [CrossRef]
14. Moreira, L.; Soares, C.G. Neural network model for estimation of hull bending moment and shear force of ships in waves. *Ocean Eng.* **2020**, *206*, 107347. [CrossRef]
15. MSC Software Corporation. MSC NASTRAN 2018 Quick Reference Guide. 2018. Available online: www.mscsoftware.com (accessed on 22 September 2023).
16. Park, H.-S.; Choi, S.-H.; Lee, Y.-S. A Study on Vibration Analysis Method Using the Global Structural Analysis Model. *J. Soc. Nav. Arch. Korea* **2007**, *44*, 314–322. [CrossRef]
17. Pesterev, A.V.; Tan, C.A.; Bergman, L.A. A New Method for Calculating Bending Moment and Shear Force in Moving Load Problems. *J. Appl. Mech.* **2001**, *68*, 252–259. [CrossRef]
18. Rahman, M.K.; Chowdhury, M. Estimation of Ultimate Longitudinal Bending Moment of Ships and Box Girders. *J. Ship Res.* **1996**, *40*, 244–257. [CrossRef]
19. Recupero, A.; D'aveni, A.; Ghersi, A. Bending Moment–Shear Force Interaction Domains for Prestressed Concrete Beams. *J. Struct. Eng.* **2005**, *131*, 1413–1421. [CrossRef]
20. Valtonen, V.; Bond, J.; Hindley, R. Improved method for non-linear FE analysis of polar class ship primary structures. *Mar. Struct.* **2020**, *74*, 102825. [CrossRef]
21. Wang, C.; Wu, J.; Wang, D. Design similar scale model of a 10,000 TEU container ship through combined ultimate longitudinal bending and torsion analysis. *Appl. Ocean Res.* **2019**, *88*, 1–14. [CrossRef]
22. Waskito, K.T.; Kashiwagi, M.; Iwashita, H.; Hinatsu, M. Prediction of nonlinear vertical bending moment using measured pressure distribution on ship hull. *Appl. Ocean Res.* **2020**, *101*, 102261. [CrossRef]
23. Xu, J.; Haddara, M. Estimation of wave-induced ship hull bending moment from ship motion measurements. *Mar. Struct.* **2001**, *14*, 593–610. [CrossRef]
24. Xu, S.; Liu, B.; Garbatov, Y.; Wu, W.; Soares, C.G. Experimental and numerical analysis of ultimate strength of inland catamaran subjected to vertical bending moment. *Ocean Eng.* **2019**, *188*, 106320. [CrossRef]
25. Yun, R.H.; Jang, B.S.; Du Kim, J.D. Improvement of the bending behavior of a flexible riser: Part I—Nonlinear bending behavior considering the shear deformation of polymer layers. *Appl. Ocean Res.* **2020**, *101*, 102249. [CrossRef]

Journal of
*Marine Science
and Engineering*

MDPI

Article

A Study on the Lateral Ultimate Strength and Collapse Modes of Doubly Curved Stiffened Plates

Guangyu Guo [1,2], Jinju Cui [1,2,*] and Deyu Wang [1,2]

1 State Key Laboratory of Ocean Engineering, Shanghai Jiao Tong University, Shanghai 200240, China; naoce.ggy@sjtu.edu.cn (G.G.); dywang@sjtu.edu.cn (D.W.)
2 Institute of Marine Equipment, Shanghai Jiao Tong University, Shanghai 200240, China
* Correspondence: jinjucui@sjtu.edu.cn

Abstract: Bows and stems are often subjected to wave slamming loads. Stiffened plates with curvatures in both longitudinal and transversal directions are the basic members of these structures. As a result, it is important to investigate the lateral ultimate strength of the doubly curved stiffened plates. In this study, the non-linear finite element method (NFEM) is selected to investigate the collapse modes and lateral ultimate strengths of the doubly curved stiffened plates. Additionally, the influences of curvature and geometrical properties on the collapse modes and ultimate strengths of doubly curved stiffened plates are investigated. In the NFEM analysis, a series of numerical simulations, covering different aspect ratios, curvatures, and structural scantlings are performed. Different collapse modes of doubly curved stiffened plates under lateral loading cases are observed. The relationships between the collapse modes and the geometrical properties are then discussed based on the numerical results. Moreover, the results also show that larger curvatures along the stiffeners and stronger stiffeners contribute more to the lateral ultimate strengths. Subsequently, an empirical formula is derived and verified for predicting the lateral ultimate strength of the doubly curved stiffened plates. The results of the empirical formula match well with numerical calculations.

Keywords: stiffened plates; longitudinal and transversal curvatures; lateral ultimate strength; collapse modes; empirical formula

Citation: Guo, G.; Cui, J.; Wang, D. A Study on the Lateral Ultimate Strength and Collapse Modes of Doubly Curved Stiffened Plates. *J. Mar. Sci. Eng.* **2023**, *11*, 2315. https://doi.org/10.3390/jmse11122315

Academic Editor: Joško Parunov

Received: 5 November 2023
Revised: 30 November 2023
Accepted: 5 December 2023
Published: 7 December 2023

1. Introduction

Complicated curved surfaces exist in ship structures, e.g., the bow, stem, and hull surfaces exhibit varying curvatures. Consequently, stiffened plates with curvatures in both longitudinal and transversal directions are employed in these structures. In recent years, there has been increasing interest from both industry and academia in the structural ultimate strength and reliability of bows and stems.

Ultimate strengths of bows and stems are very important in the evaluation of ship structures, especially in ship–ice collision and wave slamming cases. Yang et al. [1] researched the dynamic behaviors of the large container ship's bow structures subjected to slamming pressures; a safety margin evaluating the safety performance of large container ships under a slamming pressure coefficient was presented. Shabani et al. [2] investigated the slamming loads and kinematics during bow entry events and derived an advanced central bow designing method. In such cases, lateral loads with large amplitudes are applied to doubly curved stiffened plates which constitute structures. The plates face the risk of global buckling, which can result in the collapse of the entire structure. Therefore, it is crucial to conduct further research on the ultimate strength issues associated with such scenarios. Moreover, doubly curved stiffened plates are the basic units of ship structures with complicated curved surfaces, of which the collapse mechanisms are not exactly the same as with flat stiffened plates or singly curved stiffened plates. Related research into doubly curved stiffened plates has seldom been reported.

Research on the ultimate strengths and collapse mechanisms of flat stiffened plates serves as a valuable reference for studying curved stiffened plates. Numerous empirical formulae have been developed to conveniently obtain the ultimate strength of the plates or stiffened plates. Liu and Zhang [3] analyzed the influence of the rotational stiffness, aspect ratio, and initial deflection on the ultimate strengths of plates subjected to axial compressions, and a modified empirical formula containing the above factors was proposed. Hayward and Lehmann [4] presented a new formula of the load-bearing capacity of plates under biaxial loads, which can capture the influences of the plate slenderness and aspect ratio. Kim et al. [5] reviewed the well-known formula, predicting the ultimate strength of stiffened plates by comparing it with analytical and FEA results, and proposed the ultimate strength empirical formula without considering the fluctuation behavior. Li et al. [6,7] presented an adapted algorithm to predict the stiffened plates' collapse progress, which extends the predicting capacity of the elastic stiffness and ultimate and post-ultimate features. The ultimate strength of stiffened plates subjected to the combined longitudinal compressive stress and lateral pressure has also been studied, with corresponding numerical analysis, empirical formulae, and experiments. Yao et al. [8] explored the effect of the boundary conditions on the ultimate strength of stiffened panels under the combined action of longitudinal and lateral loads.

Many investigations on collapse behaviors of plane stiffened plates have also been performed, through which several typical collapse modes, deformation mechanisms, and some recommending modeling techniques have been explored. Paik [9] explored six kinds of collapse modes for stiffened plates under different load cases and derived a series of empirical formulae for combined load cases. Ma et al. [10] explored the influence of lateral loads on ultimate strengths and collapse modes of stiffeners under combined load cases and provided a useful explanation about different collapse modes of stiffeners. Xu et al. [11] performed a series of NFEM analyses and investigated the influence of boundary conditions on collapse behaviors of the stiffened plates under different load cases. Li et al. [12,13] investigated the deformation behaviors of continuous plates under combined biaxial loads and lateral pressure. The loading components include both constant loads and cyclic loads.

By taking transversal curvature into account, many studies on singly curved stiffened plates have been conducted. Cui et al. [14] investigated the ultimate strength and collapse behaviors under different load cases of single-curved stiffened plates from cargo containers and derived a series of empirical formulae for predicting the axial ultimate strength of the single-stiffened plates. Park et al. [15] researched the fundamental buckling behaviors of cylindrically curved plates subjected to axial loading and investigated the effects of the curvature, magnitude of initial imperfection, slenderness ratio, and aspect ratio on the characteristics of the buckling and post-buckling collapse behavior of cylindrically curved plates. Seo et al. [16] researched the influence of curvatures on the buckling/post-buckling characteristics and collapse behaviors of singly curved stiffened plates under axial compression and then derived relevant formulae on axial strengths. Park et al. [17] performed a series of elastic-plastic large deflection analyses on singly curved stiffened plates to clarify the fundamental behaviors of cylindrically curved plates under axial compression and lateral pressure. A double beta formula considering secondary buckling behaviors was derived based on the results. Cho et al. [18] performed an experiment on singly curved stiffened plates and investigated the influence of the curvature effect on the ultimate strengths of the single-curved stiffened plates. Research on the collapse modes and ultimate strengths of flat or singly curved stiffened plates has laid a foundation for studying doubly curved stiffened plates. The methodologies, concepts, and ultimate strengths' influencing factors involved in these studies are instructive.

In this study, the collapse behaviors and ultimate strengths of doubly curved stiffened plates under lateral loading cases are investigated. Considering different kinds of ship structures, the influences of varying types of stiffeners, longitudinal and transversal curvatures, and geometric scantlings on the ultimate strengths and collapse modes are taken into account. Based on above influential factors studies, a prediction formula for lateral ultimate strengths of such structures has been derived and verified, resulting in good accuracy with simple form. The results from this research can be helpful for the understanding of doubly curved panels' ultimate strength and collapse behavior as well as its contribution to the ultimate strengths of ship structures with complicated curved surfaces.

2. Methodology

In this study, the geometrical parameters of doubly curved stiffeners are firstly obtained according to the data from different ships. The NFEM models of the doubly curved with different slenderness ratios are established in the next step. Then, the lateral ultimate strengths and collapse modes of the models are obtained via the NFEM. Finally, an empirical formula of the doubly curved stiffened plates' lateral ultimate strength is derived.

2.1. Description of Doubly Curved Stiffened Plate

Doubly curved stiffened plates make up the complex surfaces of bows and stems, and they contribute to the strengths and the loading capacities of the structures. Such stiffened plates have curvatures in both the longitudinal and transversal directions, and the plates form complicated curved surfaces with varying curvatures. The stiffened plates in this study are selected as single span and single bay models. According to the recommendation of ISSC (2012) [19], such a model extent is acceptable. As such, stiffened plates are supported by transversal and longitudinal primary supporting members (PSMs), and, in the case of lateral loading, the boundary conditions of the stiffened structures at the PSMs can be regarded as clamped supported. Considering the background of this study, several stiffened plates derived from the bows and stems of an icebreaker, an oil tanker, and a cargo container are selected as references models. Models with different curvatures are set to explore the influence of curvature on the ultimate strengths of doubly curved stiffened plates. The structural settings of the models are shown in Figure 1, and the original geometric parameters can be found in Table 1. And the parameters appear in this research are shown in Nomenclature Section, where units for geometry and stress are mm and MPa, respectively. The detailed structural settings of the models in this research can be found in Appendix A.

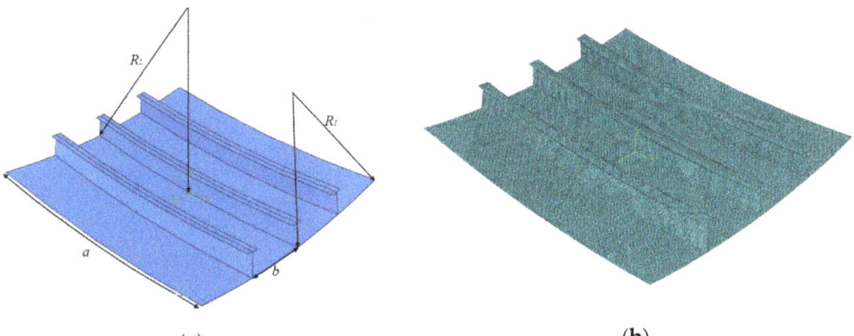

(a) (b)

Figure 1. Structural settings of typical doubly curved stiffened plates: (**a**) geometric model; (**b**) NFEM model.

As shown in Figure 1a, a denotes the longitudinal span, b denotes the spacing between adjacent stiffeners, R_L denotes the curvature radius along the stiffeners, and R_T denotes the curvature radius along the direction that is perpendicular to the stiffeners. To illustrate the curvature and its influence on the ultimate strengths, the curvatures in longitudinal and transversal directions of all the models are represented by groups of dimensionless terms $\theta_L = a/R_L$ and $\theta_T = b/R_T$, which are shown in Figure 2. All the stiffeners are equipped with a 'T' type section, which is depicted in Figure 3.

Table 1. Structural settings of the original doubly curved stiffened plates models for this study.

Ship	R_L (mm)	R_T (mm)	a (mm)	b (mm)	Stiffener	Aspect Ratio	σ_Y (MPa)	Number of Stiffeners
Icebreaker	14,523	40,306	4000	350	Tee bar, T360 × 20 + 90 × 20	11.42	355	5
Icebreaker	17,075	48,849	4000	350	Tee bar, T360 × 20 + 90 × 20	11.42	355	7
Icebreaker	5719	45,560	2375	350	Tee bar, T360 × 20 + 90 × 20	6.78	355	5
Icebreaker	5475	54,020	2375	350	Tee bar, T360 × 20 + 90 × 20	6.78	355	7
Container ship	74,243	4770	2400	650	Tee bar, T280 × 11 + 120 × 20	3.69	355	3
Container ship	13,104	7198	3044	675	Tee bar, T320 × 12 + 120 × 20	4.50	355	3
Icebreaker	11,500	59,500	2375	350	Tee bar, T360 × 20 + 90 × 20	6.78	355	5
Oil tanker	26,500	37,500	3200	800	Tee bar, T420 × 10 + 120 × 16	4	355	5
Oil tanker	41,600	27,200	3200	800	Tee bar, T420 × 10 + 120 × 16	4	355	5

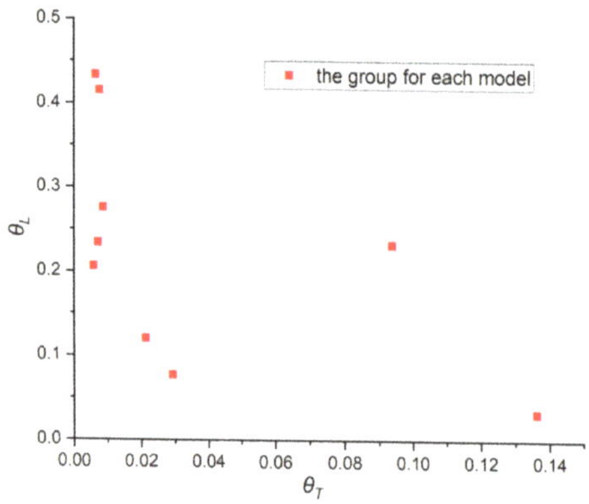

Figure 2. The curvature radiuses of the stiffened plate models.

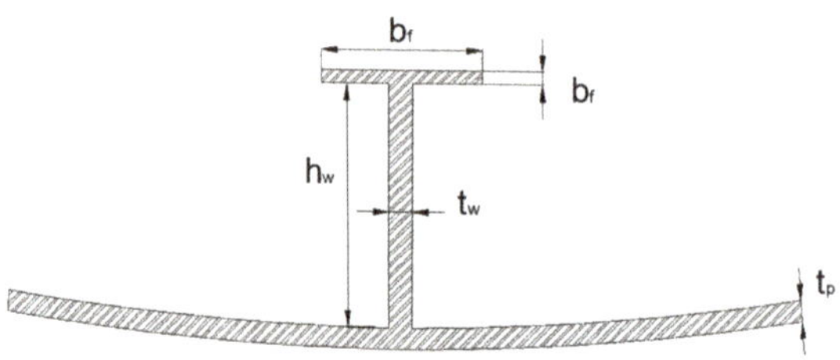

Figure 3. The dimensions of the section of the stiffener.

As there is curvature in the longitudinal or transversal direction, Seo et al. [16] investigated the singly curved stiffened plate and concluded that the ultimate strengths of such plates are related to the slenderness of flat plates with identical geometric dimensions. Therefore, the structural strengths of doubly curved stiffened plates can be evaluated by the slenderness ratios λ and β, where λ denotes the column slenderness ratio of stiffener with attached plating and β denotes the slenderness ratio of plating between stiffeners, which are expressed by Equation (1):

$$\begin{aligned} \beta &= \frac{b}{t_p}\sqrt{\frac{\sigma_Y}{E}} \\ \lambda &= \frac{a}{\pi r}\sqrt{\frac{\sigma_Y}{E}} \\ r &= \sqrt{\frac{I}{A}} \end{aligned} \quad (1)$$

To clarify the influence of initial imperfections on lateral ultimate strengths, a curved stiffened plate with transversal curvature identical to model 'R5' is established. The 'buckling' mode initial imperfection recommended by ISSC [19] and the 'thin horse' mode initial imperfection conducted by Yao et al. [20] are applied to the model, respectively. The modes of the initial deflections are depicted in Figure 4 and Equations (2)–(5). As shown in Figure 5, the lateral strengths of models with different geometrical imperfections are similar. Moreover, the lateral ultimate strength does not significantly decrease with the occurrence of the initial deflections, and the differences in the ultimate strengths between models with and without initial deflection are less than 5%. Consequently, the influence of the initial deflection on lateral ultimate strengths is actually limited. And the results from models without initial imperfections can be considered reasonable.

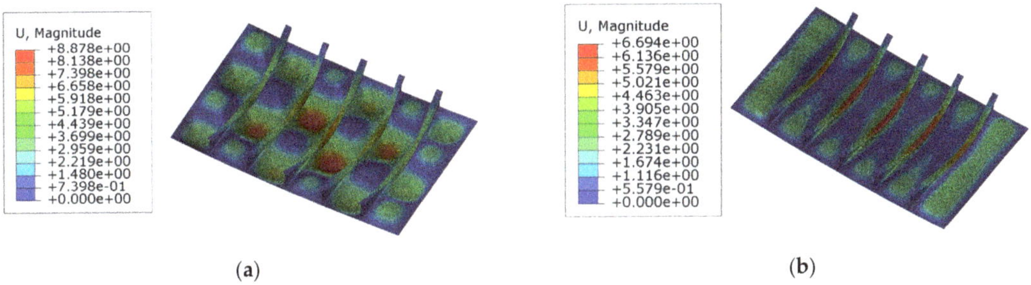

Figure 4. The modes' initial imperfections. (**a**) Buckling mode; (**b**) thin horse mode. Scale: 40.

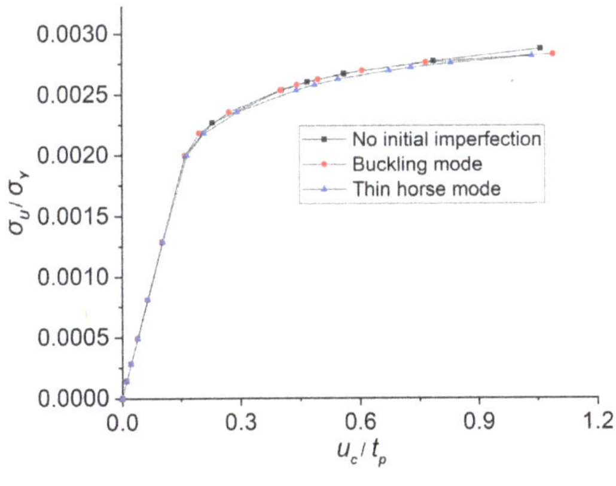

Figure 5. Influence of initial imperfections on ultimate strengths.

The column type deformation of a stiffened plate:

$$v_{oc} = B_0 \sin\left(\frac{\pi z}{a}\right) \sin\left(\frac{\pi x}{B}\right) \tag{2}$$

where B_0 denotes the magnitude of column type mode and is recommended as 0.0015a.

The initial tripping distortion of a stiffener:

$$v_{oc} = B_0 \sin\left(\frac{\pi z}{a}\right) \sin\left(\frac{\pi x}{B}\right) \tag{3}$$

The buckling mode initial deflection of local plating:

$$v_{opl} = A_0 \sin\left(\frac{m\pi z}{a}\right) \sin\left(\frac{\pi x}{b}\right) \tag{4}$$

The thin horse mode initial deflection of local plating:

$$v_{opl} = \left| \sum_{k=1}^{11} A_{0k} \sin\left(\frac{k\pi z}{a}\right) \sin\left(\frac{\pi x}{b}\right) \right| \tag{5}$$

where A_{0k} are determined according to the empirical value derived by Yao et al. [20], and is shown in Table 2, and the maximum magnitude A_0 is the same as the previous buckling mode.

Table 2. Coefficients of initial deflection magnitudes.

A_{01}/A_0	A_{02}/A_0	A_{03}/A_0	A_{04}/A_0	A_{05}/A_0	A_{06}/A_0	A_{07}/A_0	A_{08}/A_0	A_{09}/A_0	A_{10}/A_0	A_{11}/A_0
1.1439	−0.0677	0.3385	0.0316	0.1579	−0.0149	−0.0043	0.0008	0.0039	−0.0002	−0.0011

In shipyards, the manufacturing of doubly curved plates with complicated curvatures is achieved through the line heating technique. Consequently, the existing empirical formulae for initial imperfections such as 'buckling' and 'thin horse' are not suitable for such plates. Moreover, in actual conditions, residual stress will shake down after multiple loading and de-loading processes; thus, the influence of residual stress is negligible in a real ship structure. The calculations performed above have proved that the influence of the initial deflection is neglectable, and there are few data about the initial imperfections that are applicable for doubly curved stiffened plates. Therefore, the influence of the initial deflections and residual stress is not considered in this study.

2.2. NFEM Analysis

The computer code ABAQUS is considered as the numerical analyzing method in this study. The type of elements is chosen as S4R, which simulates the shell and possesses four nodes and six degrees of freedom. In the NFEM models, the nonlinearities of material and geometry are all taken into account. The geometrical nonlinearity is performed by the using the 'NLGEOM' option in the code. This option takes large deformations and the nonlinearity between displacements and strains into account. And the nonlinearity of the material is performed by considering the yielding strength in the constitutive relationship of the material. Material nonlinearity and geometric nonlinearity must be considered in research, while initial imperfection is not a mandatory input; meanwhile, there is no relevant data available for doubly curved stiffened plates, so it is not considered in this study. The property of steel is defined with the ideal elastic-plastic model (the material has no plastic strain, and the nonlinearity is realized by yielding strength, σ_Y), which is illustrated in Figure 6.

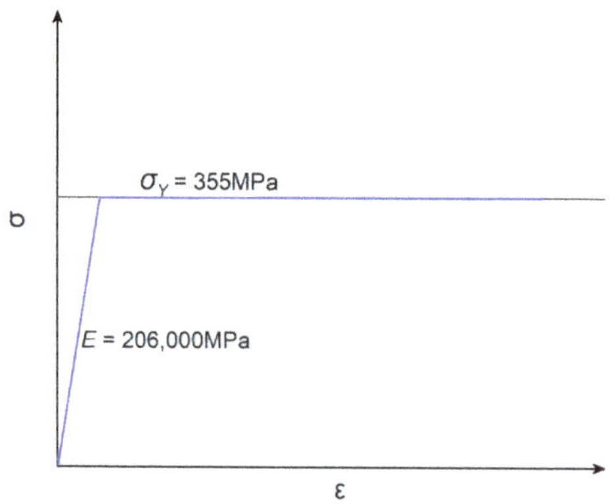

Figure 6. The ideal elastic-plastic constitutive relationship for the steel.

Additionally, the implicit static method is applied during the calculation. As the size of the mesh has an important impact on the speed and results of the calculation, to keep a balance between the accuracy and efficiency of the calculation, an appropriate size of the mesh should be confirmed. The mesh sizes are selected as 10 mm, 15 mm, 20 mm, 25 mm, and 45 mm, respectively. As shown in Figure 7, the ultimate strength exhibits a stable state when the element sizes are within 10 mm and 20 mm. As a result, an element size of 20 mm was used in the finite element model to ensure relatively accurate results and higher calculation efficiency. Moreover, to justify the reliability of the NFEM, we also performed a comparison calculation on an axial ultimate strength problem of the standard stiffened plates conducted by ISSC 2012 [19], and the relevant information and results are shown in Table 3. The error between the results of the NFEM and ISSC is within 2.5%, signifying that the NFEM employed in this study is reliable.

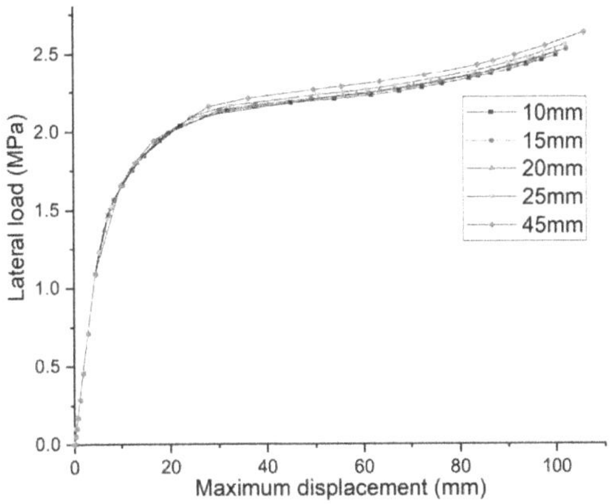

Figure 7. Results of the mesh sensitivity analysis.

Table 3. Results of the benchmark calculation.

σ_Y (MPa)	t_p (mm)	Stiffeners' Parameters (mm)	Result (MPa)	Result of ISSC (MPa)	Error
313.6	15	580 × 15/150 × 20, Tee Bar	232.52	227.05	2.4%

2.3. Boundary Conditions and Load Applications

The quantity of longitudinal and transversal primary supporting members in bow or stem structures is generally larger than those in other parts; therefore, doubly curved stiffened plates are subjected to stronger constraints than stiffened plates in other parts. Additionally, the deformations of the shell, when subjected to uniformly distributed lateral loads, are symmetric in longitudinal or transversal direction. Consequently, in the investigation of the lateral ultimate strength of the doubly curved stiffened plates, the clamped constraint is applied on the boundaries of the numerical model, restricting all the translations and rotations. The lateral loads are applied on the outer surface without stiffeners in the form of a uniformly distributed normal load. The settings of load and boundary conditions are shown in Figure 8.

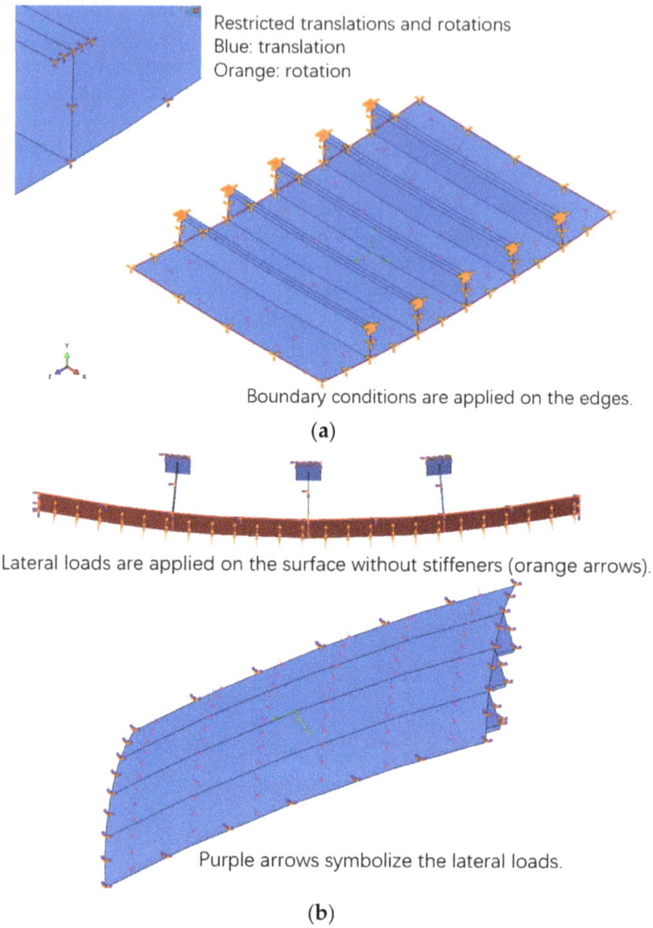

Figure 8. NFEM settings. (**a**) Boundary conditions; (**b**) lateral loading.

3. Results and Discussion

Based on the NFEM, a series of calculations have been performed on typical doubly curved stiffened plate models designed according to corresponding ships. The following results have been obtained by analyzing the target models with 144 different structural dimensions. The variations in the curvature radius, plate thickness, and stiffeners' sections are considered in the design of these numerical models. According to the structural designs of target ships and the information from Zhang [21], the column slenderness of the model ranges from 0 to 1.5 and the plate slenderness ranges from 0 to 3. To describe the geometrical properties conveniently, the stiffeners are named in the form of 'Curvature radius-Plate thickness-Stiffener'. Nine groups of curvature radiuses are represented by 'R1'-'R9'. In each group, four types of plate thickness and stiffeners' sections are selected, denoted as 'T1'–'T4' and 'S1'–'S4', respectively, and the corresponding number represents a specific type. For example, 'R2' denotes these models have the same longitudinal and transversal curvature radius as those in group 2, while 'T1' or 'S1' denotes the same setting as the original model. Similarly, the model 'R7T2S3' signifies that the plate thickness and stiffeners' properties differ from those of the original models.

3.1. The Determination of the Lateral Ultimate Strength

In this study, the lateral ultimate strength of the doubly curved stiffened plate is determined by identifying the lateral load at the inflection point of the lateral load-central displacement curve. The magnitude displacement of the central point is selected as the response of the structure in the lateral loading cases. Figure 9b illustrates the lateral load-magnitude displacement curve for each case, where four points are selected to define two straight lines. These lines represent different parts of the curve; namely, the linear loading stage before buckling occurs and the post-buckling stage with large deformations. The lateral load at the intersection of these lines is considered as the lateral ultimate load, measured in MPa. The lateral ultimate strength and the structural response are normalized by σ_U/σ_Y and w/t_p, respectively, where σ_Y denotes the yielding stress of the material and u_c denotes the normal displacement of the central point of the model.

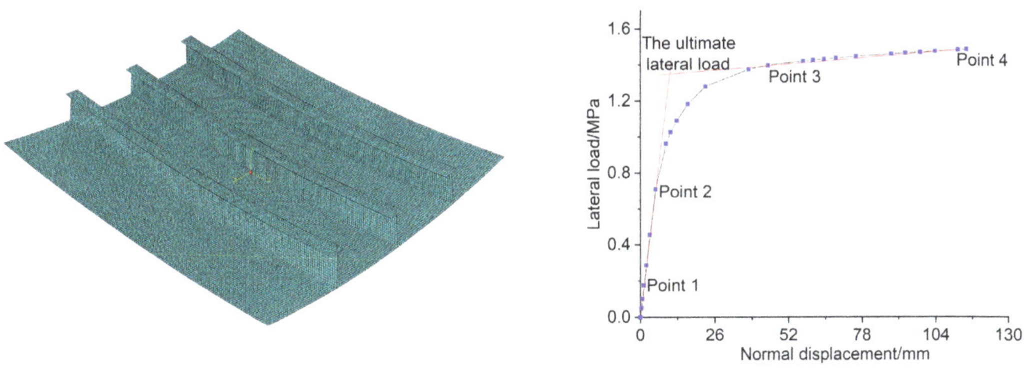

(a) (b)

Figure 9. The definition of the lateral ultimate strength. (**a**) The central point of the model (the red highlighted node); (**b**) the determination of the lateral ultimate load.

3.2. Collapse Modes and Ultimate Strength of Doubly Curved Stiffened Plates under Lateral Pressure Loading

3.2.1. The Lateral Ultimate Strength of the Doubly Curved Stiffened Plates

The lateral load—central displacement curves of model R1 are shown in Figure 10. The column slenderness ratios of model R1 range from 0.368 to 0.684, while the plate slenderness ratios range from 0.605 to 1.816. The thicknesses of plates are set as 24 mm, 14 mm, 11 mm, and 8 mm, respectively, and are labeled as 'T1'–'T4'. Initially, the displacement of the central point increases slowly as the lateral load increases, indicating that the lateral load applied to the stiffened plate is within its loading capacity. However, once the lateral load exceeds the ultimate lateral strength, even a small increment in lateral load leads to rapid growth in the normal displacement of the central point, indicating the loss of the loading ability and collapse of the stiffened plate. Figure 10 shows that when λ is set constant, the lateral ultimate strengths decrease significantly for the stiffened plates with thinner plates. Models with thickness 'T4' exhibit the lowest lateral ultimate strength. It can be inferred that for the doubly curved stiffened plates with stronger stiffeners, the thicknesses of the plates exert a significant influence on the lateral ultimate strength.

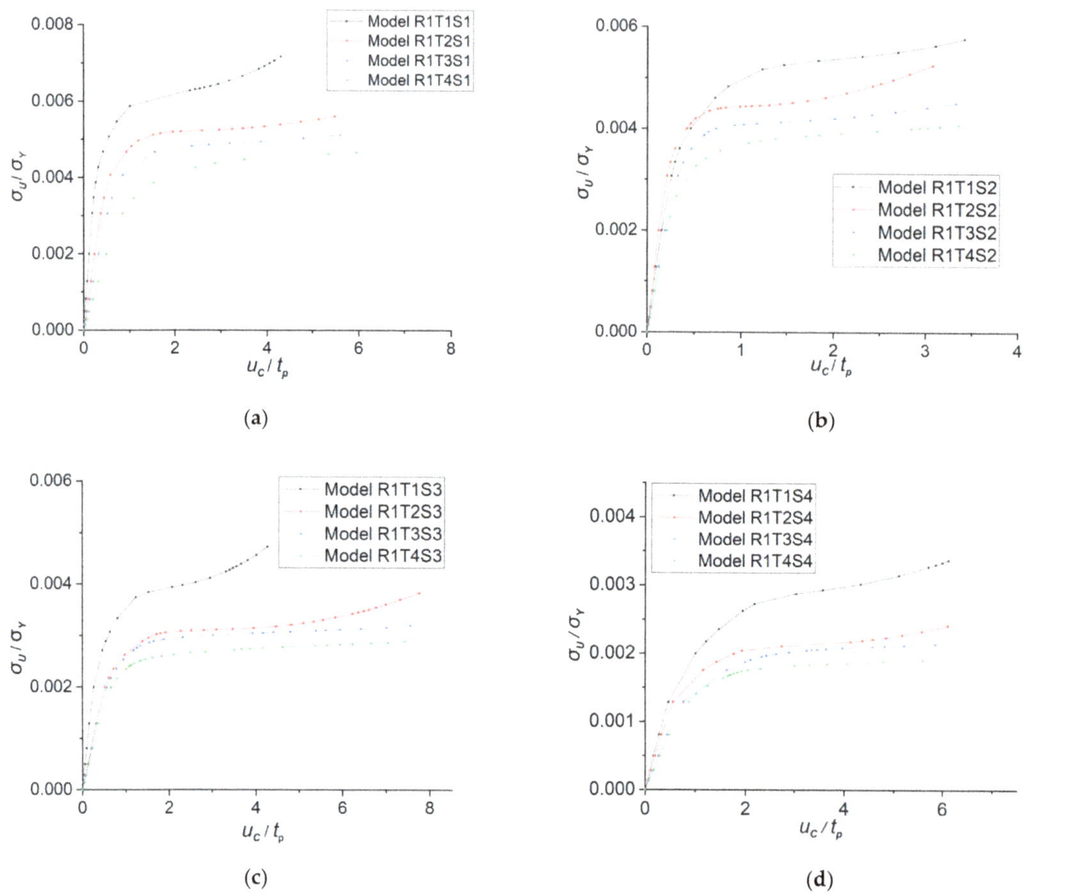

Figure 10. The lateral ultimate strength-displacement curve of model R1. (**a**) Models with stiffener 'S1'; (**b**) models with stiffener 'S2'; (**c**) models with stiffener 'S3'; (**d**) models with stiffener 'S4'.

The lateral ultimate strengths of models 'R1'–'R4' are depicted in Figure 11. The distance between adjacent stiffeners and the plate slenderness ratios of models 'R1'–'R4' are identical. Models 'R3' and 'R4' have smaller longitudinal curvature radiuses, resulting in significantly larger longitudinal curvature angles (0.415 rad for 'R3'; 0.434 rad for 'R4'; 0.275 rad for 'R1'; and 0.234 rad for 'R2'). Figure 11 shows that as the stiffeners become weaker, the lateral ultimate strengths of the doubly curved stiffened plates decrease. The relationship between lateral ultimate strength and column slenderness is inversely proportional. An interesting phenomenon is observed in cases with different plate thicknesses: the lateral bearing capacity increases significantly as the curvature angle decreases. On one hand, the curves for 'R3' and 'R4' are above those for 'R1' and 'R2'. On the other hand, the lateral bearing ability of models 'R3' and 'R4' are stronger than those of models 'R1' and 'R2' even if the column slenderness ratios are significantly larger. As the curvature angle of the stiffener increases, the shapes of the stiffened plates approach spherical shells, leading to more restrained deformation of plates and stiffeners. Consequently, the lateral bearing capacities increase significantly.

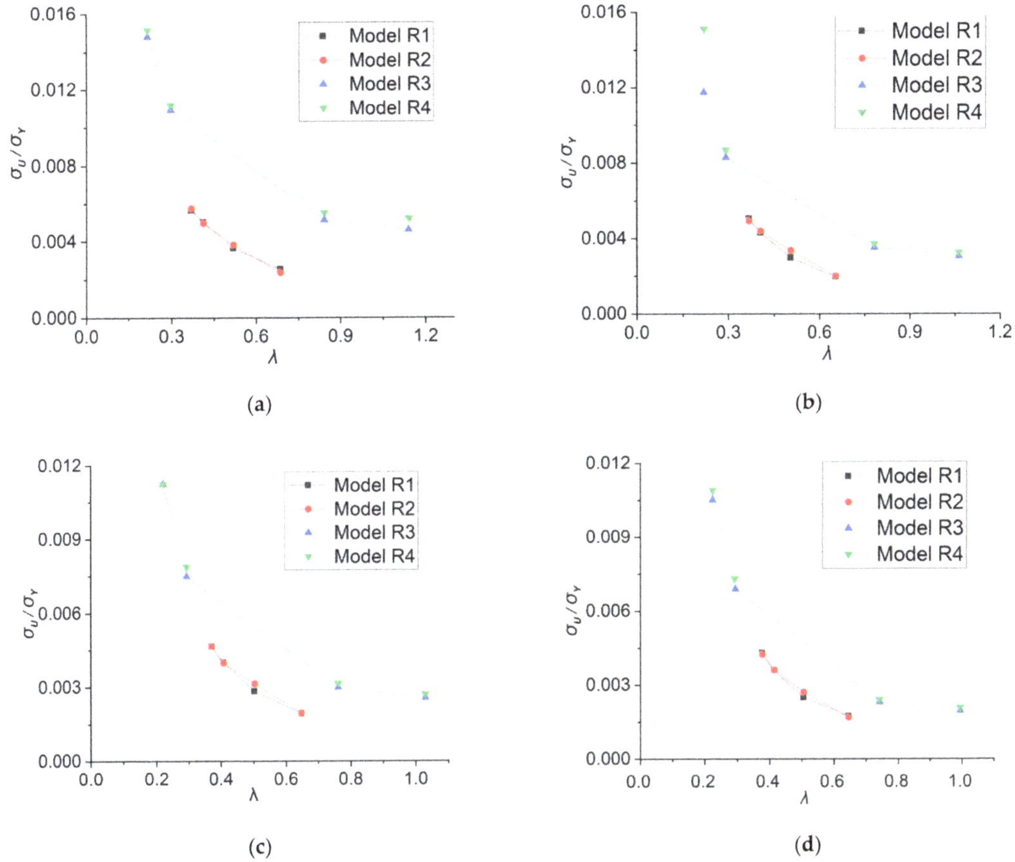

Figure 11. The lateral ultimate strengths of models R1–R4: (**a**) $\beta = 0.605$, (**b**) $\beta = 1.038$, (**c**) $\beta = 1.321$, and (**d**) $\beta = 1.816$.

In addition, the influence of low temperature has been considered in this study. Wang et al. [22] conducted a uniaxial tensile experiment on the marine steel selected for this study to investigate the effect of low-temperature conditions on the steel's properties. The results indicate that low temperature increases the elastic modulus and yielding strength, and the ideal elastic-plastic model is still applicable under these conditions. According to the outcomes of the experiment, the elastic modulus is set as 228,000 MPa and the yielding strength is set as 373 MPa (to simulate the steel at −20 °C). Figure 12 shows the lateral strengths of model 'R4T1S1' under 20 °C and −20 °C. The shape difference between the two curves is not significant, while the lateral ultimate strength of the model under −20 °C is higher. And the difference between the two cases is 4.13%. As the temperature selected in the calculation is similar to the actual working conditions of ice breakers, it can be concluded that the lateral ultimate strength of the plates will increase slightly in the actual low-temperature environment. Consequently, results from models employed in this paper are conservative.

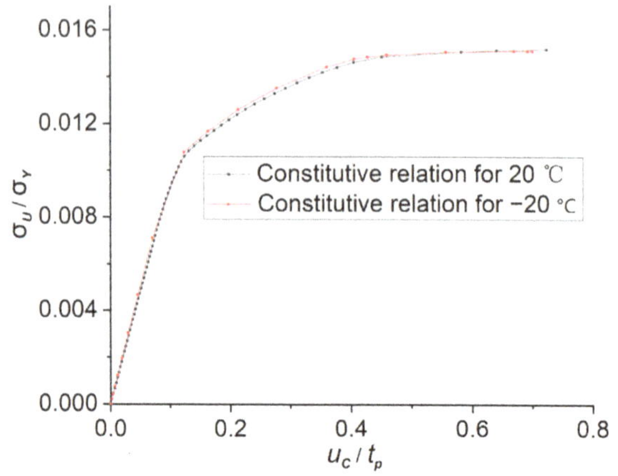

Figure 12. The lateral ultimate strengths of model R4T1S1 (at different temperatures).

The influence of the boundary condition has also been researched. The simply supported boundary condition is set on the edges of model 'R4S1'. The ultimate strengths under different boundary conditions are shown in Figure 13. There is a difference of around 10% between the results from models with different boundary conditions. From Figure 10, it can be concluded that the clamped boundary condition results in much larger ultimate strengths. However, since the members in the bows and stems are stronger, the results of the models with clamped boundary conditions are still referential.

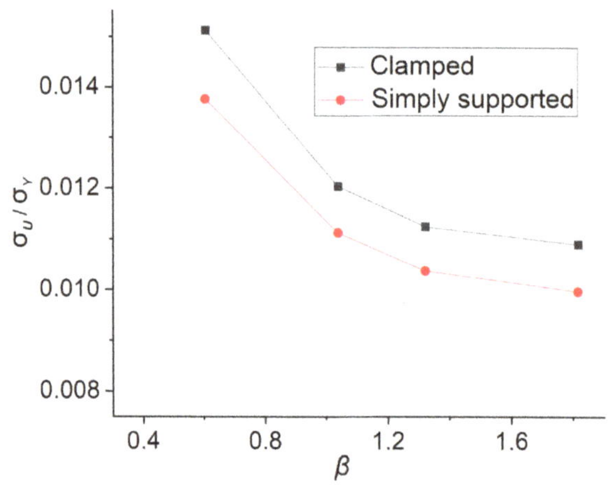

Figure 13. The lateral ultimate strengths of model 'R4S1' (with different boundary conditions).

3.2.2. Typical Collapse Modes of the Doubly Curved Stiffened Plates

The stress distribution and collapse mode of models 'R1T1S1' and 'R2T1S1' are shown in Figure 14a–d. From the picture, it can be inferred that for stiffened plates with longitudinal curvature angles around 0.2, yielding and severe deformation mainly occur in the center of the plate, the upper part of the mid-span of the stiffeners, and the areas around the endings of the stiffeners. Figure 10a,b show that under lateral loading, the yielding firstly occurs around the endings of the stiffeners, and the yielding of the plates and the mid-span of the flanges occurs suddenly after the collapse around the endings of the stiffeners. Figure 15a,b demonstrate that under lateral loading, yielding first occurs around the endings of the stiffeners. The plates and mid-span of the flanges suddenly yield after the stiffeners collapse around the endings. Such a phenomenon indicates that the plate and the stiffener buckle from the peak of the curved surface simultaneously, and the corresponding collapse mode of doubly curved stiffened plates can be described as global collapse originating from the central point.

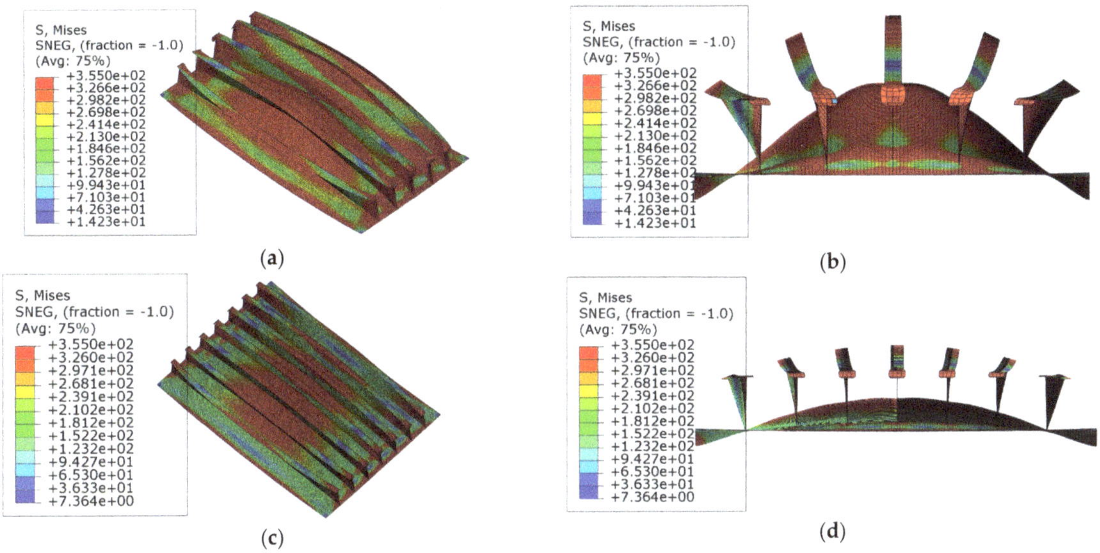

Figure 14. The collapse modes of models R1T1S1 and R2T1S1: (**a**) the deformation and stress distribution at the ultimate state of model 'R1T1S1'; (**b**) the front view of model 'R1T1S1'; (**c**) the deformation and stress distribution at the ultimate state of model 'R2T1S1'; (**d**) the front view of model 'R2T1S1'. Scale: 10.

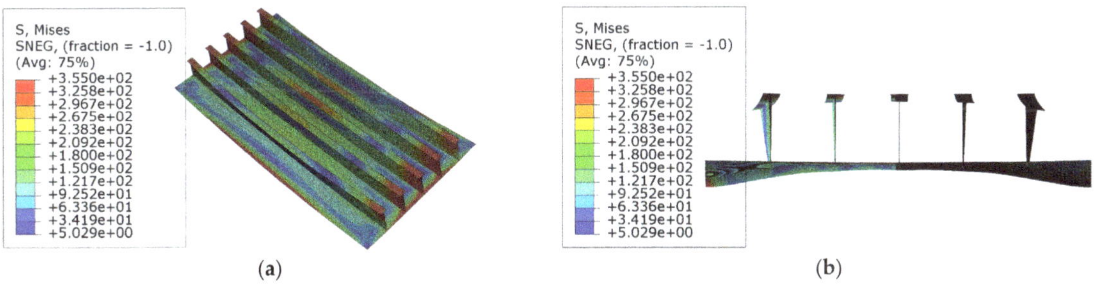

Figure 15. The deformations and stress distributions of model R1T1S1 before the ultimate state: (**a**) the global view of model 'R1T1S1'; (**b**) the front view of model 'R1T1S1'. Scale: 10.

The collapse modes and stress distributions of the models with larger curvature angles are shown in Figure 16. In Figure 16b,f, besides global deformation, the local buckling of plates is also observed. In comparing with the phenomena shown in Figure 14, the amplitude of the global collapse decreases, whereas the area of yielding in the plate expands. Figure 17 shows the stress distribution in model 'R3T1S1' and model 'R4T1S1' before reaching the ultimate state. It is evident that the global buckling and tripping of the stiffeners occur simultaneously. These phenomena signify that the plates are subjected to stronger constraints in the normal direction, leading to increased participation of the stiffened plates in the loading capacity. In model 'R3T1S1' and model 'R4T1S1', larger curvature angles enhance the lateral bearing capacity of the stiffeners; hence, stronger normal constraints are applied to the plates. The lateral bearing capacity of the stiffened plates is finally enhanced.

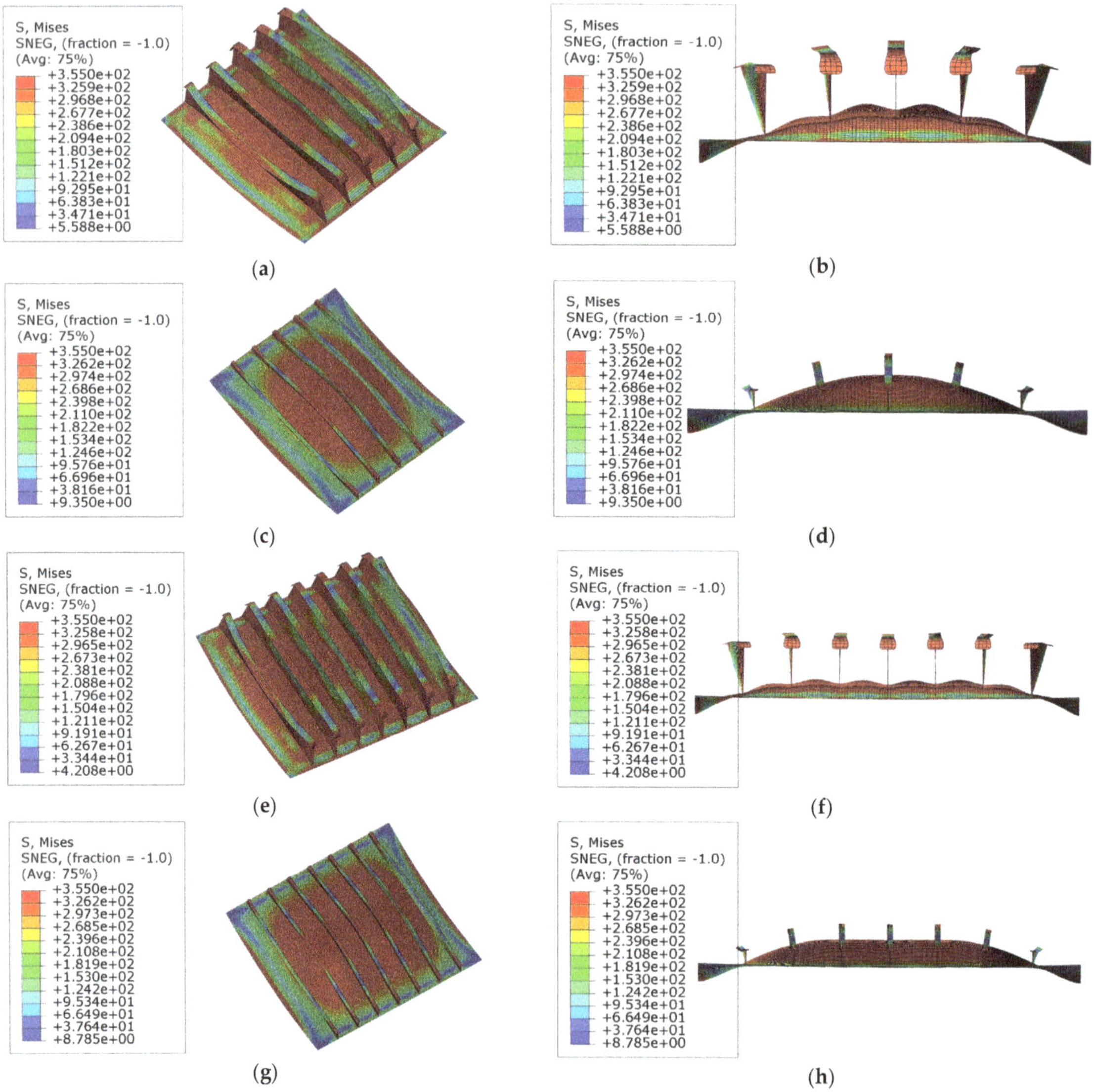

Figure 16. The collapse mode of models 'R3' and 'R4': (**a**) the deformation and stress distribution at the ultimate state of model 'R3T1S1'; (**b**) the front view of model 'R3T1S1'; (**c**) the deformation and stress distribution at the ultimate state of model 'R3T1S3'; (**d**) the front view of model 'R3T1S3'; scale: 15. (**e**) The deformation and stress distribution at the ultimate state of model 'R4T1S1'; (**f**) the front view of model 'R4T1S1'; (**g**) the deformation and stress distribution at the ultimate state of model 'R3T1S3'; (**h**) the front view of model 'R3T1S3'; scale: 15.

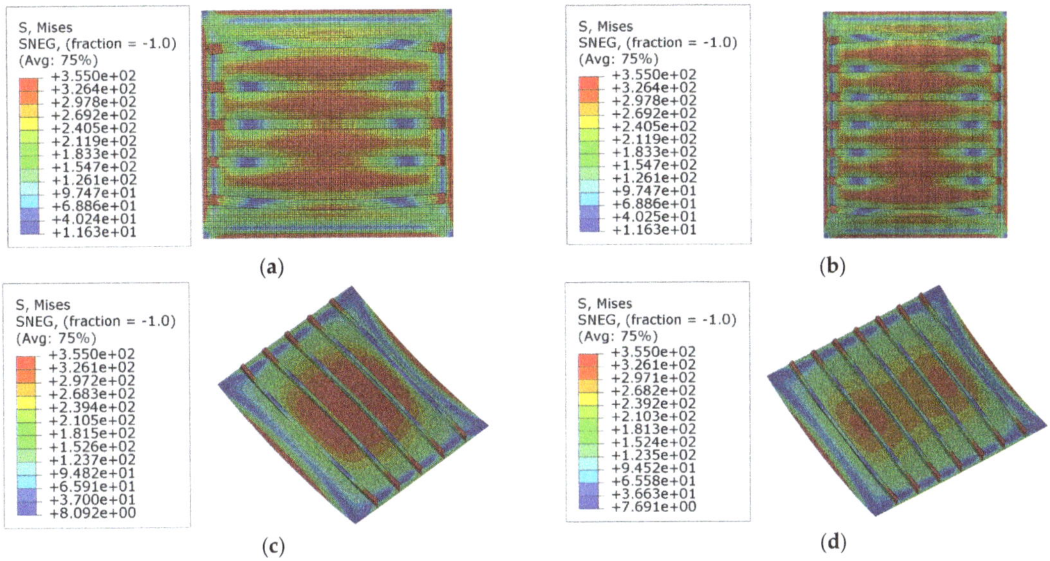

Figure 17. The stress distribution before the ultimate state in models 'R3' and 'R4': (**a**) Model 'R3T1S1'; (**b**) Model 'R4T1S1'; (**c**) Model 'R3T1S3'; (**d**) Model 'R4T1S3'; scale: 15.

The collapse modes and stress distribution in models 'R3' and 'R4' with weaker stiffeners are illustrated in Figure 16c,d,g,h. In these models, the global collapse occurs during the loading process. It is because the lateral bearing capacity of stiffeners decreases as the column slenderness ratio increases, resulting in the global buckling collapse mode. As shown in Figure 17c,d, the yielding happens in the areas around the endings of the stiffeners and the center of the plates before reaching the ultimate state. This phenomenon indicates that larger curvature angles also enhance the lateral bearing capacity of the plates, and the yielding of the plates occurs earlier, which differs from the collapse mode observed in models 'R1' and 'R2'.

The deformation behaviors and stress distributions of model 'R1S1' at the ultimate state are shown in Figure 18. It is observed that the thickness of the plate gradually decreases, leading to a more noticeable buckling of local plates. Particularly, the most severe local buckling occurs in the mid-span of the plates. As the plate slenderness ratio increases, the local buckling of the plates becomes significant, and the collapse mode converts to the combination of global buckling of the stiffened plates and the local buckling of the plates when the plate slenderness ratio becomes larger.

The deformation behaviors and stress distributions of models 'R3S1' are shown in Figures 19 and 20. In models with larger curvature angles along the stiffeners, the local buckling of plates becomes more significant as the plate slenderness ratio increases. As the plates become thinner, the local buckling of the plates becomes more severe, and the buckling of plate happens earlier. Consequently, the dominant collapse mode converts into the local buckling of the plates.

The deformation behaviors and stress distribution of models with larger slenderness ratios of plates are shown in Figure 21. In contrast to the models 'R1' and 'R2' taken from an ice breaker, these models have larger plate slenderness ratios. Another collapse mode, which is different from the models discussed above, can be observed. Figure 21b,d,f show the stress distribution and deformation behaviors of these models before reaching the ultimate state. From these figures, it can be concluded that the yielding and collapse of both the plate and stiffeners occur almost simultaneously during the lateral loading process of such stiffened plates. As shown in Figure 16a–c, the buckling of plates is quite significant at the ultimate state, and the deformations of stiffeners has also become more severe.

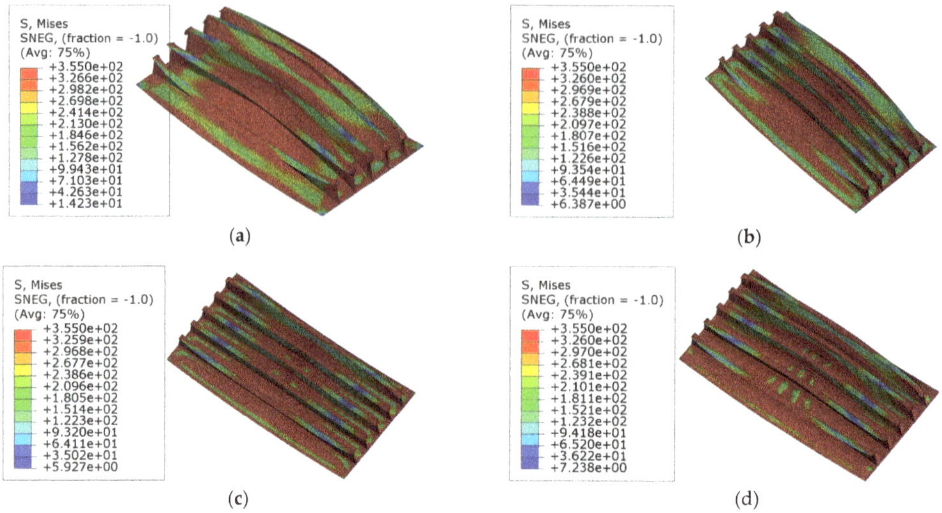

Figure 18. The collapse modes and stress distribution at the ultimate state of models 'R1S1': (**a**) Model 'R1T1S1'; (**b**) Model 'R1T2S1'; (**c**) Model 'R1T3S1'; (**d**) Model 'R1T4S1'. Scale: 10.

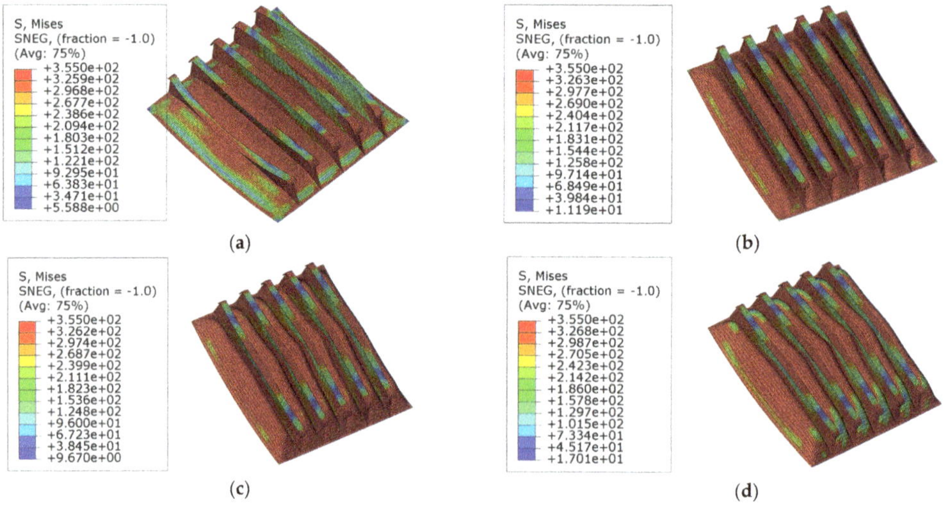

Figure 19. The collapse modes and stress distribution at the ultimate state of model 'R3S1': (**a**) Model 'R3T1S1'; (**b**) Model 'R3T2S1'; (**c**) Model 'R3T3S1'; (**d**) Model 'R3T4S1'. Scale: 15.

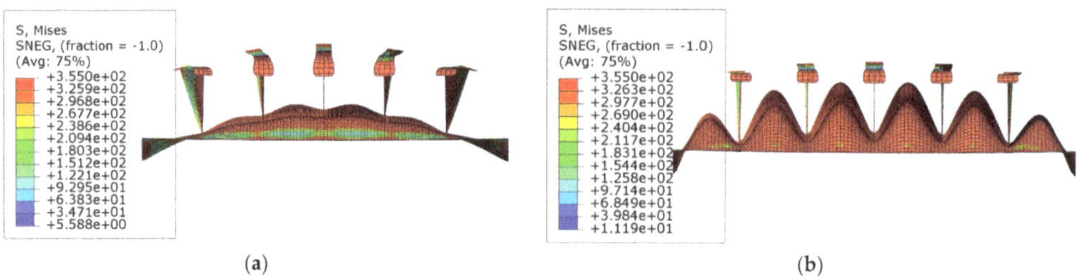

Figure 20. *Cont.*

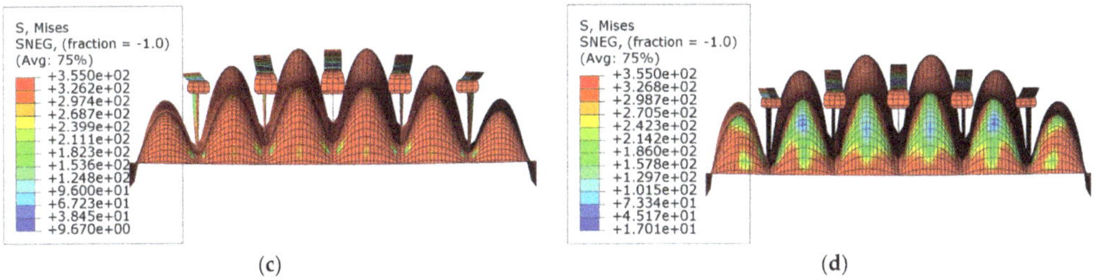

Figure 20. The front views of the collapse modes of model 'R3S1': (**a**) Model 'R3T1S1'; (**b**) Model 'R3T2S1'; (**c**) Model 'R3T3S1'; (**d**) Model 'R3T4S1'. Scale: 15.

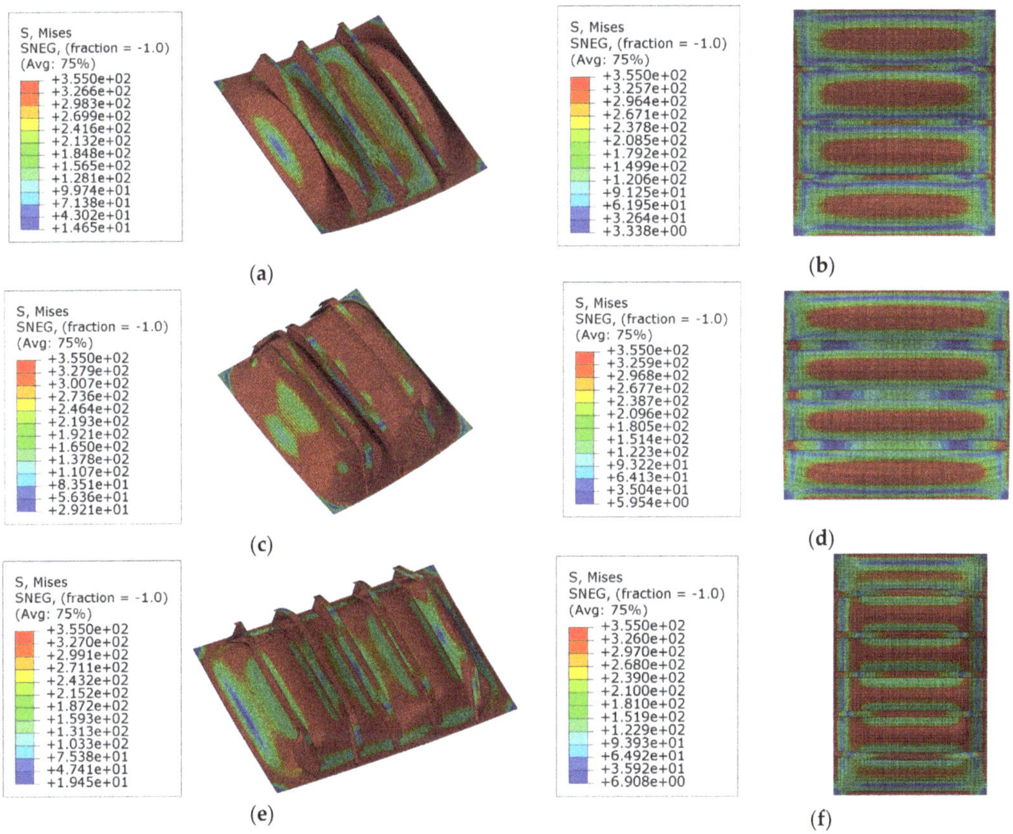

Figure 21. The collapse modes and stress distributions of model 'R5', 'R6', and 'R8': (**a**) ultimate state, model 'R5T1S1'; (**b**) before the ultimate state, model 'R5T1S1' (**c**) ultimate state, model 'R6T1S1'; (**d**) before the ultimate state, model 'R6T1S1'; (**e**) ultimate state, model 'R8T1S1'; (**f**) ultimate state, model 'R8T1S1'. Scale: 10.

4. Empirical Formula of the Lateral Ultimate Strengths

Although the NFEM codes is a powerful tool for structural strength analysis, it requires a relatively long time to calculate and process data. Therefore, in practical structure design, it is highly desirable to use an empirical formula that contains key variables and provides a straightforward prediction for the lateral ultimate strength of stiffened plates. Before the empirical formula is presented, certain preparations have been made. First of all, the design variables have been confirmed based on experiences from structural design and analysis. Next, the variables for the formula are selected. Then, an appropriate form of the formula is constructed, considering the analyzing examples and reflecting the impact of geometrical properties on ultimate strengths. Finally, a formula is derived using the minimum mean-square error method and the NFEM results.

There are various empirical formulae for predicting the axial ultimate strength of the stiffened plates, which can provide a reasonable reference for the choice of the structure of the empirical formula for the lateral ultimate strength of the doubly curved stiffened plates. Some are only related to plate slenderness (Faulkner [23] and Paik [24]), Moreover, there are two influencing factors in other empirical formulae, including the plate slenderness and the column slenderness. As mentioned in the analysis above, both the plate's slenderness ratio and the column's slenderness ratio are the key influence factors of the lateral ultimate strength of the doubly curved stiffened plates.

Zhang and Khan [21] presented an empirical formula for calculating the axial compression ultimate strength of a stiffened plate, as shown in Equation (6). Based on this form, Shi [25] derived an empirical formula for evaluating the axial strength of the U-type stiffened plates, as shown in Equation (7). From the formula, we can discover that there are three coefficients selected according to the specific situation. The coefficients in such a form can signify certain characteristic of different models and load cases and it is well worth applying in other situations.

$$\frac{\sigma_{UA}}{\sigma_Y} = \frac{1}{\beta^{0.28}} \frac{1}{(1+\lambda^{3.2})^{0.5}} \tag{6}$$

$$\frac{\sigma_{UA}}{\sigma_Y} = \frac{1}{\beta^{0.2}} \frac{1}{(1+\lambda^{20})^{0.05}} \tag{7}$$

Taking the variations in the curvatures and sections of the stiffeners into consideration, a series of doubly curved plates models are established, and the lateral ultimate strength can be calculated by employing the NFEM and the results are shown in the Appendix. The influence of the curvatures on the ultimate strength is one of the main concerns. Based on these results, using the basis function fitting method and the minimum mean-square error (MMSE) technique, the corresponding ultimate strength prediction formulae in terms of slenderness ratios λ and β and the curvature angles θ_L and θ_T are derived.

Based on the form of Equation (6), the empirical formula for predicting the lateral ultimate strength of the doubly curved stiffened is expressed as follows. To consider the effect of the curvature in longitudinal and transversal directions, the coefficients are replaced by a series of rational low-order polynomial functions. The basic form of the empirical formula is shown in Equation (8), and functions α_1, α_2, and α_3 are expressed in Equations (9)–(11), where the parameters α_1, α_2, and α_3 denote the influences of the curvature.

$$\frac{\sigma_U}{\sigma_Y} = \frac{1}{\beta^{\alpha_1}} \frac{1}{(1+\lambda^{\alpha_2})^{\alpha_3}} \tag{8}$$

$$\begin{aligned} \alpha_1 &= 2.489 - 61.76\theta_T - 24.98\theta_L + 345.4(\theta_T)^2 + 771\theta_L\theta_T + 78.29(\theta_L)^2 \\ &\quad -2398(\theta_T)^2\theta_L - 1549\theta_T(\theta_L)^2 - 71.21(\theta_L)^3 \end{aligned} \tag{9}$$

$$\begin{aligned} \alpha_2 &= 0.4812 - 0.02773\theta_T - 3.222\theta_L - 26.46(\theta_T)^2 - 95.56\theta_L\theta_T \\ &\quad +21.94(\theta_L)^2 - 959.5(\theta_T)^2\theta_L - 13.93\theta_T(\theta_L)^2 - 33.28(\theta_L)^3 \end{aligned} \tag{10}$$

$$\begin{aligned} \alpha_3 &= 7.671 + 44.03\theta_T - 2.319\theta_L - 469.6(\theta_T)^2 + 295.2\theta_L\theta_T \\ &\quad +94.74(\theta_L)^2 + 2716(\theta_T)^2\theta_L - 2848\theta_T(\theta_L)^2 - 170.4(\theta_L)^3 \end{aligned} \tag{11}$$

Comparisons with results derived from the formula and the practical NFEM are shown in Figure 22. We can find that the prediction of the formula and the NFEM results match well, with $R^2 > 0.95$. We can also find that the lateral ultimate strengths of the doubly curved stiffened plates decrease with the increase of the parameters λ and β from the picture; moreover, the difference distribution frequency is shown in Figure 23.

A stiffened plate captured from an oil tanker ($\theta_L = 0.022$ and $\theta_T = 0.13$) is calculated using the formula and the NFEM, and the comparison between the two methods is shown in Figure 24. The values obtained from both methods are similar, and the difference between them is around 5%. And the results from the formula are slightly smaller than those from the NFEM.

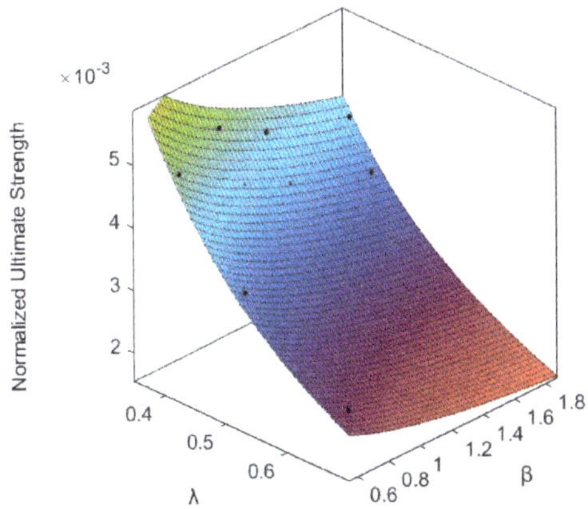

(a)

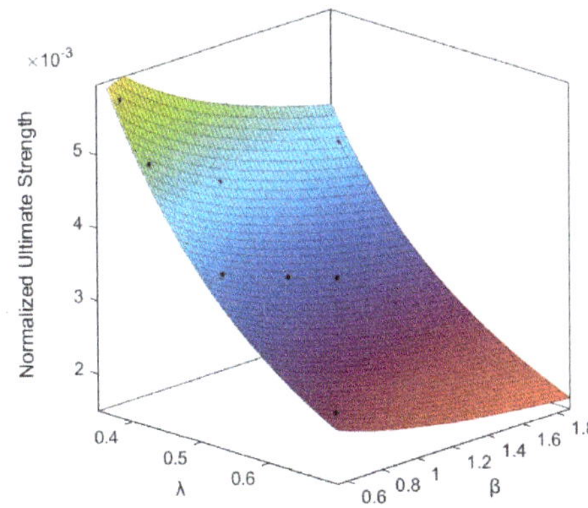

(b)

Figure 22. *Cont.*

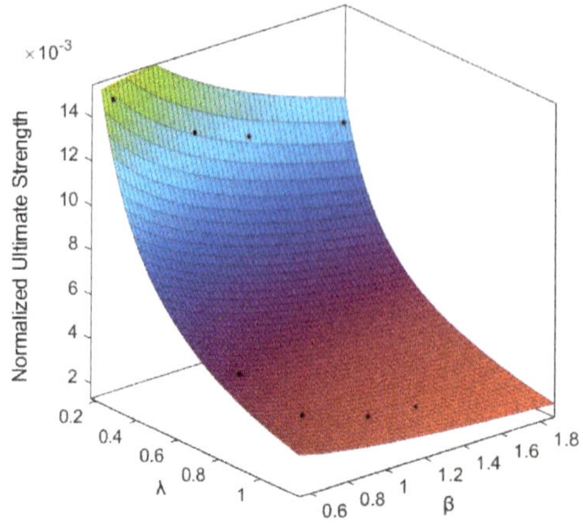

(c)

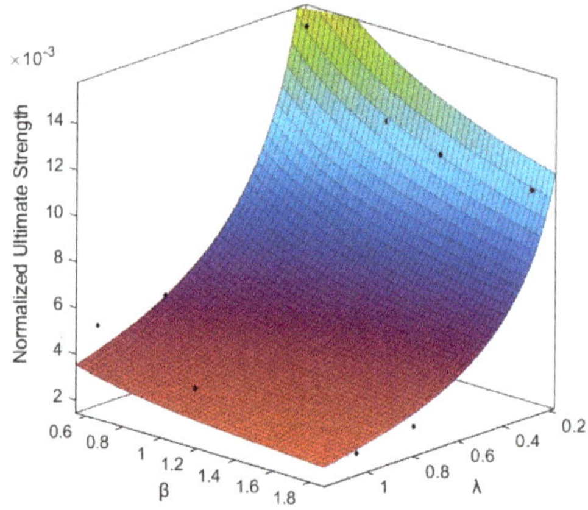

(d)

Figure 22. Formulae predictions based on the NFEM results: (**a**) Model 'R1' R^2 = 0.9894; (**b**) Model 'R2', R^2 = 0.952; (**c**) T Model 'R3', R^2 = 0.9679; (**d**) Model 'R4', R^2 = 0.9631.

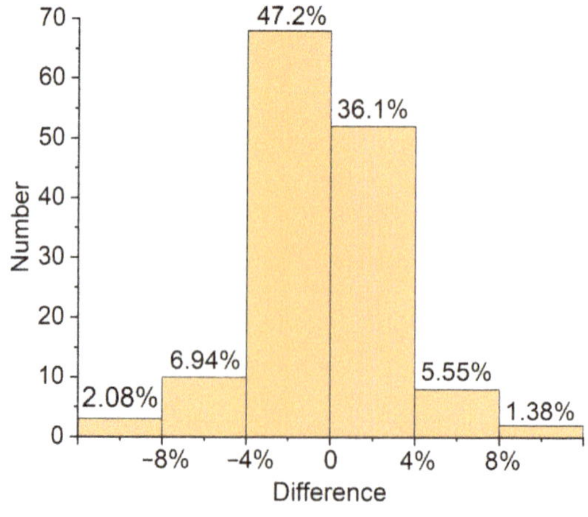

Figure 23. Difference distribution frequency.

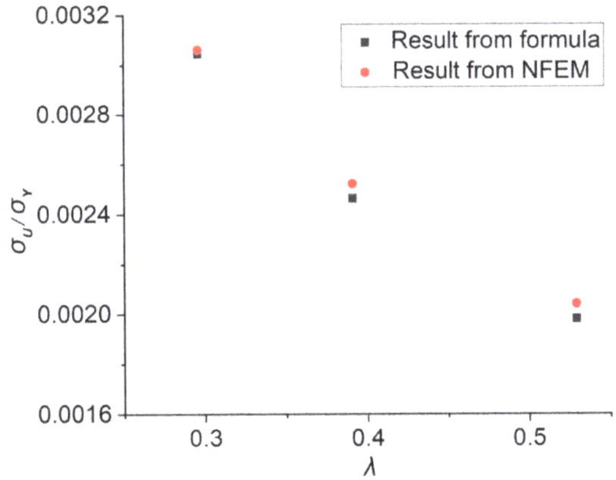

Figure 24. Comparison between the formula and the NFEM.

5. Conclusions

In this paper, the collapse modes and ultimate strengths of doubly curved stiffened plates under lateral pressure have been investigated by using the NFEM, where different groups of curvature angles have been considered and the influences of plates and stiffeners scantlings have been studied. Meanwhile, the deformation behaviors and collapse sequence of target models in lateral loading cases have been also studied. The following conclusions can be drawn:

(1) The curvature angle, along the direction of stiffeners, has significant impact on the lateral ultimate strength of the doubly curved stiffened plate. The normal bearing capacity of the stiffener enhances as the curvature of stiffeners increases. As a result, a stronger constraint is applied to the plates, which strengthens the lateral bearing capacity and the ultimate strength of the doubly curved stiffeners.

(2) In the lateral loading process, yielding firstly happens in the areas around the endings of the stiffeners. In most load cases, the stiffeners collapse before the plates. In actual designs, enhancing the strength of the stiffeners is a more effective way to improve the lateral loading capacity.

(3) The collapse modes of the doubly curved stiffened plates can be concluded as the combination of global buckling and the local buckling of the plates. If the strength of the stiffener and the plates are similar or if the column slenderness is quite large, the global buckling will become the dominant collapse mode. The local buckling of the plates is significant in the cases with strong stiffeners and weak plates.

(4) A new empirical formula Equation (4) is proposed for lateral ultimate strength prediction of doubly curved stiffeners. The proposed formula is applicable with $R^2 > 0.95$ comparing with 144 NFEM results. And the average absolute differences between the results from the formula and the NFEM is 3.7%. Over the wide range of λ (0–1.5) and β (0.6–3) considered in this paper, the agreement between the proposed formula and numerical results is very good. Additionally, the differences between the values from the formula and the NFEM on a practical case are around 5%, which can be acceptable in engineering.

Author Contributions: Conceptualization, J.C.; methodology, J.C.; software, G.G.; validation, G.G.; writing—review and editing, G.G. and J.C.; supervision, D.W.; project administration, J.C. and D.W.; funding acquisition, J.C. and D.W. All authors have read and agreed to the published version of the manuscript.

Funding: This research was funded by NSFC (No. 52371328, No. 51809167, No. 51979163 and No. U2241266) and the Fundamental Research Funds for the Central Universities.

Data Availability Statement: Data is contained within the article.

Conflicts of Interest: The authors declare no conflict of interest.

Nomenclature

Notations	Descriptions
a	Plate length
b	Plate breath
t_p	Thickness of attached plate
t_w	Web thickness of the stiffener
h_w	Web height of the stiffener
t_f	Flange thickness of the stiffener
b_f	Flange breadth of the stiffener
u_C	The magnitude displacement of the doubly curved stiffened plate's central point
E	Elastic modulus
I	The moment of inertia of the stiffener with the attached plate
A	Area of transverse section of the stiffened plate
r	Gyration radius of the stiffened plate
σ_Y	Yield stress of the material
σ_U	The normalized lateral ultimate stress of the ultimate state (Defined as the ratio of the lateral ultimate load-carrying capacity to the sectional area and the value of the yield stress.)
σ_{UA}	The normalized axial ultimate stress of the ultimate state (Defined as the ratio of the axial ultimate load-carrying capacity to the sectional area and the value of the yield stress.)
λ	Column slenderness
β	Plate slenderness
R_L	Longitudinal curve radius of the plate
R_T	Transversal curve radius of the plate
θ_L	Longitudinal curvature angle
θ_T	Transversal curvature angle
$\alpha_i (i = 1,2,3)$	Parameters in terms of curvature angles

Appendix A. Structural Dimensions, Properties, and NFEM Solutions of the Models

R_L	R_T	a	b	θ_L	θ_T	t_p	h_w	t_w	b_f	t_f	σ_Y	E	λ	β	σ_U/σ_Y
mm	mm	mm	mm	-	-	mm	mm	mm	mm	mm	MPa	MPa	-	-	-
14,523	40,306	4000	350	0.00868	0.275	24	360	20	90	20	355	206,000	0.368	0.605	0.00568
						14	360	20	90	20	355	206,000	0.366	1.038	0.00506
						11	360	20	90	20	355	206,000	0.369	1.321	0.00466
						8	360	20	90	20	355	206,000	0.375	1.816	0.00431
						24	330	20	80	18	355	206,000	0.41	0.605	0.00504
						14	330	20	80	18	355	206,000	0.404	1.038	0.00431
						11	330	20	80	18	355	206,000	0.406	1.321	0.004
						8	330	20	80	18	355	206,000	0.412	1.816	0.00361
						24	280	20	60	12	355	206,000	0.518	0.605	0.00365
						14	280	20	60	12	355	206,000	0.503	1.038	0.003
						11	280	20	60	12	355	206,000	0.501	1.321	0.00284
						8	280	20	60	12	355	206,000	0.504	1.816	0.00250
						24	220	20	50	10	355	206,000	0.684	0.605	0.00254
						14	220	20	50	10	355	206,000	0.652	1.038	0.00197
						11	220	20	50	10	355	206,000	0.646	1.321	0.00195
						8	220	20	50	10	355	206,000	0.644	1.816	0.00175

R_L	R_T	a	b	θ_L	θ_T	t_p	h_w	t_w	b_f	t_f	σ_Y	E	λ	β	σ_U/σ_Y
mm	mm	mm	mm	-	-	mm	mm	mm	mm	mm	MPa	MPa	-	-	-
17,075	48,849	4000	350	0.00717	0.234	24	360	20	90	20	355	206,000	0.368	0.605	0.00575
						14	360	20	90	20	355	206,000	0.366	1.038	0.00496
						11	360	20	90	20	355	206,000	0.369	1.321	0.00466
						8	360	20	90	20	355	206,000	0.375	1.816	0.00425
						24	330	20	80	18	355	206,000	0.41	0.605	0.00499
						14	330	20	80	18	355	206,000	0.404	1.038	0.00439
						11	330	20	80	18	355	206,000	0.406	1.321	0.00397
						8	330	20	80	18	355	206,000	0.412	1.816	0.00361
						24	280	20	60	12	355	206,000	0.518	0.605	0.0038
						14	280	20	60	12	355	206,000	0.503	1.038	0.00337
						11	280	20	60	12	355	206,000	0.501	1.321	0.00314
						8	280	20	60	12	355	206,000	0.504	1.816	0.00270
						24	220	20	50	10	355	206,000	0.684	0.605	0.00237
						14	220	20	50	10	355	206,000	0.652	1.038	0.002
						11	220	20	50	10	355	206,000	0.646	1.321	0.00195
						8	220	20	50	10	355	206,000	0.644	1.816	0.00169
5719	45,560	2375	350	0.00768	0.415	24	360	20	90	20	355	206,000	0.218	0.605	0.0148
						14	360	20	90	20	355	206,000	0.217	1.038	0.0118
						11	360	20	90	20	355	206,000	0.219	1.321	0.0113
						8	360	20	90	20	355	206,000	0.222	1.816	0.0105
						24	280	20	60	16	355	206,000	0.298	0.605	0.0109
						14	280	20	60	16	355	206,000	0.29	1.038	0.00831
						11	280	20	60	16	355	206,000	0.29	1.321	0.00749
						8	280	20	60	16	355	206,000	0.292	1.816	0.00690
						24	110	20	40	10	355	206,000	0.84	0.605	0.00515
						14	110	20	40	10	355	206,000	0.78	1.038	0.00352
						11	110	20	40	10	355	206,000	0.76	1.321	0.00298
						8	110	20	40	10	355	206,000	0.74	1.816	0.00230
						24	80	20	40	10	355	206,000	1.138	0.605	0.00462
						14	80	20	40	10	355	206,000	1.061	1.038	0.00306
						11	80	20	40	10	355	206,000	1.028	1.321	0.00256
						8	80	20	40	10	355	206,000	0.993	1.816	0.00195
5475	54,020	2375	350	0.00648	0.434	24	360	20	90	20	355	206,000	0.218	0.605	0.0151
						14	360	20	90	20	355	206,000	0.217	1.038	0.0120
						11	360	20	90	20	355	206,000	0.219	1.321	0.0112
						8	360	20	90	20	355	206,000	0.222	1.816	0.0109
						24	280	20	60	16	355	206,000	0.298	0.605	0.01115
						14	280	20	60	16	355	206,000	0.29	1.038	0.00869
						11	280	20	60	16	355	206,000	0.29	1.321	0.00789
						8	280	20	60	16	355	206,000	0.29	1.816	0.00731
						24	110	20	40	10	355	206,000	0.84	0.605	0.00552
						14	110	20	40	10	355	206,000	0.78	1.038	0.00372
						11	110	20	40	10	355	206,000	0.76	1.321	0.00314
						8	110	20	40	10	355	206,000	0.74	1.816	0.0024
						24	80	20	40	10	355	206,000	1.14	0.605	0.00522
						14	80	20	40	10	355	206,000	1.061	1.038	0.00326
						11	80	20	40	10	355	206,000	1.028	1.321	0.00270
						8	80	20	40	10	355	206,000	0.993	1.816	0.00207
74,243	4770	2400	650	0.136	0.0323	20	280	11	120	20	355	206,000	0.284	1.349	0.00450
						24	280	11	120	20	355	206,000	0.298	1.12	0.00604
						16	280	11	120	20	355	206,000	0.268	1.69	0.00316
						10	280	11	120	20	355	206,000	0.25	2.7	0.002
						20	320	11	60	15	355	206,000	0.305	1.349	0.00432
						24	320	11	60	15	355	206,000	0.319	1.12	0.00613
						16	320	11	60	15	355	206,000	0.29	1.69	0.00314
						10	320	11	60	15	355	206,000	0.267	2.7	0.00195
						20	210	11	60	15	355	206,000	0.475	1.349	0.00425
						24	210	11	60	15	355	206,000	0.499	1.12	0.00550
						16	210	11	60	15	355	206,000	0.448	1.69	0.00294
						10	210	11	60	15	355	206,000	0.404	2.7	0.00156
						20	120	11	40	10	355	206,000	1.01	1.349	0.00354
						24	120	11	40	10	355	206,000	1.064	1.12	0.00456
						16	120	11	40	10	355	206,000	0.958	1.69	0.00255
						10	120	11	40	10	355	206,000	0.842	2.7	0.00104

R_L	R_T	a	b	θ_L	θ_T	t_p	h_w	t_w	b_f	t_f	σ_Y	E	λ	β	σ_U/σ_Y
mm	mm	mm	mm	-	-	mm	mm	mm	mm	mm	MPa	MPa	-	-	-
13,104	7198	3044	675	0.0938	0.232	20	320	12	120	20	355	206,000	0.32	1.401	0.00417
						16	320	12	120	20	355	206,000	0.308	1.751	0.00337
						12	320	12	120	20	355	206,000	0.296	2.335	0.00230
						22	320	12	120	20	355	206,000	0.326	1.274	0.00470
						20	280	12	120	20	355	206,000	0.363	1.401	0.00407
						16	280	12	120	20	355	206,000	0.349	1.751	0.00327
						12	280	12	120	20	355	206,000	0.335	2.335	0.00224
						22	280	12	120	20	355	206,000	0.37	1.274	0.00457
						20	220	16	60	16	355	206,000	0.567	1.401	0.00366
						16	220	16	60	16	355	206,000	0.537	1.751	0.00271
						12	220	16	60	16	355	206,000	0.503	2.335	0.00168
						22	220	16	60	16	355	206,000	0.581	1.274	0.00447
						20	180	16	40	12	355	206,000	0.764	1.401	0.00335
						16	180	16	40	12	355	206,000	0.724	1.751	0.00253
						12	180	16	40	12	355	206,000	0.679	2.335	0.00167
						22	180	16	40	12	355	206,000	0.782	1.274	0.00373
11,500	59,500	2375	350	0.00588	0.207	24	360	20	90	20	355	206,000	0.218	0.605	0.0124
						12	360	20	90	20	355	206,000	0.218	1.21	0.0112
						10	360	20	90	20	355	206,000	0.22	1.453	0.0105
						6	360	20	90	20	355	206,000	0.227	2.422	0.0101
						24	200	20	80	10	355	206,000	0.424	0.605	0.00584
						12	200	20	80	10	355	206,000	0.402	1.21	0.00472
						10	200	20	80	10	355	206,000	0.4	1.453	0.00433
						6	200	20	80	10	355	206,000	0.403	2.422	0.00373
						24	140	20	80	10	355	206,000	0.599	0.605	0.00369
						12	140	20	80	10	355	206,000	0.557	1.21	0.00297
						10	140	20	80	10	355	206,000	0.551	1.453	0.00280
						6	140	20	80	10	355	206,000	0.548	2.422	0.00222
						24	100	20	80	10	355	206,000	0.818	0.605	0.00273
						12	100	20	80	10	355	206,000	0.756	1.21	0.00191
						10	100	20	80	10	355	206,000	0.744	1.453	0.00180
						6	100	20	80	10	355	206,000	0.73	2.422	0.00144
26,500	37,500	3200	800	0.0213	0.121	24	420	10	120	16	355	206,000	0.296	1.38	0.00305
						20	420	10	120	16	355	206,000	0.283	1.66	0.00278
						18	420	10	120	16	355	206,000	0.277	1.84	0.00263
						14	420	10	120	16	355	206,000	0.263	2.37	0.00222
						24	320	10	120	16	355	206,000	0.391	1.38	0.00249
						20	320	10	120	16	355	206,000	0.373	1.66	0.00230
						18	320	10	120	16	355	206,000	0.363	1.84	0.00217
						14	320	10	120	16	355	206,000	0.344	2.37	0.00186
						24	280	10	80	12	355	206,000	0.529	1.38	0.00192
						20	280	10	80	12	355	206,000	0.501	1.66	0.00180
						18	280	10	80	12	355	206,000	0.486	1.84	0.00174
						14	280	10	80	12	355	206,000	0.455	2.37	0.00145
						24	190	10	80	6	355	206,000	0.933	1.38	0.00146
						20	190	10	80	6	355	206,000	0.883	1.66	0.00135
						18	190	10	80	6	355	206,000	0.855	1.84	0.00126
						14	190	10	80	6	355	206,000	0.793	2.37	0.00114
41,600	27,200	3200	800	0.0294	0.0769	24	420	10	120	16	355	20,600	0.296	1.38	0.00303
						20	420	10	120	16	355	20,600	0.283	1.66	0.00270
						18	420	10	120	16	355	20,600	0.277	1.84	0.00248
						14	420	10	120	16	355	20,600	0.263	2.37	0.00221
						24	320	10	120	16	355	206,000	0.391	1.38	0.00244
						20	320	10	120	16	355	206,000	0.373	1.66	0.00229
						18	320	10	120	16	355	206,000	0.363	1.84	0.00205
						14	320	10	120	16	355	206,000	0.344	2.37	0.00175
						24	280	10	80	12	355	206,000	0.529	1.38	0.00203
						20	280	10	80	12	355	206,000	0.501	1.66	0.00174
						18	280	10	80	12	355	206,000	0.486	1.84	0.00167
						14	280	10	80	12	355	206,000	0.455	2.37	0.00136
						24	190	10	80	6	355	206,000	0.933	1.38	0.00144
						20	190	10	80	6	355	206,000	0.883	1.66	0.00130
						18	190	10	80	6	355	206,000	0.855	1.84	0.00121
						14	190	10	80	6	355	206,000	0.793	2.37	0.00112

References

1. Yang, B.; Wang, D.Y. Numerical study on the dynamic response of the large containership's bow structure under slamming pressures. *Mar. Struct.* **2018**, *61*, 524–539. [CrossRef]
2. Shabani, B.; Lavroff, J.; Davis, M.R.; Holloway, D.S.; Thomas, G.A. Slam Loads and Kinematics of Wave-Piercing Catamarans During Bow Entry Events in Head Seas. *J. Ship. Res.* **2018**, *62*, 134–155. [CrossRef]
3. Liu, C.; Zhang, S.L. A simple empirical formula for predicting the ultimate strength of ship plates with elastically restrained edges in axial compression. *Ocean Eng.* **2020**, *34*, 571–580. [CrossRef]
4. Hayward, R.; Lehmann, E. Development of a new proof of plate capacity under combined in-plane loads. *Ships Offshore Struct.* **2017**, *12*, 174–188. [CrossRef]
5. Kim, D.K.; Lim, H.L.; Yu, S.Y. A technical review on ultimate strength prediction of stiffened panels in axial compression. *Ocean Eng.* **2018**, *170*, 392–406. [CrossRef]
6. Li, S.; Kim, D.K.; Benson, S. The influence of residual stress on the ultimate strength of longitudinally compressed stiffened panels. *Ocean. Eng.* **2021**, *231*, 108839. [CrossRef]
7. Li, S.; Kim, D.K.; Benson, S. An adaptable algorithm to predict the load- shortening curves of stiffened panels in compression. *Ships Offshore Struct.* **2021**, *16*, 122–139. [CrossRef]
8. Yao, T.; Fujikubo, M.; Varghese, B.; Yamamura, K.; Niho, O. Buckling/plastic collapse strength of wide rectangular plate under combined pressure and thrust. *J. Soc. Nav. Archit. Jpn.* **1997**, *182*, 561–570. [CrossRef] [PubMed]
9. Paik, J.K.; Kim, B.J. Ultimate strength formulations for stiffened panels under combined axial load, in-plane bending and lateral pressure: A benchmark study. *Thin-Walled Struct.* **2002**, *40*, 45–83. [CrossRef]
10. Ma, H.Y.; Wang, D.Y. Lateral pressure effect on the ultimate strength of the stiffened plate subjected to combined loads. *Ocean Eng.* **2021**, *239*, 109926. [CrossRef]
11. Xu, M.C.; Yanagihara, D.; Fujikubo, M. Guedes Soares, C. Influence of boundary conditions on the collapse behaviour of stiffened panels under combined loads. *Mar. Struct.* **2013**, *34*, 205–225. [CrossRef]
12. Li, D.Y.; Chen, Z.; Chen, X.C. Numerical investigation on the ultimate strength behaviour and assessment of continuous hull plate under combined biaxial cyclic loads and lateral pressure. *Mar. Struct.* **2023**, *89*, 103408. [CrossRef]
13. Li, D.Y.; Chen, Z. Advanced empirical formulae for the ultimate strength assessment of continuous hull plate under combined biaxial compression and lateral pressure. *Eng. Struct.* **2023**, *285*, 116401. [CrossRef]
14. Cui, J.J.; Wang, D.Y. Ultimate Strength of Container Ship Bilge Panels Subjected to Axial Compression Combined with Bending and Lateral Pressure. *Int. J. Offshore Polar Eng.* **2020**, *30*, 327–339. [CrossRef]
15. Park, J.S.; Iijima, K.; Yao, T. Characteristics of buckling and ultimate strength and collapse behaviour of cylindrically curved plates subjected to axial compression. *Adv. Mater. Res.* **2008**, *33–37*, 1195–1200.
16. Seo, J.K.; Song, C.H.; Park, J.S.; Paik, J.K. Nonlinear structural behaviour and design formulae for calculating the ultimate strength of stiffened curved plates under axial compression. *Thin-Walled Struct.* **2016**, *107*, 1–17. [CrossRef]
17. Park, J.S.; Paik, J.K.; Seo, J.K. Numerical investigation and development of design formula for cylindrically curved plates on ships and offshore structures. *Thin-Walled Struct.* **2018**, *132*, 93–110. [CrossRef]
18. Cho, S.; Park, H.; Kim, H.; Seo, J. Experimental and numerical investigations on the ultimate strength of curved stiffened plates. In Proceedings of the 10th International Symposium on Practical Design of Ships and Other Floating Structures, Huston, TX, USA, 1–5 October 2007.
19. Paik, J.K.; Amlashi, H.; Boon, B.; Branner, K.; Caridis, P.; Das, P.; Fujikubo, M.; Huang, C.-H.; Josefson, L.; Kaeding, P.; et al. Report of specialist committee III.1 ultimate strength. In Proceedings of the18th International Ship and Offshore Structures Congress (ISSC), Rostock, Germany, 9–13 September 2012; pp. 289–352.
20. Yao, T.; Fujikubo, M. *Buckling and Ultimate Strength of Ship and Ship-like Floating Structure*; Elsevier: Kidlington, UK, 2016; pp. 22–30.
21. Zhang, S.M.; Khan, T. Buckling and ultimate capability of plates and stiffened panels in axial compression. *Mar. Struct.* **2009**, *22*, 791–808. [CrossRef]
22. Wang, K.; Wu, L.; Li, Y.Z.; Qin, C. Experimental study on low temperature fatigue performance of polar icebreaking ship steel. *Ocean Eng.* **2020**, *216*, 107889. [CrossRef]
23. Faulkner, D. A review of effective plating for the analysis of stiffened plating in bending and compression. *J. Ship Res.* **1975**, *19*, 1–17. [CrossRef]
24. Paik, J.K.; Thayamballi, A.K.; Lee, J.M. Effect of initial deflection shape on the ultimate strength behavior of welded steel plates under Biaxial compressive loads. *J. Ship Res.* **2004**, *48*, 45–60. [CrossRef]
25. Shi, G.J.; Gao, D.W. Ultimate strength of U-type stiffened panels for hatch covers used in ship cargo holds. *Ships Offshore Struct.* **2020**, *16*, 280–291. [CrossRef]

Journal of
Marine Science and Engineering

MDPI

Article

Enhancing the Sealing Performance of Bolted Ball Joints by Gaskets: Numerical Simulation and Experiment

Wenfeng Du [1], Jinchao Gu [1], Guilin Sheng [1,*], Guang Guo [1], Yongrun Zhao [1] and Zhijian Liu [2]

1 School of Civil and Architectural Engineering, Henan University, Kaifeng 475004, China; dwf0925@hotmail.com (W.D.); gujinchao0125@163.com (J.G.); henuguoguang@163.com (G.G.); zhaoyr_518@163.com (Y.Z.)
2 Henan Tianyuan Equipment Engineering Co., Ltd., Kaifeng 475004, China; liuzhijian315@163.com
* Correspondence: shengguilin315@163.com

Abstract: With the increasing utilization of bolted ball joint steel mesh structures in offshore floating platforms and deep-sea fish cages, the issue of seawater infiltrating the joints and members through the installation gaps of the bolted ball joint, leading to subsequent corrosion, has become increasingly prominent. This article presents an innovative method to improve the sealing performance of bolted ball joints. The approach involves creating sealed surfaces within the contact gaps between the sleeve and connecting components by adding circular grooves and sealing washers to both ends of the sleeve. Subsequently, a two-dimensional finite element analysis model of the bolted ball joint with the sealing structure was created using SOLIDWORKS 2021 and ANSYS Workbench 2022 R1. The study analyzes the sealing gasket's contact pressure at various compression levels and evaluates its performance with bubble tests for air tightness. Research results show a linear relationship between the contact pressure and compression rate, achieving sealing pressures of 2.91 MPa, 4.22 MPa, and 5.95 MPa at compression levels of 8%, 11%, and 14%, respectively. Experimental testing demonstrates that the improved bolted ball joint exhibits excellent sealing performance.

Keywords: bolted ball joint; sealing method; numerical simulation analysis; air tightness test

Citation: Du, W.; Gu, J.; Sheng, G.; Guo, G.; Zhao, Y.; Liu, Z. Enhancing the Sealing Performance of Bolted Ball Joints by Gaskets: Numerical Simulation and Experiment. *J. Mar. Sci. Eng.* **2023**, *11*, 2050. https://doi.org/10.3390/jmse11112050

Academic Editor: Cristiano Fragassa

Received: 11 October 2023
Revised: 19 October 2023
Accepted: 24 October 2023
Published: 26 October 2023

1. Introduction

The bolted ball joint steel grid structure is a spatial grid system comprised of members assembled through bolted ball joints based on specific geometric principles. This architectural framework offers advantages such as well-distributed stress, dependable connections, and convenient construction processes. Its applicability extends beyond terrestrial spatial frameworks, encompassing diverse settings like exhibition centers, sports venues, and industrial factories, while also finding progressive integration within the realm of marine engineering [1–3]. Prominent applications include floating platforms, deep-sea fish cages, and more, as depicted in Figure 1.

(a)

(b)

Figure 1. Steel grid structure with bolted ball joints: (**a**) Floating platform; (**b**) Deep-sea fish cage.

In the marine environment, the corrosion of steel materials can rapidly lead to a reduction in the cross-sectional area and load-bearing capacity of steel components. This vulnerability represents a critical weakness that undermines the safe operation of steel grid structures with bolted ball joints [4,5]. To address this issue, several corrosion protection measures for steel components in oceanic settings are employed, including weathering steel anti-corrosion, hot-dip galvanizing anti-corrosion, and coating anti-corrosion techniques [6–8]. Weathering-resistant steel anti-corrosion involves incorporating trace amounts of alloying elements into conventional steel. This process facilitates the formation of a dense oxide coating on the steel's surface, effectively preventing further corrosion and oxidation. In contrast, the approach of hot-dip galvanizing anti-corrosion entails immersing steel components into molten zinc, resulting in the formation of an iron-zinc alloy coating on the steel's surface. This layer is a barrier against corrosive agents, significantly enhancing the steel's anti-corrosion capabilities. Another strategy, coating anti-corrosion, centers around applying specialized anti-corrosion coatings onto the surfaces of steel structures. These coatings establish a protective shield, effectively isolating the steel from direct contact with the external environment. Among the aforementioned corrosion protection measures, coating anti-corrosion is extensively employed in numerous practical projects owing to its benefits of low cost, ease of repair, and straightforward construction procedures. This approach has yielded a favorable effect on the external corrosion protection of members and joints, effectively controlling corrosion on the structural exterior.

However, installation gaps between the components of the bolt ball joint and the rods, as shown in Figure 2, allow seawater to infiltrate the interior of the joints and rods, causing the corrosion of the bolt threads and the steel pipe's interior. This has emerged as the primary challenge in implementing the bolt ball joint steel grid structure in marine engineering. Given the limited internal space within the joint and the fine nature of components like bolt threads, achieving complete coverage of all surfaces with coating anti-corrosion and hot-dip galvanizing anti-corrosion is challenging. Moreover, it would compromise installation precision, rendering coating anti-corrosion and hot-dip galvanizing anti-corrosion unsuitable for addressing this issue.

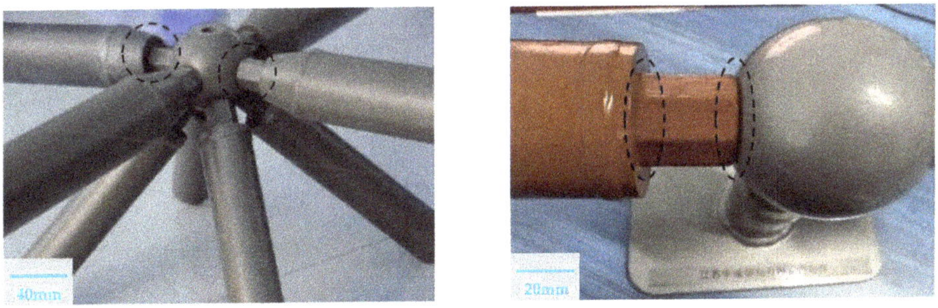

(**a**) (**b**)

Figure 2. Installation gaps between the components of the bolted ball joint: (**a**) Case 1; (**b**) Case 2.

Therefore, how to improve the joint structure and enhance the sealing performance of the joint, thereby eliminating the problem of internal corrosion caused by installation gaps, has essential research value. Huang [9] proposed a method for achieving the sealing of bolt ball joints by adding an outer layer on the surface of the bolt ball and a sealing sleeve on the periphery of the sleeve to achieve sealing protection of the bolted ball joint. However, this sealing method is complex in construction and becomes challenging to apply when there are numerous connecting rods at the joint. Zhang [10] proposed a sealed connection rod for the bolted ball joint, which seals the inside of the member by adding sealing partitions at both ends of the rod. However, this sealing method focuses solely on safeguarding the member's internal corrosion, overlooking the corrosion susceptibility of the sleeve and bolt ball components. Furthermore, the mentioned proposed sealing methods are currently in their initial stages, lacking the crucial analysis of sealing performance and experimental validation. Consequently, there exists a necessity for in-depth research to develop a pragmatic and viable sealing approach that not only enhances the sealing performance of bolted ball joints but also effectively addresses internal corrosion challenges.

To address the aforementioned challenges, this article presents a novel approach aimed at enhancing the sealing performance of bolted ball joints. The primary objectives of this approach are to effectively mitigate the risk of interstitial corrosion in bolted ball joints within the realm of offshore engineering, improve the safety and durability of bolted ball joint steel mesh structures in marine applications, and further promote their use in offshore engineering projects. This study is organized as follows: Section 1, 'offers an overview of this study. Section 2, offers a detailed explanation of the concept and construction of the sealing method. Section 3, conducts an analysis of the sealing performance of bolted ball joints with the proposed sealing structure using numerical simulation. Section 4, presents the results of tests on the sealing performance of the bolted ball joint. Section 5, draws the main conclusions of this study based on the analyses.

2. Ideas and Structural Design

To tackle the issue of seawater intrusion and corrosion in bolted ball joints arising from installation gaps, this paper explores the utilization of sealing washers to fill these installation gaps in bolted ball joints to seal the joint. Sealing washers are commonly used components in sealing applications, and their function is to occupy the spaces between two contacting surfaces, thus forming a sealing interface to prevent the permeation or leakage of gases or liquids between these surfaces.

For achieving effective joint sealing, sealing washers must be crafted from materials exhibiting commendable compressibility and resilience. Additionally, in consonance with the operational environment and objectives of the bolted ball joint, these sealing washers should also possess attributes like corrosion resistance and longevity. Common sealing materials encompass rubber, silicone, and polytetrafluoroethylene. Among these options, Nitrile Butadiene Rubber (NBR), a prominent synthetic rubber in the sealing realm, emerges as a prime contender. Noteworthy for its exceptional qualities including heightened wear resistance, robust adhesion, impressive heat resistance, resilience against aging, and superior airtightness [11,12], NBR squarely meets the sealing requirements of the bolted ball joint. Consequently, this paper selects NBR as the designated material for the sealing gasket.

The enhanced bolted ball joint featuring a sealing structure is depicted in Figure 3. This configuration primarily comprises a bolt ball, high-strength bolt, sleeves, pin, cone head, and sealing washer. Diverging from conventional bolted ball joints, the innovation introduced in bolted ball joints with sealed structures centers around the incorporation of circular grooves and sealing washers at both ends of the sleeve. These sealing washers are securely embedded within the circular grooves to ensure precise alignment and steadfastness during installation. Upon the tightening of the high-strength bolt, the sealing washer undergoes compression. This ingenious design effectively engenders a sealing interface within the contact gap between the sleeve and bolt ball and the cone head, thereby thwarting the infiltration of seawater into the joint's interior. During the installation process, the prescribed procedure entails initially positioning the sealing washer within the circular groove at each end of the sleeve. Subsequently, the sleeve and sealing washer are collectively inserted into the high-strength bolt. The subsequent step involves threading the fastening screw into the pinhole to facilitate the bolt's insertion into the threaded hole until the predefined depth is attained.

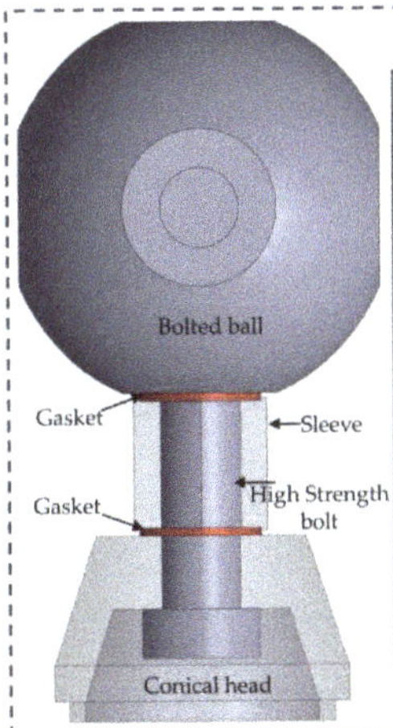

Component	Dimension
Bolted ball	100 mm diameter
Sleeve	hexagonal section inscribed circle diameter 34 mm, length 33 mm
Conical head	bottom plate thickness is 18 mm, cone tail thickness is 3.75 mm
Member	75 mm diameter, 3.75 wall thickness
Gasket	outer diameter 31 mm, inner diameter 23 mm, thickness 2 mm
High strength bolt	20 mm diameter

(**a**)

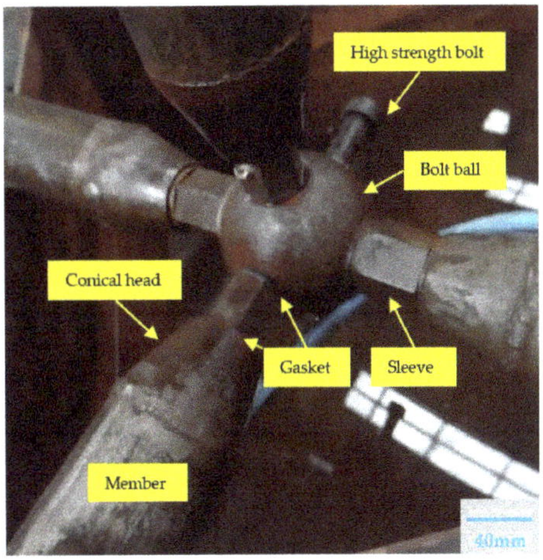

(**b**)

Figure 3. Schematic diagram of the joint structure: (**a**) Schematic diagram; (**b**) Real model.

3. Numerical Simulation

3.1. Guidelines for Sealing Evaluation

Based on the inherent nature and sealing principles of NBR materials, the contact pressure between the sealing cushion and the corresponding component's contact surface plays a pivotal role in determining the sealing efficacy of the overall structure. To establish a dependable seal, the contact pressure must surpass the pressure exerted by the sealing medium while also maintaining a specific contact length. Consequently, the criteria used to assess the structural sealing are selected as the guiding standards for evaluating the effectiveness of the sealing structure [13,14]. The criteria can be expressed as a formula:

$$P_c \geq P_m \quad (1)$$

where P_c is the effective contact pressure on the sealing contact surface, and P_m is the pressure of the gas filled inside the joint.

3.2. The Numerical Model

3.2.1. Geometric Modeling

Firstly, the joint model undergoes simplification by transforming the outer section of the sleeve from a hexagonal shape to a circular one, and the resultant simplified joint model is depicted in Figure 4a. The focal point of this study is the contact region of the seal. The analysis assumes the sleeve's centerline as the axis of symmetry, rendering the geometry and boundary conditions of the simplified nodal model axisymmetric. Consequently, a two-dimensional axisymmetric model is employed for analysis. Due to symmetry considerations, one of the seals at the sleeve's ends is selected for detailed analysis. The specific process for creating the model is as follows: Firstly, an assembly model of the bolted ball joint with the sealing structure is constructed using SOLIDWORKS. Subsequently, the stretch excision command is employed to derive a 1/4 model of the component under analysis. With the use of the isometric surface command, cross-sections of each component within the joint model are replicated. Subsequently, the joint entity is removed, yielding a two-dimensional axisymmetric geometric model of the joint, as depicted in Figure 4d. Finally, the completed joint model file is imported into ANSYS for further processing. The finite element model (FEM), post-processing, is shown in Figure 4e, and the processing procedure is detailed in subsequent Sections 3.2.2–3.2.5

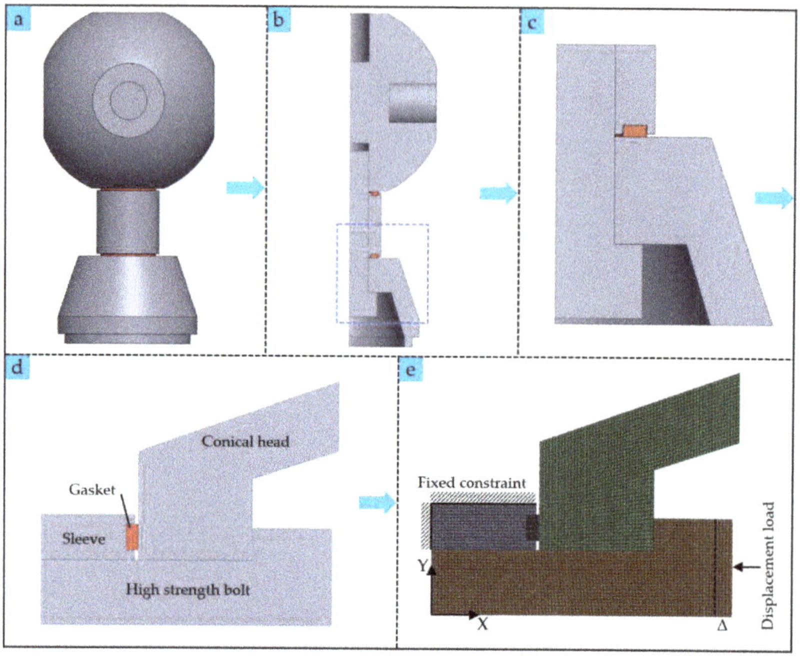

Figure 4. Modeling Creation: (**a**) The simplified model; (**b**) The 1/4 model; (**c**) The analysis section; (**d**) The two-dimensional model; (**e**) The FEM.

3.2.2. Material Constitutive Relationship

During the finite element analysis process, the components including high-strength bolts, sleeves, and cone heads are constructed using Q355 steel. Due to their minimal deformations, these components are treated as either rigid or elastic, possessing an elastic modulus of 206 GPa and a Poisson's rate of 0.3. On the other hand, the sealing ring is fashioned from NBR material, with its constitutive relationship defined through the utilization of the Mooney Rivlin 2 parameter model [15,16]. The corresponding strain energy density function is provided below:

$$W = C_{10}(I_1 - 3) + C_{01}(I_2 - 3) \quad (2)$$

where W is the strain energy, C_{10} and C_{01} are material-related constants, and I is the principal strain in the principal direction. The constants for C_{10} and C_{01} in this article are 1.87 and 0.47, respectively [11].

3.2.3. Meshing

During the finite element analysis process, the PLANE182 planar element is employed for simulation. The PLANE182 element, a quadrilateral 4-node element, possesses the capability for plasticity, hyperelasticity, stress stiffness, large deformation, and extensive strain. This enables it to provide more accurate simulations of the deformation of incompressible hyperelastic materials. Figure 4e illustrates the mesh division of the finite element model. In this diagram, the rubber ring is assigned a mesh size of 0.2 mm, while the remaining components are given a mesh size of 0.5 mm. The division results in a total of 13,456 nodes and 4292 elements. Increasing the mesh count by a factor of 1 beyond this configuration yields calculation results with an error margin below 0.45%. However, this enhancement comes at the expense of significantly increased computation time. Consequently, the mesh configuration employed in this study strikes a balance between computational efficiency and accuracy, rendering it reasonable and capable of yielding satisfactory calculation precision.

3.2.4. Contact Setting

Considering the interaction between the components of the sealing structure, the contact of the bolt with the cone head and the sleeve is set as frictional contact, and the coefficient of friction is set as 0.15. The contact of the sealing washer with the sleeve and cone head is set as frictional contact, and the coefficient of friction is set as 0.2 [15].

3.2.5. Loads and Boundary Conditions

The sleeve's left and upper boundaries are constrained in both X and Y directions, and constraints are imposed in the Y direction on the bolts and cone heads. With the use of the displacement loading mode, a displacement is applied in the X direction to the right boundary of the bolt, as visually illustrated in Figure 4e.

3.3. Numerical Simulation Results

To ensure the attainment of effective sealing performance, the typical compression rate for rectangular seals falls between 8% and 14% [17]. For the scope of this study, the model's compression spans from 0.16 mm to 0.28 mm. Leftward displacements of 0.16 mm, 0.22 mm, and 0.28 mm are then applied to the bolt's tail end to simulate corresponding seal compressions during the tightening process of the bolted ball joint. Subsequently, the deformation cloud diagrams of the sealing washer, the equivalent force cloud diagram, and the contact pressure cloud diagram are depicted in Figures 5–7.

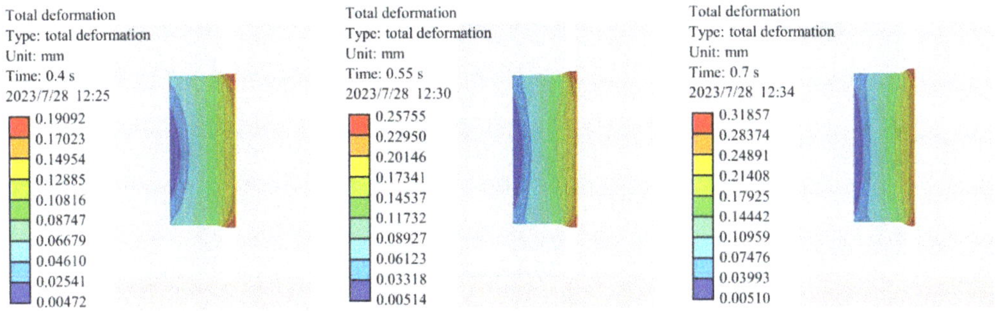

(a) (b) (c)

Figure 5. The displacement cloud diagrams of rubber rings at different compression rates: (**a**) 8%; (**b**) 11%; (**c**) 14%.

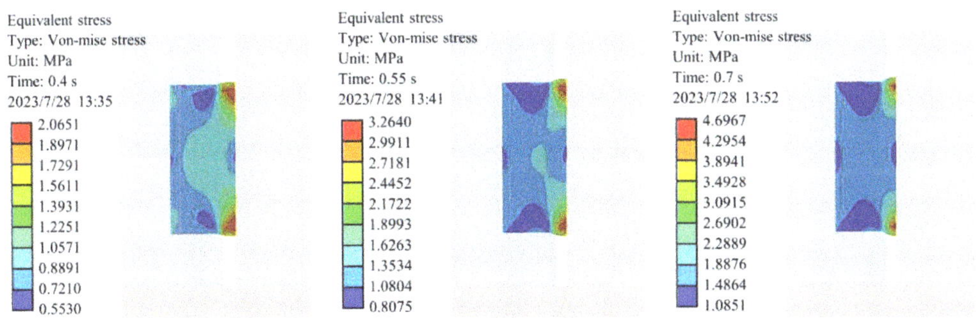

(a) (b) (c)

Figure 6. The equivalent stress cloud diagrams of rubber rings at different compression rates: (**a**) 8%; (**b**) 11%; (**c**) 14%.

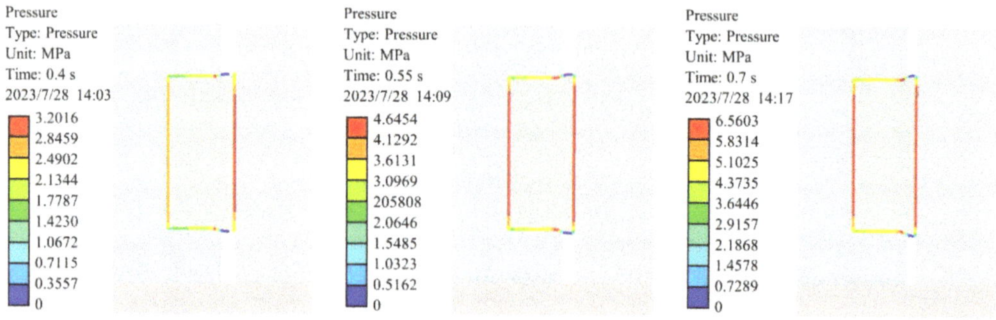

(a) (b) (c)

Figure 7. The contact pressure cloud diagrams of rubber rings at different compression rates: (**a**) 8%; (**b**) 11%; (**c**) 14%.

Based on the analysis of the deformation of the cloud map, it becomes evident that the deformation distribution of the sealing ring maintains considerable uniformity. With the gradual increase in the compression rate, the extent of the sealing ring's deformation also amplifies. It is noteworthy that the most pronounced deformation regions are localized at the upper and lower extremities of the contact surface between the sealing ring and the cone head. This phenomenon is attributed to the progressive filling of the contact gap by the sealing ring during the compression, leading to a more notable deformation in the upper and lower portions of the contact surface. This phenomenon plays a pivotal role in the sealing process, ensuring the efficacy and stability of the seal's performance.

The stress cloud diagrams reveal that the equivalent stress level of the seal escalates as the compression increases. Notably, the stress values at the upper and lower extremities of the contact surface between the seal and the cone head are notably higher, exhibiting a concentration of stress. This observation aligns with the findings presented in the deformation diagrams.

Based on the analysis of the contact pressure cloud diagram of the sealing ring, it becomes evident that the principal sealing effect manifests on the contact surfaces located on the left and right sides of the sealing ring. Notably, the sealing performance is dictated by the smaller of these two contact surfaces. Upon data extraction, the correlation between the contact pressure on the left and right sides of the sealing surface and the actual path distance under varying compression levels is established. This relationship is visually depicted in Figure 8.

As depicted in Figure 8, it becomes evident that the seal's contact pressure is most pronounced within the central region. This central section, spanning a length of approximately 3 mm and constituting 75% of the contact area, assumes the primary role in effecting sealing. With an increase in the compression of the sealing ring, the contact pressure on the sealing surface correspondingly escalates, and their relationship exhibits an approximate linearity. At 14% compression, the sealing gasket attains a sealing pressure of 5.95 MPa. This outcome substantiates the efficacy of the bolted ball joint's sealing structure within high-pressure contexts. Moreover, it underscores the sealing structure's commendable sealing performance, thereby affirming its viability in marine environments for safeguarding bolted ball joint mesh structures against corrosion.

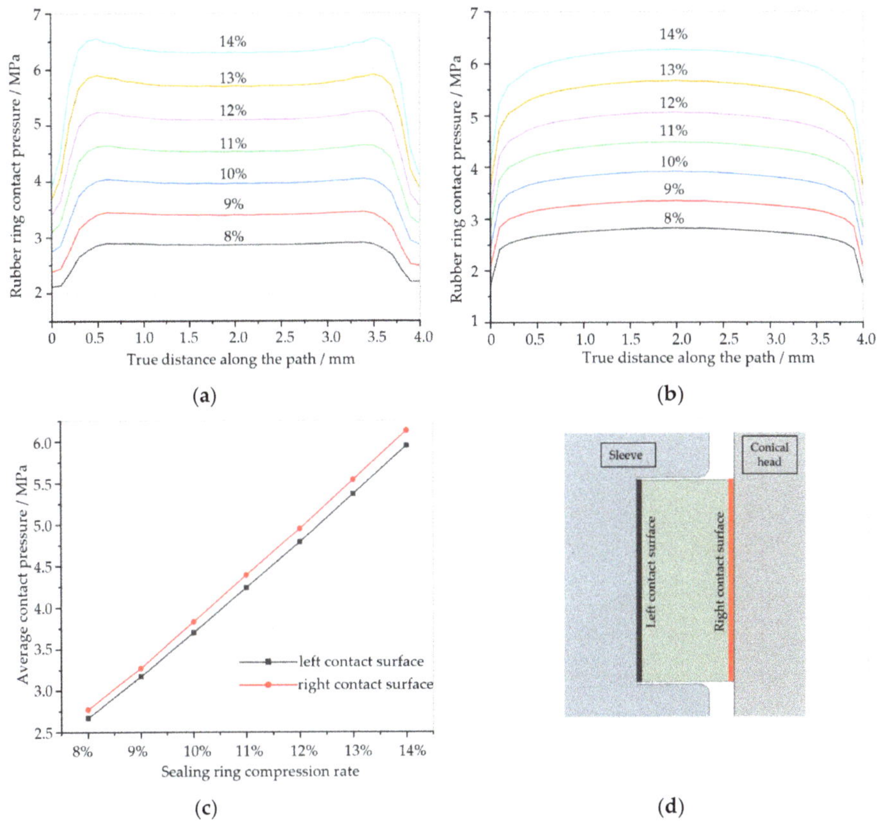

Figure 8. Contact pressure curve of the rubber ring at different compression rates: (**a**) the left contact surface; (**b**) the right contact surface; (**c**) compression rate and average contact pressure curves; (**d**) schematic representation of the contact surface.

4. Experimental Testing

4.1. Experimental Design

The chosen experimental approach employs the bubble leakage detection method. This involves introducing a specific pressure of leakage-indicating gas into the test specimen immersed in liquid. This process generates a pressure differential between the interior and exterior of the airtight cavity. If there exists a gap causing leakage within the sealing cavity, the leakage-indicating gas will flow from the higher-pressure side to the lower-pressure side through the gap. This action results in the formation of bubbles around the leakage gap. This visual occurrence aids in ascertaining the presence of a leakage phenomenon within the airtight cavity [18]. This method offers a swift and intuitive evaluation of the specimen's sealing quality with heightened sensitivity. In this test, air serves as the leakage-indicating gas, while water functions as the displaying liquid.

The experiment utilizes bolted ball joint specimens with the specifications and dimensions depicted in Figure 3. The components of the bolted ball joint are made from Q235 steel, and the sealing ring is crafted from NBR material. To facilitate the inflation of the bolted ball joint's internal pressure, a sealing plate is affixed to the end of the member, accompanied by the installation of an inflatable valve, as depicted in Figure 9a. Subsequently, an air compressor pump is engaged, connected to the member's interior to instate inflation and pressurization. This process serves the purpose of assessing the sealing pressure and detecting any potential leaks within the sealing structure. To ensure heightened reliability in the test outcomes, simultaneous pressure experiments are conducted on the four members. The essential instruments and equipment for this experiment include an air compressor, a pressure gauge, a vernier caliper, and a torque spanner. Specifically, the air compressor delivers high-pressure air inside the sealing rod, utilizing the XSZG-30A model screw air compressor by the XINLEI brand. This model consistently generates high-pressure air, reaching up to 0.8 MPa. The pressure gauge displays the air pressure output, with the experiment utilizing a pressure gauge with a range of 1.6 MPa and an accuracy rating of 1.6, ensuring precise measurement of the air pressure value. The vernier caliper measures the distance between the cone head and the bolt ball, enabling the calculation of the compression rate of the seal. In this experiment, digital display vernier calipers with a range of 0–150 mm guarantee accurate measurements. Lastly, the torque spanner plays a crucial role in tightening the bolted ball joint in accordance with specified values. The experiment employs a torque spanner with a range of 50–1000 N·m, a divisional value of 0.1, and a precision level of ±2% as indicated by the numerical display on the torque spanner. The experimental test principle and necessary equipment are elaborated in Figure 9b and Table 1, respectively.

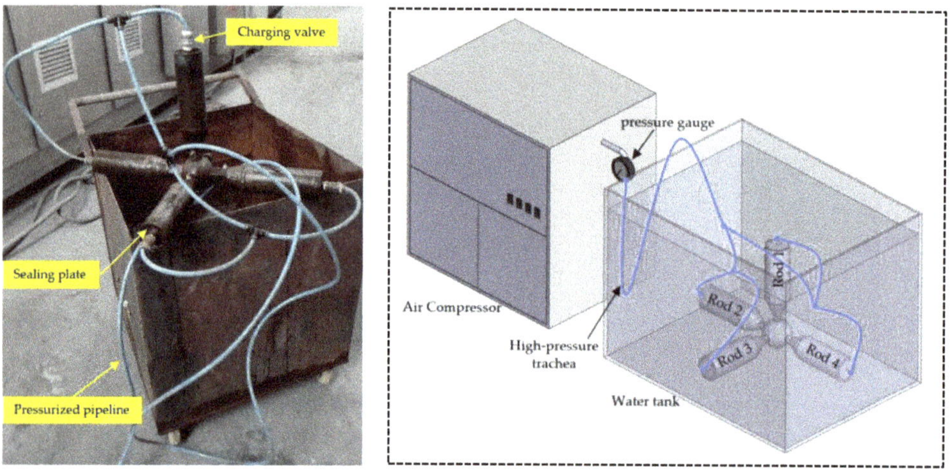

(**a**) (**b**)

Figure 9. Airtightness testing experiment: (**a**) specimen; (**b**) schematic diagram.

Table 1. Instrument and equipment for testing the sealing performance.

Device Name	Model	Amount	Performance Parameters
Air Compressor	XSZG-30A	1	Maximum output pressure 0.8 MPa (8 atmospheric pressure)
Pressure gauge	Y-100	1	Range 1.6 MPa, accuracy 1.6 grade
Dial calipers	LJ800-001	1	Range 0–150 mm
Torque spanner	WT8-30	1	Torque range 50–1000 N-m, graduation value 0.1, accuracy ±2%

4.2. Experiment Process

The initial step involves the assembly of the components of the bolted ball joint, with the application of a bolt torque during assembly adhering to the specified standard value of 137 N·m [19]. Subsequently, the distance between the cone head and the bolt ball is measured using vernier calipers, yielding a measurement of 36.50 mm, and the average compression rate of the two seals is determined to be 12.57%. Following this, the inflatable device is connected according to the test scheme and principles illustrated in Figure 9. After the connection is completed, the specimen is fully immersed in water. The air compressor is then activated, initiating a step-by-step increase in high-pressure air output, with 0.1 MPa increments. Each pressure level is maintained for 5 min, with an inspection for any signs of leakage at the seal. If no abnormalities are observed, the pressure is continuously raised until a leakage is detected or the maximum output pressure of the air compressor pump is reached.

Furthermore, upon immersing the test piece in water, bubbles might emerge at the sealing point. If these bubbles do not reappear after wiping or poking, it can be inferred that the initial bubble formation site is devoid of leakage apertures. Conversely, a consistent and persistent emission of bubbles would indicate the presence of a leakage hole at the point of bubble origination [18].

4.3. Experimental Results and Analysis

Two sets of nodal models were subjected to experimentation: the conventional bolted ball joint and the modified bolted ball joint. During the initial inflation stage of the conventional bolted ball joint, air bubbles emerged, and their quantity increased with rising inflation pressure, as depicted in Figure 10a. Notably, the conventional bolted ball joint exhibited no sealing properties whatsoever. In contrast, the modified bolted ball joint displayed impeccable performance, exhibiting no leakage and maintaining effective sealing throughout the initial inflation and pressurization stages until it reached the maximum inflation pressure of 0.8 MPa from the air compressor, as illustrated in Figure 10b.

(a) (b)

Figure 10. Airtightness testing experiment of the bolted ball joint: (**a**) No sealing setting; (**b**) With sealing setting.

Owing to power limitations in the employed inflation equipment, the experimental gas output pressure was constrained to 0.8 MPa—a water pressure value equivalent to a depth of 81.63 m in seawater. In practical applications of deep-sea fish cages, a typical depth of around 20 m is employed. This result significantly bolsters the groundwork for implementing bolted ball joints in offshore environments, ensuring their dependable adaptation to the challenges presented by seawater conditions.

Additionally, the sealing method lacks experimental validation under higher gas pressure due to the limitations of the bubble leak detection method and the specific inflation equipment used. Future research should focus on conducting refined experiments to confirm the sealing method's efficacy and validate its compatibility with the numerical simulation approach. It is worth noting that this study was conducted as a preliminary phase of a larger project focusing on deep-sea fish-farming nets utilizing the bolted ball joint.

5. Conclusions

This paper introduces a methodology to enhance the sealing performance of bolted ball joints, offering a solution for the corrosion protection of grid structures with bolted ball joints in marine environments. Numerical simulation analyses and sealing performance tests were conducted on the bolted ball joint equipped with the sealing structure. The primary conclusions are outlined below:

(1) According to the numerical simulation results, it is evident that the sealing pressure rises proportionally with the augmentation of the compression rate, illustrating a clear linear correlation between the two factors. The sealing gasket achieves impressive sealing pressures of 2.91 MPa, 4.22 MPa, and 5.95 MPa at compression rates of 8%, 11%, and 14%, respectively.

(2) Concerning the NBR seals of the specifications employed in this study, the higher contact pressure is predominantly concentrated in the intermediate region. The primary sealing function is effectively performed by the central area of the contact surface, encompassing a length of approximately 3 mm and accounting for 75% of the total contact area.

(3) Airtightness testing experiments were conducted on both the conventional and modified bolted ball joints using the bubble leakage detection method. The outcomes revealed that the conventional bolted ball joint exhibited no sealing whatsoever, while the bolted ball joint incorporating the sealing structure demonstrated impeccable sealing performance.

(4) The sealing method for bolted ball joints provides a cost-effective solution with simple installation and maintenance. Retaining the traditional appearance of the joint, it guarantees impeccable internal sealing. This prevents seawater intrusion and corrosion, aligning perfectly with the application demands of the bolted ball joint in the realm of marine engineering.

(5) The sealing method facilitates the successful adaptation of steel grid structures, extensively employed in terrestrial settings, for utilization in marine engineering applications, encompassing offshore floating platforms and deep-sea fish cages. Analytical and computational findings underscore that this approach effectively ensures optimal sealing performance even at depths of up to 607.14 m in seawater.

Author Contributions: Conceptualization, W.D. and G.S.; Methodology, J.G. and G.S.; Software, G.G.; Validation, W.D., Z.L. and Y.Z.; Writing—original draft, J.G.; Writing—review & editing, W.D.; Visualization, Y.Z.; Project administration, Z.L.; Funding acquisition, W.D. All authors have read and agreed to the published version of the manuscript.

Funding: This research was funded by the National Natural Science Foundation of China, grant numbers U1704141 and 52178172.

Institutional Review Board Statement: Not applicable.

Informed Consent Statement: Not applicable.

Data Availability Statement: Data will be made available on request.

Conflicts of Interest: The authors declare no conflict of interest.

References

1. Cao, Z.; Du, P.; Chen, Z. The stability and stressed skin effect analyses of an 80 m diameter single-layer latticed dome with bolt-ball joints. *Int. J. Steel Struct.* **2016**, *16*, 279–288. [CrossRef]
2. Liu, W.B.; Xi, S.L.; Cheng, H.S. Analysis of the structure style selection for ocean engineering floating structure. *Adv. Mater. Res.* **2011**, *163*, 501–506. [CrossRef]
3. Amaechi, C.V.; Reda, A.; Butler, H.O.; Ja'e, I.A.; An, C. Review on Fixed and Floating Offshore Structures. Part I: Types of Platforms with Some Applications. *J. Mar. Sci. Eng.* **2022**, *10*, 1074. [CrossRef]
4. Aksu, R.; Uguz, R.O.; Erdogan, M. Investigation, Modeling and Design of a Cathodic Protection System for Hull Structures in Marine Environment. *ECS Trans.* **2016**, *72*, 163. [CrossRef]

5. Si, Q.; Tang, Y.; Zong, L.; Liu, H.; Kang, B. Experimental Study and Numerical Simulation of the Tensile Properties of Corroded Bolt-Sphere Joints. *Buildings* **2022**, *12*, 1989. [CrossRef]
6. Zhao, X.D.; Xi, G.F.; Yang, J. Comparative Study on Corrosion of Mild Steel in Natural and Simulated Marine Environment. *Appl. Mech. Mater.* **2011**, *66*, 1828–1831. [CrossRef]
7. Peng, L.; Stewart, M.G.; Melchers, R.E. Corrosion and capacity prediction of marine steel infrastructure under a changing environment. *Struct. Infrastruct. Eng.* **2017**, *13*, 988–1001. [CrossRef]
8. Zeng, L.Q.; Liu, M.W.; Abi, E. Erosion characteristics of viscoelastic anticorrosive coatings for steel structures under sand flow. *Constr. Build. Mater.* **2020**, *258*, 120360. [CrossRef]
9. Huang, F.; Hong, Z.H.; Huang, P. A Sealing Structure for Bolted Ball Joints. Jiangxi Province. CN 217682657U, 28 October 2022.
10. Zhang, R.L. Sealed Connecting Rods for Bolted Ball Joint Grid Structure. Tianjin. CN 2571842, 10 September 2003.
11. Li, J.X.; Liu, P.F.; Wang, S.B.; Leng, J.X. Finite Element Analysis of O-ring Sealing Performance of Manned Submersible Viewports. *J. Fail. Anal. Prev.* **2020**, *20*, 1628–1637. [CrossRef]
12. Zhou, C.; Chen, G.; Liu, P. Finite Element Analysis of Sealing Performance of Rubber D-Ring Seal in High-Pressure Hydrogen Storage Vessel. *J. Fail. Anal. Prev.* **2018**, *18*, 846–855. [CrossRef]
13. *GB/T 33509-2017*; General Specification for Mechanical Seals. China Machinery Industry Federation: Beijing, China, 2017.
14. Zhao, Z.F.; Shi, X.; Hua, Y.L. Performance analysis and test on the sealing structure of repair clamp for subsea pipeline. *Ocean. Eng.* **2023**, *41*, 101–109.
15. Bhaumik, S.; Kumaraswamy, A.; Guruprasad, S. Investigation of friction in rectangular Nitrile-Butadiene Rubber (NBR) hydraulic rod seals for defense applications. *J. Mech. Sci. Technol.* **2015**, *29*, 4793–4799. [CrossRef]
16. Song, X.; Huang, S.; Hui, H. Analysis of sealing performance of a kind of profiled rubber gasket used in the radial contact seal structure. *Proc. Inst. Mech. Eng. Part E J. Process Mech. Eng.* **2021**, *235*, 857–862. [CrossRef]
17. Jing, T.; Weimin, Y.; Yumei, D. Finite Element Analysis of Rectangular Rubber Seals. *Lubr. Eng.* **2007**, *2*, 36–39.
18. Biram, J.; Barrows, G. Bubble tests for gas tightness. *Vacuum* **1964**, *14*, 221–226. [CrossRef]
19. *GB/16939-2016*; High Strength Bolts for Joints of Space Grid Structures. China Machinery Industry Federation: Beijing, China, 2016.

Journal of
Marine Science and Engineering

Article

A Method for Predicting the Load Interaction between Reinforced Thermoplastic Pipe and Sandy Soil Based on Model Testing

Chuan Wang [1,2,3,*], **Lianghai Liu** [1,2,3], **Ya Zhang** [1,2,3] **and Min Lou** [1,2,3]

[1] College of Mechanical and Electronic Engineering, China University of Petroleum (East China), Qingdao 266580, China
[2] Key Laboratory of Unconventional Oil & Gas Development, China University of Petroleum (East China), Ministry of Education, Qingdao 266580, China
[3] National Engineering Laboratory of Offshore Geophysical and Exploration Equipment, China University of Petroleum (East China), Qingdao 266580, China
* Correspondence: wangchuan@upc.edu.cn

Abstract: This study aims to investigate the interaction between reinforced thermoplastic pipes (RTPs) and sandy soil. The mechanical properties of sandy soil in the South China Sea region were determined through shear tests to obtain fundamental data. Subsequently, a specialized experimental setup was designed and assembled to study the pipe–soil interaction, specifically measuring the lateral soil resistance of flexible pipes at varying burial depths. Data analysis revealed the relationship between soil resistance, lateral displacement, and initial burial depth. To simulate the mechanical behavior of the pipe–soil interaction, the coupled Eulerian–Lagrangian (CEL) method was employed for numerical simulations. The research findings indicate that the lateral soil resistance is influenced by the uplift height and accumulation width of the soil ahead of the pipe. Within a lateral displacement range of 0.5 times the pipe diameter (0.5*D*), the lateral soil resistance rapidly increases, resulting in a soil uplift along the circumferential direction of the pipe. This process not only enhances the load-bearing capacity of the pipe but also increases the accumulated soil resistance, consequently expanding the soil failure zone. Furthermore, the ultimate soil resistance exhibits an increasing trend with an increasing burial depth. Once the pipe reaches a certain burial depth, the uplift height of the soil reaches a critical state. To address the grid distortion caused by soil deformation, numerical simulations based on the CEL method effectively modeled the pipe–soil interaction forces under significant lateral displacements, exhibiting good agreement with the experimental results. This study provides a solution for investigating soil resistance in submarine pipelines, thereby contributing significantly to the design and performance prediction of underwater pipelines.

Keywords: lateral soil resistance; CEL method; reinforced thermoplastic pipes (RTPs); pipe–soil interaction

Citation: Wang, C.; Liu, L.; Zhang, Y.; Lou, M. A Method for Predicting the Load Interaction between Reinforced Thermoplastic Pipe and Sandy Soil Based on Model Testing. *J. Mar. Sci. Eng.* **2023**, *11*, 2353. https://doi.org/10.3390/jmse11122353

Academic Editor: Dejan Brkić

Received: 4 November 2023
Revised: 11 December 2023
Accepted: 11 December 2023
Published: 13 December 2023

1. Introduction

Submarine oil and gas pipelines serve as the lifelines for offshore oil and gas production systems, providing the most direct and efficient means of transportation [1]. With the increasingly complex exploitation environment and the growing utilization of fiber materials, the application of RTPs based on thermoplastic materials has gained traction in the field of offshore engineering [2–5]. Submarine pipelines, which are partially embedded in the seabed, are subjected to lateral soil resistance during significant lateral movements. This resistance plays a crucial role in the lateral buckling of pipelines under thermal stress [6,7]. Therefore, studying the interaction between RTPs and soil is of paramount importance to ensure on-bottom stability.

In order to achieve an optimal design solution, it is crucial to accurately model the interaction between a pipe and the soil during large-amplitude lateral movements. In recent years, researchers have focused on studying the pipe–soil interaction for pipes that are shallowly embedded in the seabed with the aid of centrifuge testing. Several models have been proposed specifically for shallowly embedded pipes, considering different soil types such as calcareous sands [8–12] and clays [12–15]. DNV [16] proposed a small-displacement, three-stage soil resistance model. Bruton et al. [17] conducted lateral large pipeline displacement experiments and analyzed the nonlinear characteristics of soil resistance. They categorized the pipe–soil interaction into the breakthrough, unstable, and residual motion stages. Cheuk, White, and Bolton [18] conducted extensive plane strain model tests to measure the response of a partially embedded pipe segment in soft clay subjected to large-amplitude cyclic movements. The tests aimed to simulate the thermal expansion and contraction experienced at a pipeline bend. Colicchio [19] et al. proposed an innovative model for analyzing pipe–soil interaction using the level set technique for a two-phase mixture of sand and water. Dingle, White, and Gaudin [20] researched the lateral breakthrough resistance and residual resistance of soft clay. During the lateral buckling process, significant lateral displacements ranging from approximately 5 to 20 times the pipe diameter may occur. Wang [15,21] developed a dynamic soil resistance model that varies with the buried depth of the pipeline, based on a series of model tests using sand samples from the Bohai Gulf. In fact, the time-varying soil resistances during the pipe–soil interaction are strongly interconnected in both vertical and lateral directions. However, there is currently limited research attention devoted to this issue, and the coupling mechanism requires further exploration.

Wang [22] developed a new formula for predicting residual resistance based on the failure mechanism of pipe–soil interaction during lateral movement. Kong [7] explored the breakout resistances of a pipe in all possible directions to determine the most vulnerable plane where transverse displacement is likely to occur, along with the minimum breakout resistance. Liu [21] conducted large-scale model tests on Bohai sand and proposed bi-linear pipe–soil interaction models normalized by pipe weight and diameter based on these test results. Dong [23] developed a coupled large-deformation finite element (LDFE) framework with a small strain (RITSS) and discussed the effects of the interaction rate and drainage condition on the p–y curve, excess pore pressure generation and dissipation, and failure mechanisms. Liu and Ortega [24] proposed a closed-form linear–elastic solution to analyze pipes that were restrained using both thrust blocks and restrained joints. The solution took into consideration the passive soil resistance, as well as the contribution of friction resistance resulting from the increased normal pressure at the pipe–soil interface. Macaro, Utili, and Martin [25] utilized the three-dimensional distinct-element method to investigate the behavior of a partially embedded pipe segment in sand, and the simulation approach was validated against experimental results. Wu [26] presented a detailed numerical investigation to study the interaction between large-diameter pipes and soil under three typical working conditions. The study aimed to capture the structural mechanical behaviors associated with pipe deformation, soil displacement, and soil pressure. Additionally, a novel method was introduced to investigate soil–pipe interaction in the lateral and vertical directions. This method involved mounting a soil box on a six-degrees-of-freedom (DOF) system. The test results provided evidence of how different loading types affect soil–pipe interaction and demonstrated a strong coupling of the pipe response in the lateral and vertical directions. Wang [27] proposed a formula for predicting the maximum lateral soil resistance based on tests that examined pipe–soil interaction with different initial pipe embedments and weights. Rezadoost and Nassiraei [28] conducted a comprehensive study on the criteria for modeling fiber-reinforced polymer (FRP) using the finite element method. The accuracy of these models was rigorously verified through meticulous comparisons with experimental results.

There has been extensive research on the interaction between steel pipes and soil in this field. However, research on the soil resistance of RTPs moving on the surface of sandy soil remains largely unexplored. Additionally, existing large-displacement lateral testing devices have various limitations in terms of efficiency. Therefore, there is a need to enhance the research on the interaction between reinforced thermoplastic (RTPs) pipes and soil.

The aim of this study is to address a research gap by designing and manufacturing a test apparatus for investigating the horizontal soil resistance of flexible pipes in response to different initial settlement depths. This apparatus enables the detection of the lateral soil resistance characteristics of pipelines under varying burial conditions. The primary objective is to reveal the correlation between typical resistance and lateral movement at different stages. Moreover, the study aims to resolve the issue of grid distortion caused by soil deformation by employing the CEL method to simulate the interaction between pipes and soil. The effectiveness of the simulation was validated through a comparison with the experimental results.

2. Experimental System and Content

2.1. Experimental Configuration

Typically, pipelines are assumed to be infinitely long, and axial displacements are disregarded, allowing the lateral pipe–soil interaction problem to be simplified as a plane strain problem [29]. Based on this conception, a testing system for lateral pipe–soil interaction was designed to replicate significant lateral movements of the pipe above the seabed, as depicted in Figure 1.

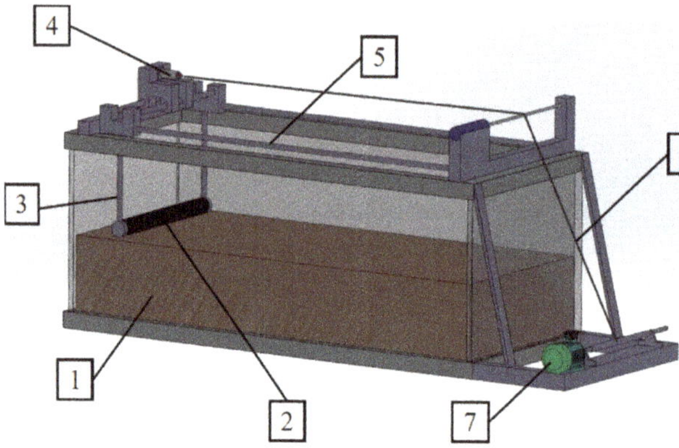

Figure 1. Sketch of the experimental device. 1—soil box; 2—flexible pipe; 3—anti-roll linkage mechanism; 4—tension–compression load sensor; 5—slide (with wiping lubricant); 6—steel wire; and 7—electric motor.

(1) Experimental system: In a rectangular soil box, two parallel tracks with a length of 1.5 m were built, and a connection structure with a flexible pipeline was designed to connect the pipe to the motive force transmission device and install it on the soil box. The motive force transmission device was connected to the pressure sensor, and the pipeline was driven horizontally in the soil sample to obtain the relationship curve between the pipe displacement and the soil resistance force.

The design of the connecting components between the pipeline and the transmission structure is shown in Figure 2. The connecting mechanism consists of the following components:

- A connecting rod: It provides the structural connection between different parts of the mechanism;
- A connecting outer circle: Its diameter is equal to the outer diameter of the pipeline. It ensures the connection between the mechanism and the outer surface of the pipeline;
- A connecting inner embedded circle: Its diameter is equal to the inner diameter of the pipeline. It ensures the connection between the mechanism and the inner surface of the pipeline;

- An anti-rotation latch: It is embedded in the gap of the pipeline to prevent circumferential movement during horizontal motion of the pipeline.

(2) Experimental equipment: The main instruments and components used in this experiment included the following: a soil box size whose length × width × depth = 1.5 m × 1.0 m × 1.0 m, an LTR-1 tension–compression load, sensor, a sliding rail, a pipeline anti-roll and translation device, a motor, etc.

(3) Experimental samples: Flexible, non-metallic pipes with a diameter of 76 mm and sandy soil from the seabed.

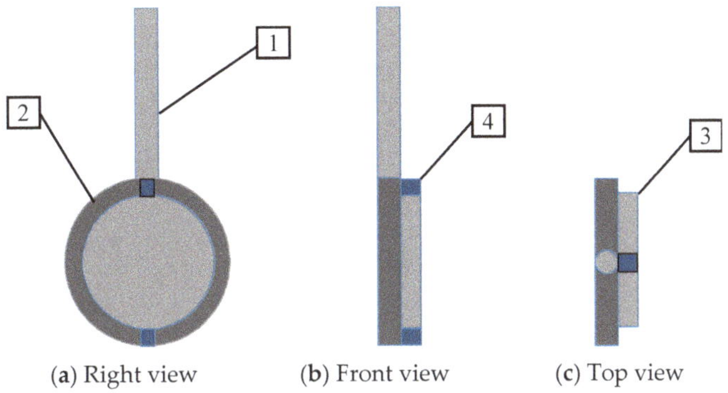

Figure 2. Schematic diagram of the pipe connection. 1—connecting rod, 2—connecting outer circle, 3—connecting inner embedded circle, and 4—anti-rotation latch.

2.2. Experiment Content

This experiment included three parts: (1) carry out soil shear tests to determine the friction angle and shear strength of the soil; (2) carry out pipe–soil interaction experiments to identify the relationship between horizontal pipe displacement and soil resistance; (3) analyze the experimental results based on the scientific research literature and obtain comparative data on the differences in the soil resistance loads of flexible, non-metallic pipes and steel pipes.

2.3. Tests Pipe

The RTPs examined in this study were made up of a polyethylene (PE) matrix with fiberglass reinforcement, as depicted in Figure 3. The reinforcement comprised fiberglass embedded within the polymer matrix, which was spirally wound around the inner PE liner at a specific angle. The layers were fused together through heating, producing a filament-wound fiber-reinforced composite pipe.

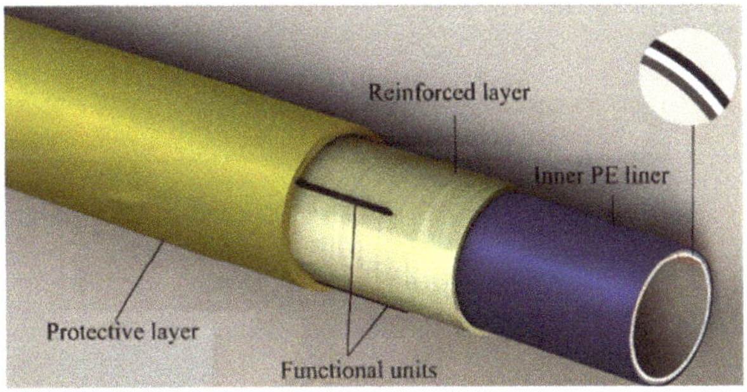

Figure 3. Structural diagram of an RTP.

2.4. Test Soil

The South China Sea is rich in oil and gas storage and transportation, and submarine pipelines have a relatively broad application space. To ensure the safety of submarine pipelines, the target oil and gas sea area of seabed sand was selected as the experimental model, which had a wide particle size distribution. Sandy soil is highly permeable and rapidly drains, resulting in the strength of a dry sample being comparable to that of a saturated sample under drained conditions. Thus, in the interest of simplification, we employed dry sand to construct the test soil bed.

2.5. Experimental Process

To study pipe–soil interactions during large displacements in sand, we conducted two series of tests, a shear strength experiment with soil tests and large-amplitude motion tests.

The main purpose of the shear strength experiment was to identify the mechanical parameters of the experimental soil samples to predict the lateral soil resistance in numerical simulations. The objective of the lateral large displacement experiment was to determine the lateral soil resistance experienced by the pipeline under different initial burial depth conditions.

1. Soil shear strength test:

The purpose of the soil shear strength test was to study the shear strength and internal friction angle of the soil sample. A constant-normal-pressure shear test was carried out at seven normal pressures. After the normal displacement was stable, the shear was started, and the shear rate was controlled at 0.8 mm/min. The target parameters were identified by processing the experimental data, as shown in Figure 4.

Figure 4. Soil shear test.

2. Pipe–soil interaction experiment:

It has been shown that the impedance of sandy soil to pipelines does not differ significantly between fully wet and fully dry states. Therefore, dry sandy soil was used in this experiment. The experimental system was set up as shown in Figure 5. The surface of the sandy soil was scraped flat, and lubricating oil was applied to the smooth track to reduce the influence of track friction on the experimental results.

The vertical position of the track was adjusted to ensure that the pipeline had initial depths of $0.1D$, $0.2D$, and $0.3D$, where $0.1D$ means 0.1 times the pipe diameter.

The electric motor was started, and the steel wire rope was tightened via the transmission rod to drive the truss structure and the test pipeline to move laterally on the surface of the sandy soil through the auxiliary pulley system. In order to better observe the change in soil resistance, the pipeline always maintained a stable and relatively low speed value during lateral movement. Table 1 shows the experiment parameters.

Figure 5. Lateral pipe–soil interaction testing system.

Table 1. Experiment parameters.

Property	Value
Pipe diameter (mm)	76
Internal friction angle (degrees)	To be tested
Cohesive strength (Pa)	To be tested
Soil density (kg·m^{-3})	1560
Horizontal displacement (mm)	760 (10 times the pipe diameter)
Initial embedment	0.1D, 0.2D, and 0.3D

3. Numerical Method

3.1. CEL Theoretical Approach

The lateral displacement of pipes can cause soil extrusion, which poses challenges for numerical calculations due to material deformation and grid distortion. In order to avoid computational failures caused by conventional methods, this study employed the coupled Eulerian–Lagrangian (CEL) method to address the issue of large deformations in pipe–soil interaction. The CEL method utilizes the Eulerian description of solid mechanics problems. The following are the principles underlying the CEL method.

The Lagrangian formulation was employed to describe the conservation equations for mass, momentum, and energy, which could be expressed as follows:

$$\frac{d\rho}{dt} + \rho\nabla\cdot\nu = 0 \tag{1}$$

$$\rho\frac{d\mathbf{v}}{dt} = \nabla\cdot\boldsymbol{\sigma} + \rho\mathbf{b} \tag{2}$$

$$\frac{de_I}{dt} = \boldsymbol{\sigma} : \dot{\boldsymbol{\varepsilon}} \tag{3}$$

$$\frac{d\psi(x,t)}{dt} = \frac{\partial\psi(x,t)}{\partial t} + \mathbf{v}\cdot(\nabla\psi(x,t)) \tag{4}$$

In these equations, ρ represents the material density, $\mathbf{v}$ represents the material particle velocity, $\boldsymbol{\sigma}$ represents the Cauchy stress, $\mathbf{b}$ represents body forces, e_I represents the specific internal energy per unit volume, $\dot{\boldsymbol{\varepsilon}}$ represents the rate of strain, ":" denotes the double contraction of tensors, ∇ represents the gradient operator, and $\nabla\cdot\nu$ represents the divergence of the velocity vector.$\psi(x,t)$ represents the field function.

The Lagrangian control equations were converted into Eulerian-form control equations:

$$\frac{\partial\rho}{\partial t} + \nabla\cdot(\rho\nu) = 0 \tag{5}$$

$$\frac{\partial \rho \mathbf{v}}{\partial t} = \nabla \cdot (\rho \mathbf{v} \otimes \mathbf{v}) = \nabla \cdot \boldsymbol{\sigma} + \rho \mathbf{b} \tag{6}$$

$$\frac{\partial e_I}{\partial t} + \nabla \cdot (e_I \mathbf{v}) = \boldsymbol{\sigma} : \dot{\boldsymbol{\varepsilon}} \tag{7}$$

$$\frac{\partial \psi}{\partial t} + \nabla \cdot \boldsymbol{\psi}(\psi, v, x, t) = \mathbf{S} \tag{8}$$

Among them, the term $\boldsymbol{\psi}$ represents the flux function accounting for convective effects, and $\mathbf{S}$ represents the source term.

In the CEL method, the equation mentioned above was decomposed into two separate equations, and these two equations were solved sequentially in each numerical iteration step.

$$\frac{\partial \psi}{\partial t} = \mathbf{S} \tag{9}$$

$$\frac{\partial \psi}{\partial t} + \nabla \cdot \boldsymbol{\psi}(\psi, v, x, t) = 0 \tag{10}$$

Using an explicit time integration scheme, the velocity and displacement solutions of the physical problem were computed and updated using the central difference method.

$$\mathbf{v}^{n+1/2} = \mathbf{v}^{n-1/2} + \Delta t \mathbf{M}^{-1} \left\{ \mathbf{F}^{n}_{external} - \int \mathbf{B}^{T} \bar{\sigma} d\Omega \right\} \tag{11}$$

$$\mathbf{x}^{n+1} = \mathbf{x}^{n} + \Delta t \mathbf{v}^{n+1/2} \tag{12}$$

In these equations, $\mathbf{M}$ represents the mass diagonal matrix, $\mathbf{F}^{n}_{external}$ represents the external force vector, $\mathbf{B}$ represents the strain matrix, σ represents the average stress within the element, n represents the nth time increment step, Ω represents the computational domain, and Δt represents the duration of the nth time increment step.

3.2. *Modeling Criteria*

a. Constitutive Model

The Mohr–Coulomb model was employed to the constitutive relationship for the soil in this study. The soil domain and air domain were discretized using Eulerian elements, while the pipe model was represented using Lagrangian elements. The structure mesh and fluid mesh were coupled using the CEL contact algorithm to enable the numerical simulation of large lateral displacements in the pipe–soil system.

b. Pipe–Soil Interaction

This approach was based on the time-domain dynamic analysis method to investigate the pipe–soil interaction. In order to prevent the settlement of the soil under its own weight and ensure the relative stability of the soil properties, the loading was applied through internal forces between the soil layers. This loading method allowed the formation of an initial stress field in the soil that satisfied the given conditions and achieved stress equilibrium.

c. Modeling Dimension Verification

Considering the influence of the grid scale on computational results, it is necessary to investigate the sensitivity of the grid size in the soil–pipe interaction region. Taking the self-weight settlement of the pipeline as a case study, this research analyzed the temporal variation curve of the soil–pipe interaction force and compared the effects of different grid sizes on the unit length of soil reaction. Three scenarios were selected for calculation: a large grid (8 mm), a medium grid (5 mm), and a small grid (3 mm). The results indicate that the soil–pipe interaction forces converged and exhibited relatively consistent mean values in all three scenarios, as per Figure 6. However, the large grid (8 mm) showed significant fluctuations during the stable phase, while the medium and small grids exhibited smoother variations. The computational cost of the small grid was approximately 4.5 times higher than that of the medium grid. Therefore, to obtain accurate results with a relatively small computational workload, it is recommended to set the soil grid size to 5 mm and gradually transition towards the surrounding boundaries.

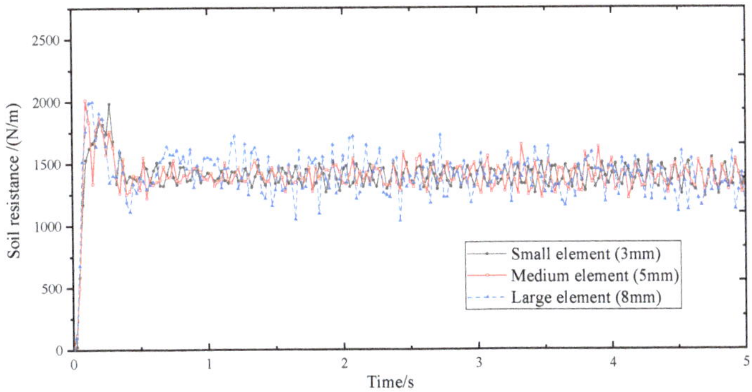

Figure 6. Analysis of grid sensitivity in the CEL method.

4. Results and Discussion

4.1. The Results of the Soil Shear Strength Test

The force readings of the measuring ring under different normal pressures are shown in Table 2, and the shear strength and vertical pressure relationship curve is shown in Figure 7, which identified the friction angle and cohesive force of the soil sample.

Table 2. Shear strength experimental data.

No.	Normal Pressure (kPa)	Load Coefficient (kPa·mm^{-1})	Load Measurement (mm)	Shear Pressure (kPa)
1	100	1.572	34.72	54.58
2	150	1.558	70.69	110.13
3	200	1.572	83.18	130.76
4	250	1.585	98.69	156.42
5	300	1.585	122.29	193.83
6	350	1.558	141.50	220.46
7	400	1.558	166.90	260.03

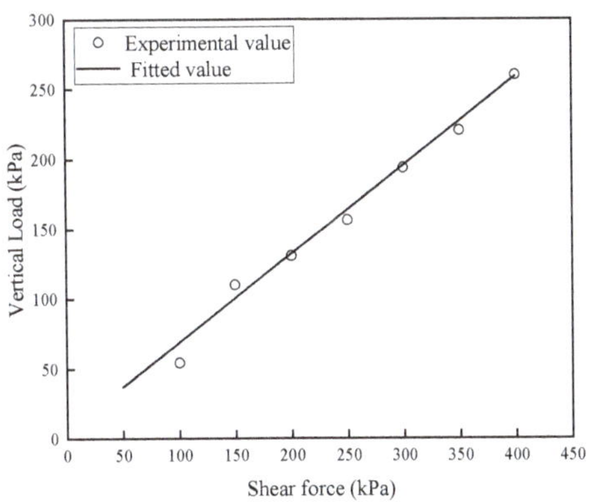

Figure 7. Shear force–vertical load relationship curve.

The experimental data were fitted using the linear fitting method. The intercept of the curve at the y-axis for the lateral displacement of the pipeline at different depths was found to be 0.158, with a slope of k = 0.6429. The cohesive force was found to be 0.158 kPa, and the internal friction angle was found to be 32.75°.

4.2. Pipe-Soil Interaction Experiment Results

(1) Experimental phenomena:

Based on the analysis of experimental observations shown in Figure 8, the following conclusions can be drawn. At the initial settlement position, the pipeline maintained its position stability under the influence of soil resistance (Figure 8a). Under lateral displacement, the soil resistance increased with the rise of the soil bulging height. With small displacements (Figure 8b), the soil resistance exhibited a typical softening response. The soil bulged around the pipeline, leading to soil damage, uplift, and accumulation on the movement side, resulting in an increase in the soil height.

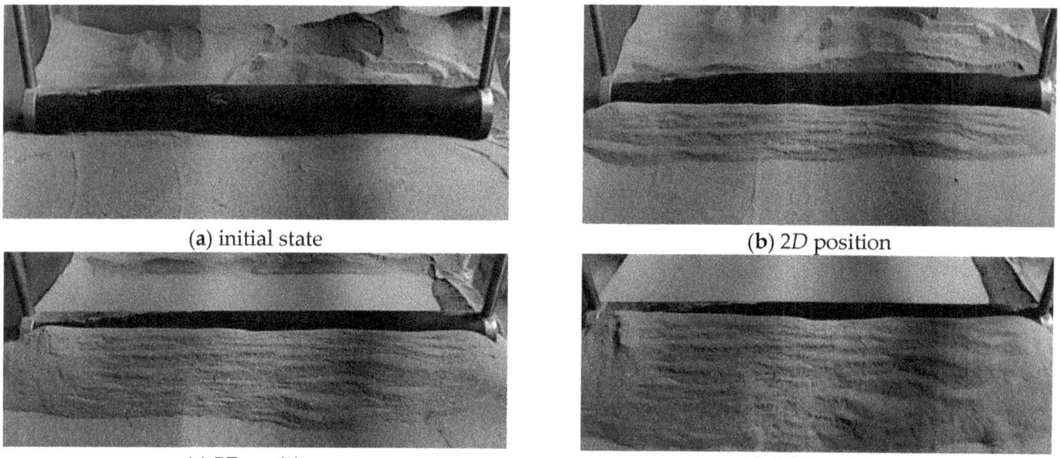

(**a**) initial state (**b**) 2D position (**c**) 5D position (**d**) 10D position

Figure 8. Soil failure at lateral displacements in the 3D initial embedment.

As the lateral displacement increased (Figure 8c), the accumulation of the soil expanded the failure range, enhancing the resistance to the lateral displacement of the simulated pipeline. When the lateral displacement further increased (Figure 8d), the accumulated soil on top of the pipeline could not be sustained, leading to detachment. The degree of detachment was dependent on the cohesive strength of the soil layer.

(2) Experimental data

Figure 9 shows the curve of lateral soil resistance measured during the experiment for a 76-mm-diameter pipeline at different initial depths and horizontal displacements.

From the analysis of the collected experimental data, it could be observed that, at a fixed burial depth, during the initial stage where the lateral displacement was within the range of 0.5D, the lateral soil resistance rapidly increased and gradually increased with the increase in the horizontal displacement of the pipeline. This is because, during the initial stage, the pipeline remained within the range of the in-place stability capacity. However, when the displacement exceeded 0.5D, the pipeline transitioned into a phase of overall large deformation. As the lateral displacement of the pipeline increased, the front soil sample gradually bulged, causing the soil resistance to increase gradually with the displacement as the pipeline moves laterally. This increase in the soil resistance was closely related to the gradual rise in the height of the front soil.

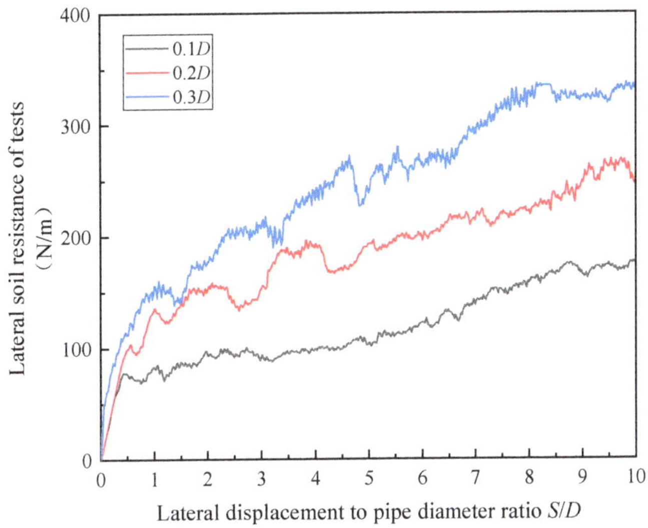

Figure 9. Lateral soil resistance curve of the tests.

Based on the analysis of the experimental results from pipelines with different initial burial depths, it could be observed that, as the initial settlement depth of the pipeline increased, the rate of increase in the soil resistance during the pipeline initiation stage was higher, and the final lateral soil resistance at 10*D* displacement was also greater. The soil resistance values obtained from the experiments are shown in Table 3, and the peak soil resistance between adjacent test conditions decreased.

Table 3. Soil resistance measured via tests.

No.	Initial Embedment	Soil Resistance/N	Difference/N
1	0.1*D*	160.38	——
2	0.2*D*	252.69	92.30
3	0.3*D*	315.77	63.07

In order to obtain a more accurate numerical simulation method for the interaction between pipes and the soil, this study employed the CEL method to numerically simulate the resistance of pipe–soil interaction. The simulation was conducted using an initial burial depth of 0.5*D* as an example. The deformation of the soil is presented in Figure 10. The results indicate that, at a lateral displacement of 0.5*D* (Figure 10a), there was a noticeable uplift of the soil, with a significant increase in height compared to its initial horizontal position. When the displacement reached 2*D* (Figure 10b), the height of the soil was approximately equal to the pipe diameter. As the displacement gradually increased to 4*D* (Figure 10c), the soil continued to accumulate above the pipe, and the horizontal width of the uplifted soil also increased, consistent with the experimental observations. When the lateral displacement reached 7*D* (Figure 10d), the accumulated soil above the pipe underwent failure, resulting in soil detachment towards the opposite movement side. Subsequently, the soil's interaction force reached a stable state.

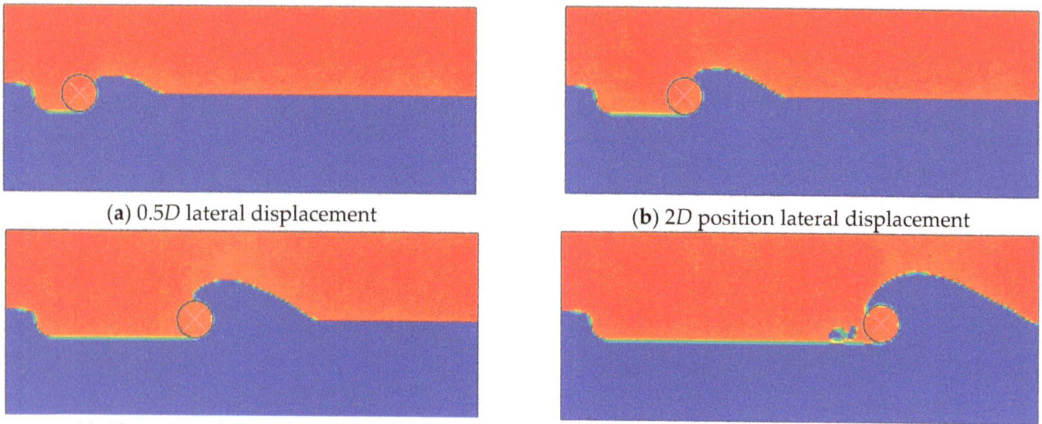

(**a**) 0.5*D* lateral displacement (**b**) 2*D* position lateral displacement

(**c**) 4*D* position lateral displacement (**d**) 7*D* position lateral displacement

Figure 10. Soil failure with lateral displacement in 0.5*D* via a numerical simulation.

Figure 11 presents schematic diagrams of the lateral and axial pipe–soil interaction at a lateral displacement of 7*D* in the model experiments. From the figure, it can be observed that the height of the soil bulging in the direction of the pipeline displacement gradually increased with the increase in lateral displacement. The height of the soil bulging approached approximately 1.5 times the diameter of the pipeline. As shown in Figure 11a, the distribution of the soil bulging height along the length direction was not uniformly consistent, which was attributed to the initial non-uniformity of the soil layer height in the experimental setup. During the model experiments, soil detachment began to occur at the 7*D* position (Figure 11b), which was consistent with the numerical simulation results and the observed phenomenon.

(**a**) 7*D* position lateral displacement (test) (**b**) 7*D* position lateral displacement (test)

Figure 11. Height of soil bulging.

Figure 12 illustrates the soil resistance curves at different burial depths, based on the CEL method. The results show that the ultimate soil resistance increased in a sequential manner with the increase in burial depth. The numerical simulation results demonstrate that, within the lateral displacement range of the initial 0.5*D*, the soil resistance increased sharply. In the initial stage, the pipeline was positioned within the trench, and under external forces, it was required to overcome the lateral soil resistance, resulting in lateral displacement. Therefore, the soil resistance rapidly increased from zero, reaching its peak growth rate within approximately 0.2*D*. As the lateral displacement continued to increase, the soil resistance force continued to rise, but at a diminishing growth rate.

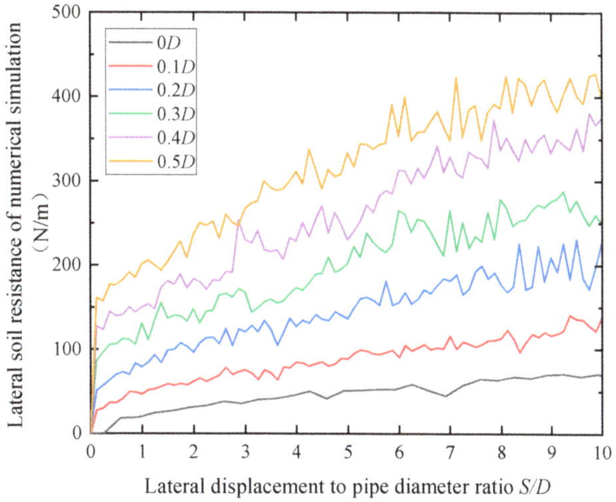

Figure 12. Lateral soil resistance curve of the simulation.

For the conditions of 0*D*, 0.1*D*, 0.2*D*, 0.3*D*, and 0.4*D*, the soil resistance continued to increase with the displacement, and it exhibits an upward trend even at the 10*D* position. Under the 0.5*D* condition, a distinct phase of cyclically stable soil resistance was observed. This is because only under the 0.5*D* condition did the soil reach the critical state of soil detachment failure. A comparison between the calculated soil resistance results based on the CEL method and the experimental data is shown in Figure 13. It is evident that the experimental soil resistance values are higher than the numerical simulation results. Table 4 presents the corresponding numerical calculation results of the simulation and experimental data, indicating a discrepancy that decreased with increasing burial depth, reaching approximately 10%. This difference was attributed to the influence of frictional resistance in the sliding device mechanism, which decreased in proportion as the burial depth increased. The computational results demonstrate that the proposed CEL method in this study achieves good accuracy in predicting pipe–soil interaction forces and effectively addresses the issue of distorted finite element meshes, thereby accurately representing the deformation and flow characteristics of the soil.

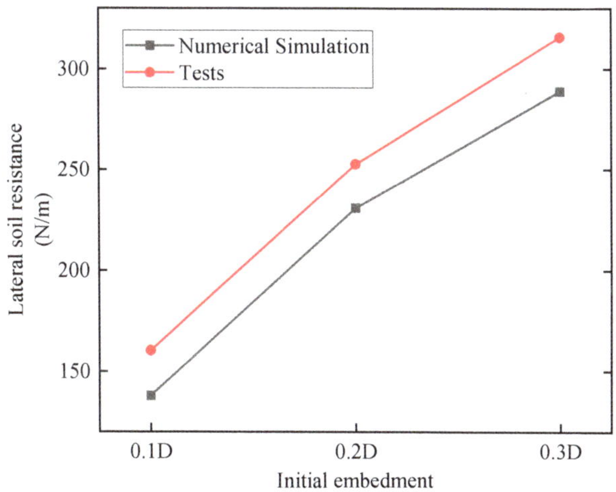

Figure 13. Comparison between the simulation of and experiment on soil resistance.

Table 4. Difference between the simulation of and experiment on soil resistance.

No.	Initial Embedment	Numerical Simulation/N	Tests/N	Difference/N
1	0.1*D*	138	160.38	13.9%
2	0.2*D*	231	252.69	8.6%
3	0.3*D*	289	315.77	8.5%

5. Conclusions

This study designed and conducted enhanced experimental tests on the mechanical interaction between reinforced flexible pipes and sandy soil. The magnitudes of the soil resistance acting on the pipes were measured for both exposure and different burial depths. A simulation method for lateral soil resistance based on the CEL method was developed. The conclusions are as follows:

(1) Within the lateral displacement range of 0.5*D*, the lateral soil resistance increases rapidly. As the lateral displacement of the pipe increases, the soil climbs along the circumferential direction of the pipe, increasing the load-bearing capacity of the pipe and the accumulated soil resistance. The accumulation of soil leads to a larger range of soil failure.

(2) With an increase in the burial depth, the ultimate soil resistance exhibits an increasing sequence, and the final uplifted height of the soil reaches a critical state. A larger initial burial depth allows for a quicker attainment of the critical soil resistance state. The differences in the final soil resistance among different burial depths decrease gradually, and this conclusion was validated through data calculations.

(3) The CEL method successfully addresses the issue of grid distortion caused by soil deformation and effectively simulates the pipe–soil interaction forces under large lateral displacements. The results of the simulation align well with the experimental findings, demonstrating the effectiveness of this method from both qualitative and quantitative perspectives.

Author Contributions: Conceptualization, C.W.; methodology, C.W.; software, L.L.; validation, Y.Z. and L.L.; formal analysis, C.W.; investigation, M.L.; data curation, C.W.; writing—original draft preparation, Y.Z. and L.L.; writing—review and editing, C.W.; supervision, M.L.; project administration, C.W.; funding acquisition, M.L. All authors have read and agreed to the published version of the manuscript.

Funding: This research was funded by the following: the National Natural Science Foundation of China, grant number U2006226; the National Key Research and Development Program of China, grant number 2016YFC0303800; Fundamental Research Funds for the Central Universities, grant number 20CX02320A; and the Opening Fund of National Engineering Laboratory of Offshore Geophysical and Exploration Equipment.

Institutional Review Board Statement: Not applicable.

Informed Consent Statement: Not applicable.

Data Availability Statement: Data are contained within the article.

Conflicts of Interest: The authors declare no conflict of interest.

References

1. Bai, Y.; Liu, T. Mechanical behavior of metallic strip flexible pipe subjected to tension. *Compos. Struct.* **2017**, *170*, 1–10. [CrossRef]
2. Bai, Y.; Tang, J. Collapse of reinforced thermoplastic pipe (RTP) under combined external pressure and bending moment. *Ocean Eng.* **2015**, *94*, 10–18. [CrossRef]
3. Wang, Y.Y.; Lou, M.; Tong, B.; Wang, S. Mechanical properties study of reinforced thermoplastic pipes under a tensile load. *China Ocean Eng.* **2020**, *34*, 806–816. [CrossRef]
4. Bai, Y.; Wang, Y.; Cheng, P. Analysis of Reinforced Thermoplastic Pipe (RTP) under Axial Loads. *Iliffe* **2012**, 708–724. [CrossRef]

5. Lou, M.; Wang, Y.; Tong, B. Effect of temperature on tensile properties of reinforced thermoplastic pipes. *Compos. Struct.* **2020**, *241*, 112119. [CrossRef]
6. Li, C.F.; Liu, R. Experimental and theoretical studies on lateral buckling of submarine pipelines. *Mar. Struct.* **2021**, *78*, 102983. [CrossRef]
7. Kong, D.Q.; Feng, L.Y.; Zhu, B. Assessment model of pipe soil interaction during large-amplitude lateral displacements for deep-water pipelines. *Comput. Geotech.* **2020**, *117*, 103220. [CrossRef]
8. Tian, Y.H.; Cassidy, M.J.; Gaudin, C. Advancing pipe-soil interaction models in calcareous sand. *Appl. Ocean Res.* **2010**, *32*, 284–297. [CrossRef]
9. Calvetti, F.; di Prisco, C.; Nova, R. Experimental and numerical analysis of soil-pipe interaction. *J. Geotech. Geoenviron. Eng.* **2004**, *130*, 1292–1299. [CrossRef]
10. Zhang, J.G.; Stewart, D.P.; Randolph, M.F. Modeling of shallowly embedded offshore pipelines in calcareous sand. *J. Geotech. Geoenviron. Eng.* **2002**, *128*, 363–371. [CrossRef]
11. Tian, Y.H.; Cassidy, M.J. Modeling of Pipe-Soil Interaction and Its Application in Numerical Simulation. *Int. J. Geomech.* **2008**, *8*, 213–229. [CrossRef]
12. White, D.J.; Dingle, H.R.C. The mechanism of steady friction between seabed pipelines and clay soils. *Geotechnique* **2011**, *61*, 1035–1041. [CrossRef]
13. Hodder, M.S.; Cassidy, M.J. A plasticity model for predicting the vertical and lateral behaviour of pipelines in clay soils. *Geotechnique* **2010**, *60*, 247–263. [CrossRef]
14. Bienen, B.; Cassidy, M.J. Advances in the three-dimensional fluid-structure-soil interaction analysis of offshore jack-up structures. *Mar. Struct.* **2006**, *19*, 110–140. [CrossRef]
15. Wang, L.; Liu, R. The effect of a berm on the lateral resistance of a shallow pipeline buried in sand. *Ocean Eng.* **2016**, *121*, 13–23. [CrossRef]
16. Det Norske Veritas. *On-Bottom Stability Design of Submarine Pipelines*; Det Norske Veritas Group: Oslo, Norway, 2021.
17. Bruton, D.; White, D. Pipe/Soil Interaction Behavior During Lateral Buckling. *Sex. Abuse J. Res. Treat.* **2006**, *21*, 57–75. [CrossRef]
18. Cheuk, C.Y.; White, D.J.; Bolton, M.D. Large-scale modelling of soil-pipe interaction during large amplitude cyclic movements of partially embedded pipelines. *Can. Geotech. J.* **2007**, *44*, 977–996. [CrossRef]
19. Colicchio, G.; Colagrossi, A. Pipe-soil interaction: An evaluation of a numerical model. In Proceedings of the 26th International Conference on Offshore Mechanics and Arctic Engineering (OMAE 2007), San Diego, CA, USA, 10 June 2007.
20. Dingle, H.R.C.; White, D.J.; Gaudin, C. Mechanisms of pipe embedment and lateral breakout on soft clay. *Can. Geotech. J.* **2008**, *45*, 636–652. [CrossRef]
21. Liu, R.; Liu, W. A soil resistance model for subsea pipeline global lateral buckling analysis. *Rock Soil Mech.* **2015**, *36*, 2433–2441.
22. Wang, Z.K.; Tang, Y.G. Model Test for Lateral Soil Resistance of Partially Embedded Subsea Pipelines on Sand during Large-Amplitude Lateral Movement. *J. Coast. Res.* **2017**, *33*, 607–618. [CrossRef]
23. Dong, X.Y.; Zhang, W.C. Large deformation coupled analysis of embedded pipeline-Soil lateral interaction. *Mar. Struct.* **2021**, *78*, 102971. [CrossRef]
24. Liu, M.; Ortega, R. Thrust Restraint of Buried Continuous Pressure Pipe Considering Pipe-Soil Interaction. *J. Pipel. Syst. Eng. Pract.* **2021**, *12*, 04021039. [CrossRef]
25. Macaro, G.; Utili, S.; Martin, C.M. DEM simulations of transverse pipe-soil interaction on sand. *Geotechnique* **2021**, *71*, 189–204. [CrossRef]
26. Wu, H.G.; Yu, J.H. Pipe-Soil Interaction and Sensitivity Study of Large-Diameter Buried Steel Pipes. *KSCE J. Civ. Eng.* **2021**, *25*, 793–804. [CrossRef]
27. Wang, L.; Wang, Y.F. Physical model tests of lateral pipe-soil interaction including the pipe trajectory in sand. *Eur. J. Environ. Civ. Eng.* **2022**, *26*, 1962–1976. [CrossRef]
28. Rezadoost, P.; Nassiraei, H. Identification of the most suitable probability distributions for ultimate strength of FRP-strengthened X-shaped tubular joints under axial loads. *Ocean Eng.* **2023**, *290*, 116292. [CrossRef]
29. Merifield, R.; White, D.J.; Randolph, M.F. The ultimate undrained resistance of partially embedded pipelines. *Geotechnique* **2008**, *58*, 461–470. [CrossRef]

Journal of
Marine Science and Engineering

MDPI

Article

Design and Study of Mechanical Cutting Mechanism for Submarine Cable Burial Machine

Zhou Yu [1,†], Zhangyong Jin [1,†], Kaichuang Wang [1], Chunyue Zhang [1] and Jiawang Chen [2,3,*]

[1] Ocean College, Zhejiang University, Zhoushan 316000, China; oceanman@zju.edu.cn (Z.Y.); jinzhangyong@zju.edu.cn (Z.J.); kc.wang@zju.edu.cn (K.W.); 22134068@zju.edu.cn (C.Z.)
[2] Donghai Laboratory, Zhoushan 316021, China
[3] Ocean Research Center of Zhoushan, Zhejiang University, Zhoushan 316021, China
[*] Correspondence: chenjiawang@donghailab.com or arwang@zju.edu.cn
[†] These authors contributed equally to this work.

Abstract: This paper aims to explore the current state of research on submarine cable burial machines and proposes a novel mechanical burial machine design employing a chain-type structure based on a combination of theoretical considerations and practical requirements. Through theoretical analysis, simulation, and experimental studies, the cutting process of the mechanical burial machine is investigated in detail, with special attention given to the seabed conditions in the Northeast Asia region. Starting from the dynamic process of the blade–soil interaction and the working mechanism of the blade–soil system, an accurate and reliable model for the seabed rock–soil stress is established, along with a fast computation method. A destructive analysis of rock–soil mechanical cutting is performed, elucidating the influences of the cutting depth, cutting angle, and chain blade cutting speed on cutting resistance. This paper provides reference parameters for the design of chain-type trenching devices under different seabed conditions, and adjustments are made based on simulation experiments and actual soil trenching test results. These analyses contribute practical and reliable guidance for the design and optimization of submarine cable burial machines.

Keywords: submarine cable; burial; mechanical cutting mechanism; cutting mechanism; chain-type; simulation tests

Citation: Yu, Z.; Jin, Z.; Wang, K.; Zhang, C.; Chen, J. Design and Study of Mechanical Cutting Mechanism for Submarine Cable Burial Machine. *J. Mar. Sci. Eng.* **2023**, *11*, 2371. https://doi.org/10.3390/jmse11122371

Academic Editor: José António Correia

Received: 25 October 2023
Revised: 1 December 2023
Accepted: 8 December 2023
Published: 15 December 2023

1. Introduction

Submarine cable burial machines are primarily used for trenching operations on the seabed; submarine cables are buried beneath the seafloor to protect them from external forces. The origins of submarine cable burial machines can be traced back to the 1940s and 1950s, with the jetting trenching method being employed in nearshore areas. By the 1970s, as the water depths increased and excavation conditions became harsher, the development of plough-type and mechanical burial machines was prompted. In the late 1990s to the early 21st century, three main types of burial machines emerged: jetting, mechanical, and plough-type. The characteristics of these different types of submarine cable burial machines are shown in Table 1 [1–5].

Different types of seabed cable laying machines are shown in Figure 1 [1–5].

Mechanical cable laying machines originated in the 1970s and witnessed the rapid development of deep-water mechanical cable laying technology and equipment in the early 21st century. These machines are primarily used for cutting through strong seabed soils such as rocks. Mechanical cutting devices like chain saws and disc cutters are employed to crush or even liquefy hard soils. Mechanical cable laying machines combine mechanical cutting with jetting or dredging, making them suitable for various seabed soil types. They overcome the limitations of jetting machines in cutting hard soils and the need for high-powered towing vessels for plough-type machines. Broadly categorized, they include chain-type, disc-type, and column-type mechanical cable laying machines based on their operational principles [6,7]. Among these, the chain-type machine is simple in structure, easy to assemble, creates stable trench profiles, and uses a multi-blade chain drive system to achieve high-speed, low-depth soil cutting. This design allows for continuous excavation at a slower pace, making it versatile for applications like laying underground drainage pipelines, telecommunications, and cable installations. It accommodates varying trench widths and depths, rendering it the most widely adopted mechanical cable laying machine today.

Table 1. Characteristics of different types of submarine cable burial machines.

Category	Characteristics
Jetting	Advantages: Simple structure, low cost, easy operation. Disadvantages: Cannot excavate all types of soil.
Ploughing	Advantages: Low failure rate, fast trenching speed, capable of excavating various soil types. Disadvantages: Expensive, requires a high standard for the mother ship, cannot operate intermittently.
Mechanical	Advantages: Capable of excavating various soil types, suitable for deep-water operations. Disadvantages: Complex structure, low operational efficiency.

(a) (b) (c)

Figure 1. Different types of seabed cable laying machines. (**a**) Jetting-type seabed cable laying machine. (**b**) Plow-type seabed cable laying machine. (**c**) Mechanical-type seabed cable laying machine.

This paper primarily focuses on a chain-type structure submarine cable mechanical burial machine. Through a case study of the cutting process, this paper conducts a comparative analysis between the simulation results and experimental data. This study explores the impact of various cutting parameters on traction resistance, providing reference parameters for the optimization design of chain-type trenching machines under different working conditions.

2. Analysis of Mechanical Soil Cutting Mechanism

Soil cutting is the simplest and most typical soil processing method. The ultimate goal of analyzing soil cutting is to reduce traction resistance and save energy by optimizing mechanical structures and operational parameters. Research methods include physical experimentation and numerical simulation [8–12]. The study of soil dynamics began in the 1920s, with Soviet scholars systematically studying the plowing process of soil using mathematical methods. Nichols ML in the United States studied the operation of burial machinery, but almost all research results are based on physical experiments [13–17].

This paper focuses on the marine bottom sediment characteristics in the Northeast Asia sea area. Through a theoretical, simulation, and experimental analysis of the dynamic process of blade cutting soil and the blade–soil working mechanism, it establishes an accurate and reliable force model of submarine rock and soil, as well as a fast calculation method. This paper completes the destructive analysis of rock and soil mechanical cutting, determines the effects of cutting depth, cutting angle, and chain blade cutting speed on cutting resistance. Utilizing the analysis of this paper, different chain trenching parameters for different bottom sediment conditions are determined as design references in the design process of submarine cable laying machines.

The trenching process using a chain mechanism is illustrated in Figure 2. The hydraulic control system tilts the working arm of the trenching device at a certain angle. The chain blades located on the side of the tightened chain move upwards under the action of the chain. These blades then exert cutting force on the soil, resulting in soil excavation. The excavated soil, separated by the cutting teeth of the chain blades, is conveyed to the tensioning sprocket and then unloaded to both sides of the trench through a helical soil conveyor. As the trenching machine advances, the soil is removed, creating the trench. After unloading the soil, the chain blades move downward along the upper side of the trenching device frame. Once they bypass the driven sprocket, they resume cutting the soil, and this cyclic process allows for continuous excavation [18].

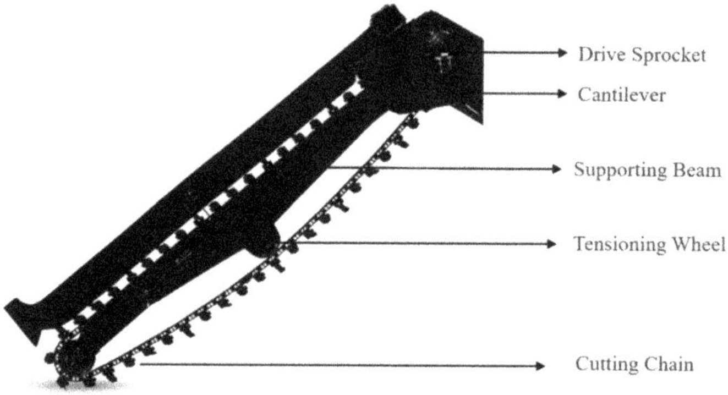

Figure 2. Chain trenching mechanism.

The kinematic relationships are derived based on kinematic and geometric principles, as shown in Figure 3 [19].

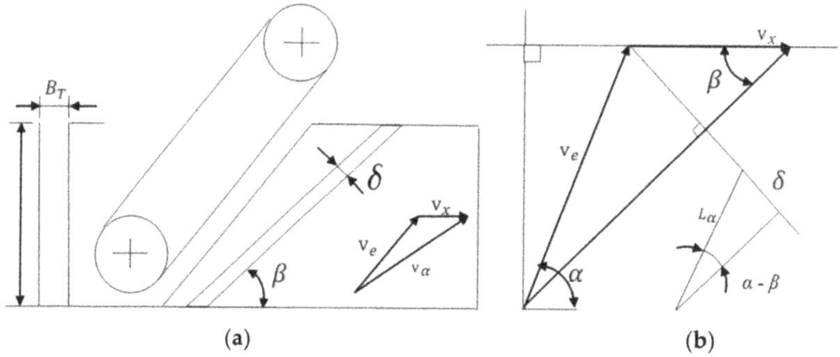

Figure 3. Kinematic relationships. (**a**) Chain Blade Thickness Vector Diagram. (**b**) Velocity Schematic.

$$v_x = P_i / \left(3.6 \times 10^3 B_T H_T\right) \quad (1)$$

$$v_a = \sqrt{v_e^2 + v_x^2 + 2\cos a v_e v_x} \quad (2)$$

where:

P_i is the theoretical productivity;
v_x is the horizontal working speed of the trenching machine;
B_T is the trench width;
H_T is the trench depth;
v_a is the absolute movement speed of the trenching machine chain;
v_e is the relative movement speed of the trenching machine chain.

α is the angle between v_e and v_x, which is the angle between the trenching chain blade and the direction of horizontal movement. The size of angle α directly affects the filling degree of the soil between the trenching chain blades. According to mechanical soil mechanics regulations [18], α is generally taken between 48° and 65°, and for the trenching chain design, α is chosen as 50°.

From Figure 3, the vector inclination of the absolute velocity v_a and the expression for the chain blade cutting thickness can be obtained as follows:

$$\tan\beta = \frac{v_e \sin\alpha}{v_e \cos\alpha + v_x} \tag{3}$$

$$\delta = nL_c \sin(\alpha - \beta) \tag{4}$$

where:

v_x is the horizontal movement speed of the trenching machine;
L_c is the pitch distance of the trenching chain;
n is the number of blades cutting the soil simultaneously by the trenching machine;
δ is the cutting thickness. Based on existing experimental results, a reasonable cutting thickness under normal conditions is [20]

$$\delta = \frac{b_c}{10 \sim 15} \tag{5}$$

where b_c is the width of the chain blade.

From Equation (4), we have the following:

$$\cot\beta = \frac{v_e \sin\alpha + v_x}{v_e \cos\alpha} = \tan\alpha + \frac{v_x}{v_e} \times \frac{1}{\cos\alpha} \tag{6}$$

Since in the actual trenching process, $v_x \ll v_e$, the above equation can be simplified as

$$\cot\beta \approx \tan\alpha \tag{7}$$

Namely, $\alpha \approx \frac{\pi}{2} - \beta$ (approximately orthogonal).

The relationship between the motion speed of the trenching machine and the cutting thickness of the chain blades is shown in Figure 4.

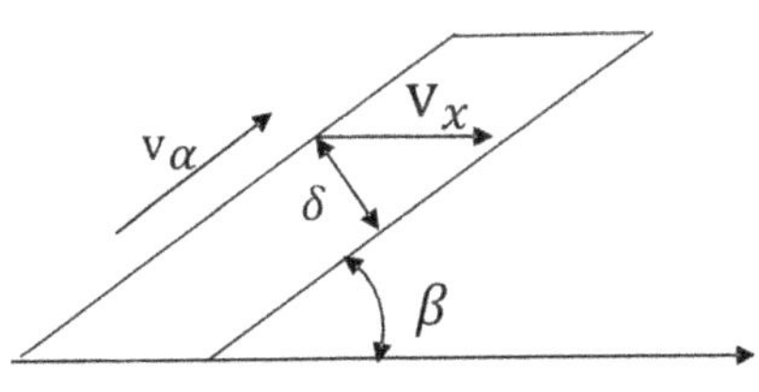

Figure 4. The relationship between the excavation speed of the mechanical soil cutting operation and the cutting thickness.

Based on this graph, the relationship between the two can be derived as follows:

$$\delta = v_x \times \sin\beta \times 1000 \tag{8}$$

For the initially selected chain blade width and pitch, it is necessary to verify their capacity for soil conveyance and removal. The correctly chosen values should meet the following conditions [21]:

$$P \leq P_i \frac{1}{K_P} \Delta \tag{9}$$

where:

P represents the actual cutting chain productivity;
P_i is the productivity calculated based on the chain blade's soil-discharging capacity and the productivity for continuous operation under the "calculation conditions";
K_P is the soil loosening coefficient;
Δ is the soil spreading coefficient related to the chain speed, as shown in Table 2.

Table 2. The coefficients of dispersion.

v_e	0.1	1	1.5	2
Δ	0.97	0.92	0.85	0.75

As the coefficient of dispersion is directly related to the chain blade's movement speed, it is necessary to investigate the movement speed of the chain blade. During trenching, the chain blade is subjected to the driving force and soil resistance, which can lead to the wear and oscillation of the blade edge and chain link bearings. In order to ensure the quality of the operation and avoid excessive vibration, there must be certain limitations on the chain blade's movement speed. According to the experiments in [22], the former Soviet scholars recommended a chain blade linear velocity of 1 to 2 m/s. In this paper, the mechanical cutting components are selected with a chain blade movement speed of 1.5 m/s, resulting in a corresponding soil dispersion coefficient of 0.85.

The calculation of the soil unloading capacity of the chain blade can be determined by comparing the chain's horizontal inclination angle with the soil's natural repose angle.

When $\alpha < \varphi_r + arctanh_c/L_c$,

$$P_i = 3600 b_c h_c v_e [1 - \tan(\alpha - \varphi_r) L_c / 2h_c] \tag{10}$$

When $\alpha > \varphi_r + arctanh_c/L_c$,

$$P_i = 3600 b_c h_c v_e \frac{h_c}{2L_C} \cot(\alpha - \varphi_r) \tag{11}$$

When $\alpha < \varphi_r$,

$$P_i = 3600 b_c h_c v_e \tag{12}$$

where:

h_c represents the height of the chain blade in the trenching mechanism;
φ_r is the angle of natural repose.

The angle α between the absolute motion velocity of the chain trencher blade v_a and the horizontal motion velocity v_x of the trencher is determined based on the experimental results from the article titled "Operator and Machine Models for Dynamic Simulation of Construction Machinery". The chain blade height is generally chosen within the range of 0.1 to 0.15 m. Considering the potential deformation and breakage of the blade when encountering obstacles at excessive heights, a smaller chain blade height is preferred while ensuring effective cutting. In this study, h_c is selected as 10 cm; L_c is 25 cm.

As Max(α) = 65°, the range of values for $\varphi_r + arctanh_c/L_c$ is as follows:

Substituting into Equation (10) yields

$$b \geq 9.2 \text{ mm} \tag{13}$$

By combining Equation (9), we can obtain

$$\pi'_p = 3600 b_c h_c v_e [1 - L_c/(2h_c)\tan(\alpha - \varphi_r)] \tag{14}$$

According to the requirements of the trenching operations, in order to ensure that the soil cut by the blade can be smoothly carried out of the trench, it is necessary for the width of the chain blade to be $\geq$18.4 mm.

The thickness of the upper soil layer cut in the vertical plane along the advancing direction of the blade is referred to as the cutting depth S. The chain-type trenching machine uses the rotation of the chain to make the blades on the chain rotate and cut the soil. The cutting depth S of the chain blade can be calculated using the following equation:

$$S = \frac{2\pi}{n\omega} v_x \tag{15}$$

where:

ω represents the angular velocity of the chain axis;
n stands for the number of trenching blades simultaneously engaged in cutting the soil.

Different types of soil exhibit distinct failure mechanisms under specific conditions, and the optimal shear angle varies for different types of failure mechanisms, leading to variations in the cutting efficiency of the tools. However, based on existing research, the magnitude of cutting forces for various failure mechanisms can be calculated using the same formula. In this paper, the influence of parameters on mechanical trenching mechanisms will be studied from the perspective of cutting resistance.

During the operation of a chain-type trencher, each chain blade cuts soil into specific shapes, forming the basis for calculating excavation resistance and productivity. The shape of the cut soil chips is influenced by the structure of the trencher's chain blade. For circular arc chain blades, the formula for calculating the cutting resistance of soil is as follows [23]:

$$F'_c = C_r z \delta (b_c K_B + Z_r K_s)(1 + 0.256 lg v_e) K_\varphi \tag{16}$$

where:

C_{rz} represents the number of impacts for hardness measurement;
b_c is the cutting width of the trenching blade;
δ is the cutting thickness of the trenching blade;
K_φ is the influence coefficient of the cutting angle, φ;
Z_r stands for the lateral cutting quantity of the trenching blade;
K_B is the influence coefficient of the trenching blade's soil processing;
K_S is the proportion coefficient of the trenching blade's lateral soil cutting.

The closed-side cutting quantity of the trenching blade is determined according to the trenching blade's working conditions. Free cutting: $Z_r = 0$; semi-closed cutting: $Z_r = 1$; and separate cutting: $Z_r = 2$.

The calculated cutting resistance of a single trenching blade is

$$F = 462.36 \text{ N} \tag{17}$$

The number of trenching blades simultaneously engaged with the soil in the longitudinal direction of the arc-shaped trenching blade is denoted as n in the formula.

$$n = \frac{H_T}{\sin\alpha L_c} \tag{18}$$

where:

α represents the angle between the relative velocity of the trenching chain blade and the horizontal velocity of movement;
H_T denotes the trenching depth;

L_c stands for the trenching chain pitch.

The total cutting resistance of the trenching arm, F_k, can be expressed as follows:

$$F_k = nF \tag{19}$$

According to relevant studies in the literature, the total cutting power of the trenching machine is

$$P_n = F_k v_a \times 10^3 \tag{20}$$

The power consumed during the cutting and transporting of soil is

$$P_r = \frac{(H/2 + H_0)P_i\gamma_s K_b}{3.6 \times 10^6} \tag{21}$$

where:

H_0 is the width between two blades;
K_b is the coefficient of potential soil particle clogging between the chain blade and the trench sidewall (set as 1 in this paper);
γ_S represents the soil bulk density.

The frictional power consumption generated by the conveyed soil and the trench soil is

$$P_f = P_r \mu_f ctg\gamma \tag{22}$$

where μ_f is the coefficient of friction between soils.

The power consumed by the helical soil conveyor moving soil to the side of the trench is given by [24]:

$$P_h = (C_0 \pi_T \gamma_s L_B)/3.6 \times 10^6 \tag{23}$$

where:

C_0 is the experimentally determined soil resistance coefficient of the helical soil conveyor, typically ranging from 4 to 5 for most soils;
L_B is the maximum distance for moving soil along the trench edge, usually taken as 0.45 to 0.5.

The total power required for the trenching device to cut and transport soil is given by the following:

$$P_t' = \left(P_n + P_r + P_f + P_c\right)/\eta_1\eta_2 \tag{24}$$

where:

η_1 represents the chain transmission efficiency;
η_2 denotes the efficiency of the equipment's travel transmission.

As for the forward resistance of the burial machine, its resistance graph is illustrated in Figure 5 below:

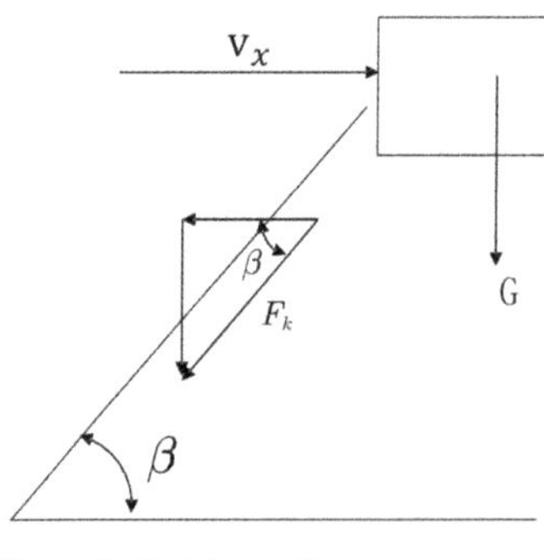

Figure 5. Resistance diagram.

The formula for forward resistance is as follows:

$$F_a = (G + F_t \sin \beta)\mu + F_t \cos \beta \tag{25}$$

where:
G stands for the weight of the trenching machine;
μ represents the coefficient of rolling resistance.

The above analysis indicates that the parameters influencing power consumption mainly include the working speed of horizontal movement of the trenching machine, denoted as v_e, the absolute motion speed of the trenching machine's chain cutter, denoted as v_a, and the relative motion speed of the chain cutter to the trenching machine, denoted as v_γ.

3. Mechanical Cutting Parameter Simulation Analysis

In order to effectively analyze the mechanical performance of the mechanical trenching device in complex underwater environments, it is necessary to establish a dynamic model for the mechanical trenching and burial machine. To efficiently study the process of chain-type trencher fracturing the underwater soil under the influence of flow fields, it is essential to develop a simulation model for chain blade soil cutting under the action of the flow field. This paper employs the LS-DYNA R11.2.2 software to carry out a numerical simulation of the cutting of soil by circular arc blades. The analysis includes the examination of cutting resistance and cutting power consumption, leading to conclusions regarding the variations in the cutting speed and cutting depth.

3.1. Model Establishment

In this study, the LS-DYNA dynamic module is employed for virtual simulation analysis. The soil model is discretized using the traditional explicit dynamic method, constraints are defined, and loads are applied to the soil and blades based on actual conditions. Due to the complexity of soil cutting processes, which involve rapid changes in stress and strain, including deformation characteristics such as elasticity, plasticity, fracture, and soil failure, the numerical simulation of the cutting process assumes the following conditions:

(1) During the blade cutting process, the forward speed and chain velocity are assumed to be constant, i.e., the direction and magnitude of the blade's absolute velocity remain unchanged.
(2) The soil model established in LS-DYNA has consistent stiffness and moisture content and a uniform soil structure.
(3) In the cutting simulation process, it is assumed that the blade always moves within the same plane.
(4) During the cutting process, the soil is assumed to remain stationary, and the combined velocity of the machine's forward movement and chain velocity is in the direction of the blade's motion.

To enhance the accuracy of the soil cutting simulation results, it is necessary to not only select a suitable soil model that reasonably reflects the actual situation and establish constitutive relationships that accurately represent the dynamic properties of the soil, but also to choose appropriate constitutive relationships and failure criteria for the analysis of high-speed soil cutting problems.

In this paper, the MAT147D material model of LS-DYNA is adopted. This material model employs a modified plasticity model. Table 3 shows the parameter for soil.

Table 3. Soil parameters of clayey sand soil.

Parameters	Value
density	2082.3 kgm^{-3}
relative density/specific gravity of soil	2.68
bulk modulus	35 MPa
shear modulus	20 MPa
cohesion force	30 KPa
angle of friction	24°
water content	34%

The cutting tool is a crucial factor in any mechanical trenching system. It penetrates the soil under a certain thrust and torque provided by the mechanical trenching system. The cutting tools transfer the machine's energy to the soil, enabling the cutting process. Therefore, the geometric shape and wear characteristics of cutting tools significantly affect the energy transferred to the soil and the achievable penetration rate.

The primary target operating areas are the Northeast Asian Sea and the Zhoushan Sea, where the seabed substrate is mainly composed of cohesive soil. Based on the principle of selecting cable routes, it is essential to avoid areas with rocky seabed as much as possible. Hence, this paper selects the shape of the cutting tool as an arc-shaped trenching blade, the structure of which is illustrated in Figure 6.

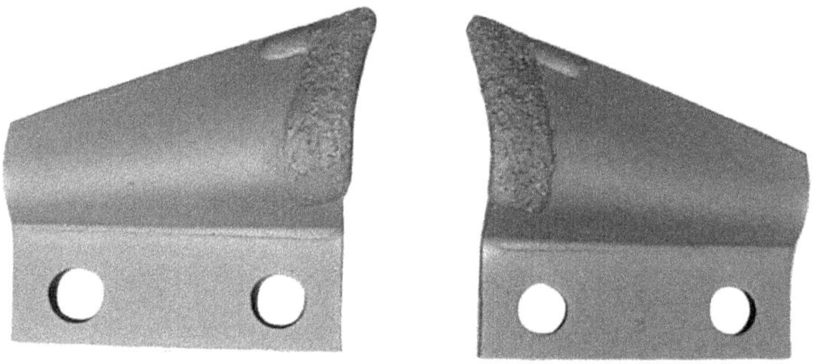

Figure 6. Arc-shaped trenching blade.

In the simulation, the material properties of the arc-shaped cutting tool are set as shown in Table 4.

Table 4. Material properties of circular arc cutting tool.

Elastic Modulus/GPa	Density/$g \cdot cm^{-3}$	Poisson's Ratio
207	7.81	0.35

Figure 7 shows the overall grid division of the finite element model for cutting soil with a chain blade. The contact type between the chain blade and the soil is automatic single-sided contact. Moreover, the chain blade is the target component, and the soil is the contact component. In the model, the forward velocity of the blade is constant, and the direction is simplified to be perpendicular to the soil module. A specific load curve is applied to the blade node group to produce a constant velocity of motion in the x direction, with zero displacement constraints in the y and z directions.

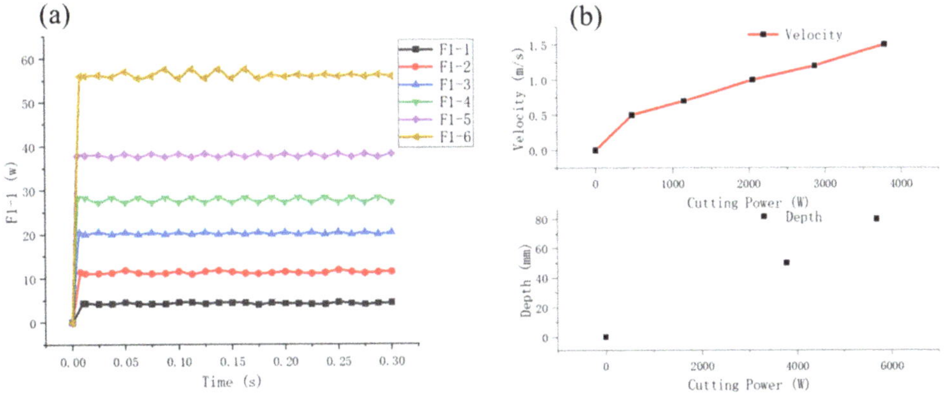

Figure 12. Cutting power under different working conditions. (**a**) Power change with time with different working conditions. (**b**) Cutting power required for different cutting speeds and cutting depths.

(c) Effect of Cutting Speed

From Figure 12a,b, It can be observed that when the speed is less than 0.6 m/s, the cutting power increases rapidly. When the speed is greater than 0.6 m/s, the cutting speed increases linearly. Therefore, when selecting the cutting speed, there is no need to worry about the power increasing geometrically with the cutting speed. From Figure 12, it can be seen that as the cutting depth increases, there is a linear increase in the cutting power. Therefore, from the perspective of numerical simulation, the influences of speed and depth on the cutting power are similar, as they are both linear.

4. Analysis of Chain Blade Cutting Experiment

Experimental testing of chain blade soil cutting not only allows for control over the soil properties and conditions to be obtained, but also facilitates the convenient adjustment of the movement state of working components, enabling the testing of parameters such as the interaction between the trenching blade and the soil. In order to further investigate the cutting mechanism and process of the arc-shaped trenching blade, this study employs arc-shaped cutting tools and utilizes a soil trenching test method to examine the relationships between the cutting speed, cutting depth, and cutting angle of the arc-shaped tool. This experiment establishes quantitative relationships between various factors and power, constructs relevant mathematical models, and utilizes cutting resistance as the evaluation index. Building upon the virtual simulation model of blade soil cutting established in Section 3, experimental research is conducted, and a regression analysis is performed on the experimental data.

4.1. Design of Experimental Platform

In the trenching process of a chain-type mechanical trencher, the detachment of soil blocks from the ground relies primarily on the cutting action of the chain blades. Within the overall power consumption of the mechanical arm, cutting resistance accounts for a significant portion of the power consumed. The rational selection of the tool and cutting parameters can reduce the resistance of the trenching machine, decrease stress on the chain blades, enhance the operational efficiency, and serve as a key design criterion for the mechanical trenching mechanism.

This experiment aims to identify the patterns of variation in the cutting resistance of the trenching apparatus, validate the accuracy of the simulation results, facilitate the optimization of the overall structure of the trenching apparatus, reduce the cutting resistance, and improve the machine's operational efficiency. The main variables of the experiment include the angle of the cutting chain blade, cutting speed, cutting depth, and the influence of different cutting chain blades on the cutting resistance. The mechanical cutting test platform consists of a height adjustment module, an angle adjustment module, and a sliding guide rail platform, with its structural components shown in Figure 13.

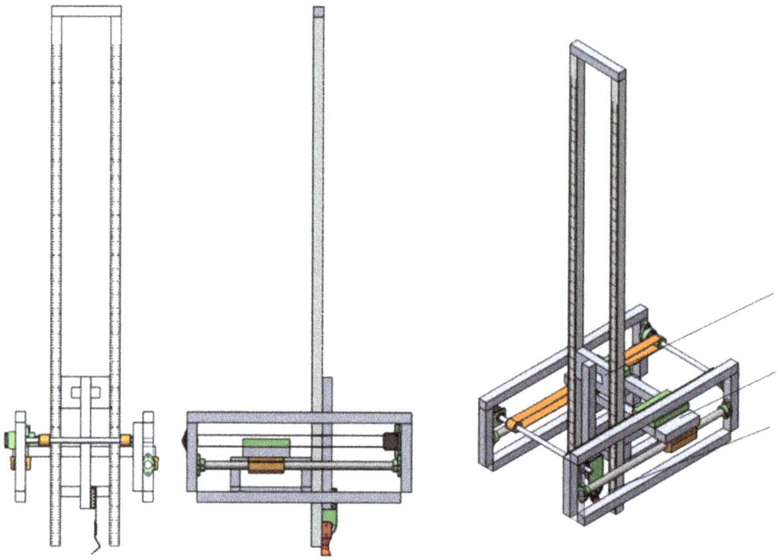

Figure 13. Composition of the mechanical trenching platform.

(a) Cutting Angle Adjustment

The angle adjustment module of the experimental platform is illustrated in Figure 14. The chain blade is hinged to the dial indicator through a pivot joint, and its angle is secured using aluminum profiles. The smallest scale on the dial indicator is 1°. This bracket allows for the adjustment of the cutting angle of the chain blade, with adjustment parameters set at 0°, 30°, and 45°.

(b) Height Adjustment

The height adjustment module is shown in Figure 15. It provides a dimension reference based on the image collected through the line diameter. During the actual jetting experiment, the bottoms of two scale rulers (at the 0 mark) always make contact with the mud surface. Meanwhile, the plane of the scale ruler is parallel to the plane of the chain blade tip. By capturing the position of the blade tip, the corresponding cutting depth can be obtained. The adjustment parameters are set at 5 cm and 10 cm.

(c) Drive System

The experimental platform is constructed using 4040 aluminum profiles, as illustrated in Figure 16. This choice of aluminum profiles facilitates the assembly of the platform. The lightweight nature of aluminum profiles reduces the vibration and deflection generated on the sliding track during the experiment, while ensuring the required rigidity. The working platform is mounted on the guide rails, with a transmission device and slider installed on the upper part of the platform. This arrangement enables the platform to move along the optical axis.

Figure 14. Cutting angle adjustment module of the experimental platform.

Figure 15. Height adjustment module.

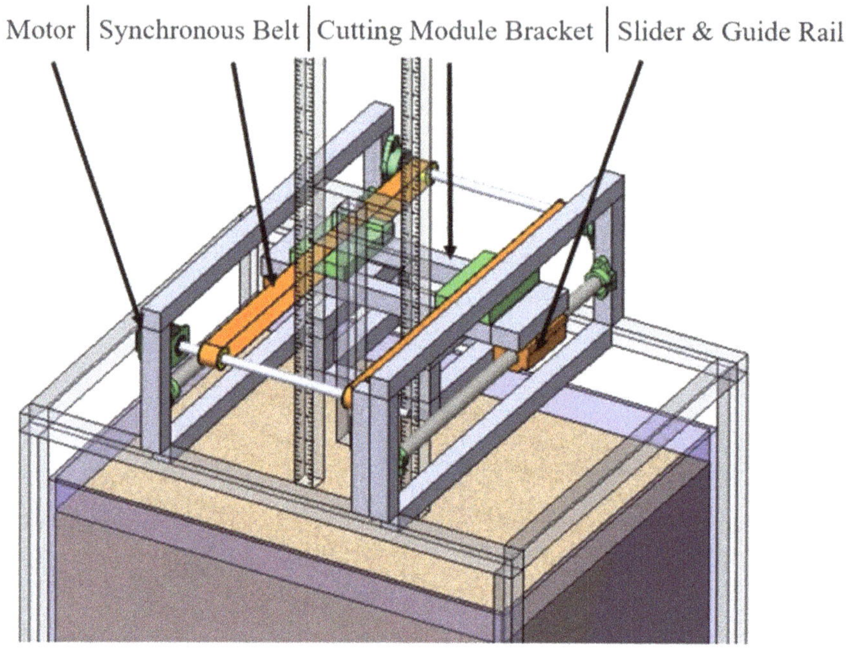

Figure 16. Drive system.

The drive system employs a transmission configuration consisting of a stepper motor and a synchronous belt. The synchronous belt pulley is fixed to the tool holder, and the movement of the tool holder is driven through the synchronous belt pulley. The tool holder is secured to four guide rail sliders. In practical applications, the horizontal motion of the mechanical trenching operation is the same as that of the jet trenching operation. Therefore, the drive system is designed to be consistent with the equipment used in the jet trenching test platform.

The drive system employs a transmission configuration consisting of a stepper motor and a synchronous belt. The synchronous belt pulley is fixed to the nozzle holder, and the movement of the nozzle holder is driven through the synchronous belt pulley. The nozzle holder is secured to four guide rail sliders. In practical engineering, the trenching operation of the buried machine is a relatively slow process, and the driving speed must be controlled within a certain range. Meanwhile, trenching operations require overcoming significant resistance, necessitating a higher driving force. The design of the drive system is based on the selection of the motor according to the driving force and speed. The reference range for the movement speed is 30 to 70 mm/s, based on existing flushing platform speeds. The ST86 series stepper motor is chosen, which enables a platform movement speed of 0 to 118.4 mm/s.

(d) Three-Dimensional Force Sensor

In this thesis, the main purpose of the mechanical trenching test is to verify the relationship between the chain cutter cutting parameters and cutting resistance. As the chain cutter moves, it experiences spatial forces. Therefore, a three-dimensional force sensor is used in this test to measure the magnitude of the spatial forces acting on the chain cutter. The arrangement of the three-dimensional force sensor is shown in Figure 17.

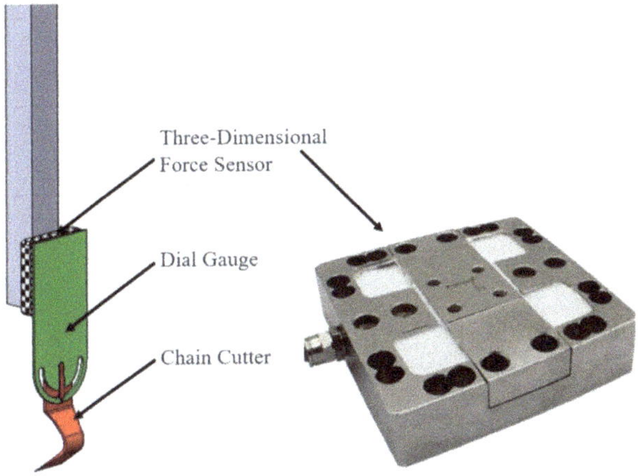

Figure 17. Three-dimensional force sensor arrangement.

The design employs the Freud DYDW-002 three-dimensional force sensor with a range of 500 N for the XY axes and a range of 2000 N for the Z axis. The sampling sensitivity can reach 1 mV/V, equipped with a 0–5 mV analog output interface. The calibration of the sensor is required prior to the experiment.

(e) Experimental Conditions

This experiment involves a four-factor orthogonal test, with cutting tests conducted according to the designed arrangement. The experimental conditions are shown in Tables 6–9. Each set of conditions underwent three repeated tests. Among them, the P1-1, P2-2, P3-2, and P4-1 conditions represent the same set of conditions.

Table 6. Variable experimental conditions for cutting speed.

Operating Condition	Cutting Speed	Cutting Depth	Blade Thickness	Cutting Angle
P1-1	0.3 m/s	50 mm	6 mm	0°
P1-2	0.5 m/s	50 mm	6 mm	0°
P1-3	0.7 m/s	50 mm	6 mm	0°

Table 7. Cutting depth variable conditions.

Operating Condition	Cutting Speed	Cutting Depth	Blade Thickness	Cutting Angle
P2-1	0.3 m/s	30 mm	6 mm	0°
P2-2	0.3 m/s	50 mm	6 mm	0°
P2-3	0.3 m/s	80 mm	6 mm	0°

Table 8. Chain blade thickness variable conditions.

Operating Condition	Cutting Speed	Cutting Depth	Blade Thickness	Cutting Angle
P3-1	0.3 m/s	50 mm	4 mm	0°
P3-2	0.3 m/s	50 mm	6 mm	0°
P3-3	0.3 m/s	50 mm	8 mm	0°

Table 9. Cutting angle variable conditions.

Operating Condition	Cutting Speed	Cutting Depth	Blade Thickness	Cutting Angle
P4-1	0.3 m/s	50 mm	6 mm	0°
P4-2	0.3 m/s	50 mm	6 mm	30°
P4-3	0.3 m/s	50 mm	6 mm	45°

4.2. Experimental Analysis

The cutting process of the circular arc trenching tool in the simulated substrate conditions during the experiment is illustrated in Figure 18.

Figure 18. Typical cutting situation of circular arc trenching tool.

To reduce errors, in this experiment, under specific conditions, the cutting resistance was repeatedly tested for various experimental conditions. The results were averaged to determine the average cutting resistance of the blade under different factors. Excluding abnormally high or low data points, the intermediate data were analyzed for the experimental results. The experimental outcomes are presented in Table 10.

Table 10. Experimental data.

Operating Condition	Cutting Speed	Cutting Depth	Blade Thickness	Cutting Angle	Average Cutting Directional Resistance
P1-1	0.3 m/s	50 mm	6 mm	0°	274.7 N
P1-2	0.5 m/s	50 mm	6 mm	0°	332 N
P1-3	0.7 m/s	50 mm	6 mm	0°	385.3 N
P2-1	0.3 m/s	30 mm	6 mm	0°	208.3 N
P2-2	0.3 m/s	50 mm	6 mm	0°	274.7 N
P2-3	0.3 m/s	80 mm	6 mm	0°	327.3 N
P3-1	0.3 m/s	50 mm	4 mm	0°	263 N
P3-2	0.3 m/s	50 mm	6 mm	0°	274.7 N
P3-3	0.3 m/s	50 mm	8 mm	0°	285 N
P4-1	0.3 m/s	50 mm	6 mm	0°	274.7 N
P4-2	0.3 m/s	50 mm	6 mm	30°	289 N
P4-3	0.3 m/s	50 mm	6 mm	45°	300.3 N

In this experiment, data analysis software was employed for experimental optimization, design, and analysis. The collected cutting resistance was used as the experimental indicator. The experimental results are presented in Table 11.

Table 11. Experimental results analysis.

Operating Condition	A Cutting Speed	B Cutting Depth	C Blade Thickness	D Cutting Angle	Average Cutting Directional Resistance
P1-1	1	2	2	1	274.7 N
P1-2	2	2	2	1	332 N
P1-3	3	2	2	1	385.3 N
P2-1	1	1	2	1	208.3 N
P2-3	1	3	2	1	327.3 N
P3-1	1	2	1	1	263 N
P3-3	1	2	3	1	285 N
P4-2	1	2	2	2	289 N
P4-3	1	2	2	3	300.3 N
K1	2047.6	208.3	263	2175.6	
K2	333	327.3	2216.9	289	
K3	385.3	2229.3	285	300.3	
k1	292.5143	208.3	263	310.8	
k2	333	327.3	316.7	289	
k3	385.3	318.4714	285	300.3	
Range R	92.78571	119	53.7	21.8	
Main-Sub		B > A > C > D			

From Table 10, it can be observed that the extreme values of cutting depth, cutting speed, blade thickness, and cutting angle decrease sequentially. According to the analysis of the orthogonal experimental data results, larger values of the range R correspond to a more significant influence of the respective factors. The variation in the blade thickness has a relatively minor impact on the cutting resistance, which can be considered negligible. The primary influencing factors on the cutting resistance are identified as the cutting depth, cutting speed, and cutting angle, with the cutting depth having the most significant impact.

A comparison was made between the experimental data and theoretical calculation results in the cutting speed of 0.5 m/s and the cutting depth of 500 mm. The theoretical value is 347 N, the experimental value is 332 N (4.3% lower than the theoretical value), and the simulated value is 416 N (19.8% higher than the theoretical value). This verifies the reliability of the experimental data. Such discrepancies may arise due to various factors, such as potential deviations between experimental equipment and conditions from the intended ones.

5. Conclusions

This paper conducts a simulation analysis and experiments on the trenching chain blade of a chain-type trenching machine. This study tests the variation in traction resistance under different trenching conditions and quantifies the impacts of various cutting parameters on traction resistance. Through an analysis of orthogonal experimental data, it was determined that changes in the cutting angle have negligible effects on the cutting resistance. The primary factors influencing cutting resistance are the cutting depth, cutting speed, and blade thickness, with the cutting depth having the most significant impact.

When selecting cutting parameters for a chain-type trenching machine, it is advisable to appropriately reduce the cutting depth, cutting speed, and blade thickness to decrease the cutting resistance and enhance machine efficiency. Additionally, this paper compares and validates the feasibility of a simulation analysis, providing a viable approach for the optimization design of chain-type trenching machines.

Author Contributions: Conceptualization, Z.Y. and Z.J.; methodology, Z.Y, Z.J. and K.W.; software, C.Z.; validation, Z.Y., Z.J. and J.C.; formal analysis, J.C.; investigation, Z.Y.; resources, Z.Y.; data curation, Z.J.; writing—original draft preparation, Z.Y. and Z.J.; writing—review and editing, K.W.; visualization, Z.Y. and Z.J.; supervision, J.C.; project administration, J.C.; funding acquisition, J.C. All authors have read and agreed to the published version of the manuscript.

Funding: This research was funded by the Key Research and Development Program of Jiangsu Province, grant number BE2022062.

Institutional Review Board Statement: Not applicable.

Informed Consent Statement: Not applicable.

Data Availability Statement: Data are contained within the article.

Acknowledgments: The authors acknowledge the funding provided by the Key Research and Development Program of Jiangsu Province, grant number BE2022062. We would like to express our gratitude for the help and support given to us throughout the engineering process. We would also like to thank the reviewers for proofreading this article.

Conflicts of Interest: The authors declare no conflict of interest.

References

1. Nordel, D. Nordel Annual Report 2003. 2004. Available online: https://inis.iaea.org/collection/NCLCollectionStore/_Public/36/019/36019496.pdf?r=1 (accessed on 30 November 2023).
2. Song, W. Present situations and development trends of the transnational interconnected electric systems. *Electr. Power Technol. Econ.* **2009**, *21*, 62–67.
3. Allan, P.G. Selecting appropriate cable burial depths-a methodology. In Proceedings of the IBC Submarine Telecommunications Conference, Cannes, France, 1 November 1998; p. 25.
4. Li, X.; Chen, G.; Zhu, H.; Zhang, R. Quantitative risk assessment of submarine pipeline instability. *J. Loss Prev. Process Ind.* **2017**, *45*, 108–115. [CrossRef]
5. Standard, O. Submarine Pipeline Systems. DNVOS-F101, Section H, 2000, 301. Available online: http://www.opimsoft.cn/download/reference/os-f101_2013-10.pdf (accessed on 21 October 2023).
6. Griffiths, P.; Zavahir, M. Planning for New Zealands Inter-Island HVDC pole 1 replacement. *CIGRE Study Committe B4 Sess.* **2008**, *27*, 1–30.
7. Zhen, Y.; Li, X.; Cao, Y.; Zhang, S. A novel method to determine critical CTOA directly by load-displacement curve. *Eng. Fract. Mech.* **2020**, *230*, 107013. [CrossRef]
8. Okubo, S.; Fukui, K.; Chen, W. Expert system for applicability of tunnel boring machines in Japan. *Rock Mech. Rock Eng.* **2003**, *36*, 305–322. [CrossRef]
9. Sapigni, M.; Berti, M.; Bethaz, E.; Busillo, A.; Cardone, G. TBM performance estimation using rock mass classifications. *Int. J. Rock Mech. Min. Sci.* **2002**, *39*, 771–788. [CrossRef]
10. Xu, Q.; Chang, G.K. Evaluation of intelligent compaction for asphalt materials. *Autom. Constr.* **2013**, *30*, 104–112. [CrossRef]
11. Hu, W.; Huang, B.; Shu, X.; Woods, M. Utilising intelligent compaction meter values to evaluate construction quality of asphalt pavement layers. *Road Mater. Pavement Des.* **2017**, *18*, 980–991. [CrossRef]
12. Shuler, S.; Nobe, M.D. Edge Cracking in residential development hot mix asphalt pavements. *Int. J. Constr. Educ. Res.* **2007**, *3*, 179–197. [CrossRef]
13. Barthelat, F. Biomimetics for next generation materials. *Philos. Trans. R. Soc. A Math. Phys. Eng. Sci.* **2007**, *365*, 2907–2919. [CrossRef] [PubMed]
14. Bhushan, B. Biomimetics: Lessons from nature—An overview. *Philos. Trans. R. Soc. A Math. Phys. Eng. Sci.* **2009**, *367*, 1445–1486. [CrossRef] [PubMed]
15. Gao, J.; Jin, Y. Soil-Cutting Simulation and Test of Oblique Rotary Tiller. In Proceedings of the Computer and Computing Technologies in Agriculture V: 5th IFIP TC 5/SIG 5.1 Conference, CCTA 2011, Beijing, China, 29–31 October 2011; Proceedings, Part III 5. Springer: Berlin/Heidelberg, Germany, 2012; pp. 140–150.
16. Jiang, Z.; Jiandong, J.; Tao, J. Deep-tillage rotavator technology based on smoothed particle hydrodynamics simulation. *J. Mech. Eng.* **2010**, *46*, 63–69.
17. Aluko, O.B.; Chandler, H.W. Characterisation and Modelling of Brittle Fracture in Two-dimensional Soil Cutting. *Biosyst. Eng.* **2004**, *88*, 369–381. [CrossRef]
18. Chen, Y.L.; Qiu, W.; Ding, W.M.; Li, Y.N.; Liu, Y.T. Improvement and test on ditcher chain transmission system of the machine for rice-wheat cyclic planting. *Appl. Mech. Mater.* **2014**, *644*, 853–857. [CrossRef]
19. Zeng, D. *Mechanical Soil Dynamics*; Beijing Science and Technology Press: Beijing, China, 1995. (In Chinese)
20. Same Size, More Power from Newest Vermeer Rock Wheel Trencher-Utility Equipment[EB/OL]. Available online: http://www.versalifteast.com/articles/vermeertrencher.htm (accessed on 2 September 2023).

21. Grant, P.; Freeman, J.S.; Vail, R.; Huck, F.P. Reparation of a Virtual Proving Ground for Construction Equipment Simulation. In Proceedings of the 1998 ASME Design Engineering Technical Conferences, Atlanta, GA, USA, 13–16 September 1998.
22. Filla, R. Operator and Machine Models for Dynamic Simulation of Construction Machiner. Ph.D. Thesis, Linköpings Universitet, Linköping, Sweden, 2005.
23. Gupta, S.C.; Hadas, A.; Schafer, R.L. Modeling Soil Mec hanical Behavior During Compaction. *Mech. Relat. Process. Struct. Agric. Soils* **1989**, *172*, 137–152.
24. Sun, Y.; Gao, X.; Yu, D. *Agricultural Soil Mechanics*; Agriculture Press: Beijing, China, 1985. (In Chinese)

Journal of
*Marine Science
and Engineering*

MDPI

Article

Analysis of Flow-Induced Vibration Control in a Pontoon Carrier Based on a Pendulum-Tuned Mass Damper

Libin Du, Yongchao Cui, Yanqun Ma, Jie Liu * and Zezheng Liu

College of Ocean Science and Engineering, Shandong University of Science and Technology, Qingdao 266590, China; 13864322713@163.com (Y.C.)
* Correspondence: cyc2713@163.com; Tel.: +86-138-6432-2713

Abstract: The pendulum-tuned mass damper (PTMD) is a widely used vibration-damping device capable of transferring and dissipating structural vibration energy, resulting in reduced structural amplitude, and offering both structural and performance advantages. Given the susceptibility of the submerged expendable conductivity, temperature, and depth profiler (SSXCTD) buoyancy platform to flow-induced vibrations during the upwelling process, the PTMD effectively suppresses the main structure's amplitude under flow field effects. To this end, we investigated the application and design of the PTMD in the SSXCTD buoyancy platform and analyzed its vibration reduction performance. Moreover, we conducted finite volume simulations of the structure using ANSYS FLUENT fluid simulation software, providing insights into its motion under flow field effects and validating the PTMD's effectiveness in mitigating the buoyancy platform's flow-induced vibrations. Our research results demonstrate that the PTMD effectively alleviates the impact of seawater flow on the buoyancy platform, leading to a significant improvement in its operational stability. The proposed research methodology and findings serve as valuable references for the design and optimization of other expendable marine detection equipment.

Keywords: pendulum-tuned mass damper; SSXCTD buoyancy platform; flow-induced vibration; structural damping; finite volume simulation

Citation: Du, L.; Cui, Y.; Ma, Y.; Liu, J.; Liu, Z. Analysis of Flow-Induced Vibration Control in a Pontoon Carrier Based on a Pendulum-Tuned Mass Damper. *J. Mar. Sci. Eng.* **2023**, *11*, 1963. https://doi.org/10.3390/jmse11101963

Academic Editors: José António Correia and Kamal Djidjeli

Received: 29 August 2023
Revised: 28 September 2023
Accepted: 9 October 2023
Published: 11 October 2023

1. Introduction

In recent years, as ocean research and development have accelerated, traditional marine detection equipment has struggled to meet the demands of increasingly complex detection environments and requirements. Consequently, new types of ocean physical parameter measurement devices, such as the submerged expendable conductivity, temperature, and depth profiler (SSXCTD), have emerged. These devices provide oceanographers with physical parameters, including seawater temperature and salinity at various depths, enabling a more accurate understanding of the specific oceanic characteristics in a given area. This has significant implications for marine economic development, military strength building, marine environmental protection, and other applications [1]. As research progresses, higher expectations are placed on the stability of the SSXCTD's operational process and the precision of measurement results [2]. In addition, the buoyancy platform plays a crucial role in the SSXCTD's operations, encompassing functions such as protection, load-bearing, release, and data transmission. Consequently, designing a buoyancy platform capable of withstanding the flow-induced vibrations in the underwater environment is essential to enhance the stability of the SSXCTD measurement system and the accuracy of its results.

The pendulum-tuned mass damper (PTMD) is an early-developed device used for structural vibration control. It primarily consists of two parts: a structural vibration system, composed of an inertial mass, and a damping system, which is generally supported or suspended on the structure. When the structure experiences external excitation and starts vibrating, the PTMD system also vibrates accordingly. The inertial forces generated by the PTMD system act in response to the structure, tuning these inertial forces to harmonize with the main structure's vibrations. This process reduces the structural vibration response and mitigates fatigue damage. Initially, the PTMD was primarily utilized in tall buildings and bridges to control the amplitude of structural vibrations. Gradually, it was introduced into mechanical structures to control mechanical vibrations. In recent years, the PTMD has found successful applications in small and medium-sized equipment in the marine domain [3]. Hu et al. [3] incorporated electromechanical inertance systems into PTMD design and applied the integrated electromechanical inertance PTMD system to the vibration control of offshore wind turbines. In 2015, Tan et al. [4] investigated the vibration control problems of suspended pipeline systems under seismic excitations and experimentally verified the vibration suppression performance of two PTMD configurations: a spring-steel type and a single-pendulum type. Guan et al. [5] studied the application of genetic algorithms in tuning the parameters of tuned mass dampers, improving the system's damping efficiency through the optimization of system parameters. Xia et al. [6] investigated the application of PTMDs in the collision damping and vibration damping control of offshore wind turbines under combined wind and wave action and used the Euler–Lagrange equations to establish a simplified model of the wind turbine considering blade, tower coupling, and pile–soil interactions, and, through numerical simulation, the dynamic response characteristics of the wind turbine under wind, wave, and seismicity were analyzed in the absence of control. The forward and backward displacements of the nacelle of an offshore monopile wind turbine were controlled using a PTMD. The analyses showed that the incorporation of viscoelastic material baffles made the PTMD less space-intensive and more robust than the TMD.

Note that the current primary application of PTMDs in the field of marine engineering equipment technology focuses on vibration control for instruments and equipment such as wind turbines and buoys above the sea surface. Although some studies have been conducted on reducing the mechanical structure vibrations of small underwater detection equipment, like submerged expendable conductivity, temperature, and depth (SSXCTD) buoys, progress in this area remains limited. Furthermore, existing research on PTMD applications mostly centers on the vibration control of tall buildings and bridges in atmospheric airflow, leaving a gap in the analysis of vibration control for small devices like SSXCTD buoys in seawater flow fields. Therefore, this paper investigates the vibration control problem of SSXCTD buoys with the use of PTMD systems in seawater flow fields. Firstly, a dynamic model of the SSXCTD buoy is established. Through research and analysis of the model, parameters affecting the buoy's motion stability are identified. These parameters are then used for the research and design of the PTMD system when applied to the SSXCTD buoy, and the system's effectiveness is validated through finite volume simulation analysis.

2. Dynamic Model of SSXCTD Buoyancy Vehicle

2.1. SSXCTD Buoyancy Vehicle

As depicted in Figure 1, this schematic diagram illustrates the prototype of a submerged expendable conductivity, temperature, and depth (SSXCTD) instrument developed by Lockheed Martin Corporation [7]. The prototype comprises three main components: the probe, buoyancy carrier, and deck unit. The diagram clearly shows that the buoyancy carrier consists of two parts: the upper buoyancy cylinder and the lower probe housing cylinder. This structural design is commonly employed in similar, currently available products.

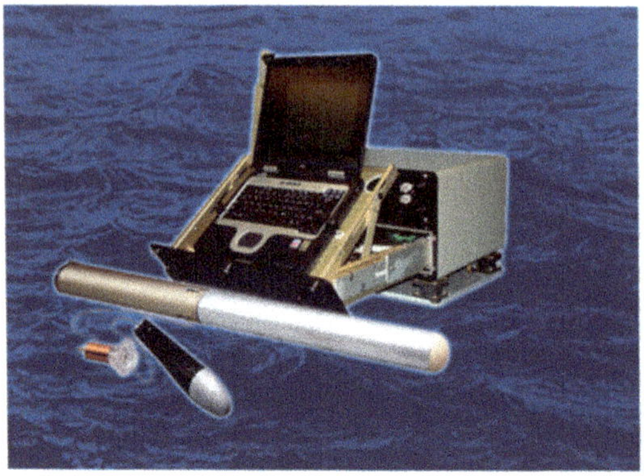

Figure 1. Lockheed Martin SSXBT principle prototype, USA.

Figure 2 illustrates the structural design of the self-developed buoyancy carrier. It comprises, from top to bottom (from right to left in the figure), the buoyancy chamber, near-surface attitude stabilization chamber, upwelling process attitude correction chamber, and probe housing chamber. The overall structure forms a cylindrical shape. In this paper, we will concentrate on this structure as the research subject and explore the design and application of the pendulum-tuned mass damper (PTMD) in small-scale oceanic detection equipment such as the SSXCTD.

Figure 2. Rendering of self-developed SSXCTD float carrier.

2.2. The Dynamical Model of the SSXCTD Buoyancy Vehicle

During the operational process, the SSXCTD buoyancy carrier functions as a complex multi-degree-of-freedom system. In the buoyancy carrier's upwelling process, the hydrodynamic forces of seawater provide the main load. To facilitate analysis and reduce the complexity of the buoyancy carrier's load, we simplified its structure to a cylindrical shape and established a six-degrees-of-freedom model system [8].

To model and analyze the motion of the buoyancy carrier, it is essential to establish relevant coordinate systems, namely, the body coordinate system and the inertial coordinate system, as shown in Figure 3. Now, let us introduce each of them.

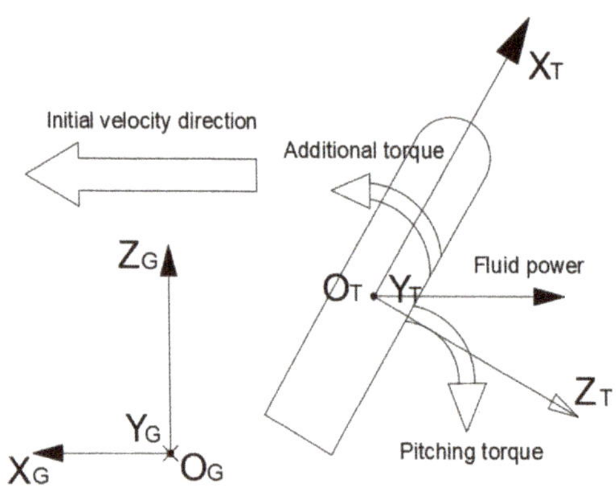

Figure 3. SSXCTD float carrier model coordinate system.

The body coordinate system $O_TX_TY_TZ_T$: The origin O_T of this coordinate system is located at the centroid of the buoyancy carrier. The positive direction of the O_TX_T axis aligns with the longitudinal direction of the buoyancy carrier, pointing towards the head of the buoyancy carrier. The O_TY_T axis is perpendicular to the O_TX_T axis and lies within the symmetry plane of the buoyancy carrier. The O_TZ_T axis is orthogonal to both the O_TX_T and O_TY_T axes, forming a right-handed Cartesian coordinate system.

The inertial coordinate system $O_GX_GY_GZ_G$: The origin O_G of this coordinate system is located at the launch point of the deployment platform. The O_GX_G axis lies within the horizontal plane of the launch point and points in the direction of the launch. The O_GZ_G axis is perpendicular to the O_GX_G axis, with its positive direction aligned along the vertical line at the launch point, pointing upward. The O_GY_G axis is orthogonal to both the O_GX_G and O_GZ_G axes, forming a right-handed Cartesian coordinate system.

After the buoyancy carrier is launched from the platform into the water, its motion can be considered as free motion. During the motion process, it experiences forces such as gravity, buoyancy, hydrodynamic forces, and added mass effects. In the inertial coordinate system, the dynamic models of the buoyancy carrier's translational and rotational motions around its center of mass can be expressed as follows [8]:

$$\begin{cases} \dot{V}_{x1} = \left[m(V_{y1}\omega_{z1} - V_{z1}\omega_{y1}) + F_{\lambda x1} + R_{x1} + F_{x1} + G_{x1}\right]/(m+\lambda_{11}) \\ \dot{V}_{y1} = \left[m(V_{z1}\omega_{x1} - V_{x1}\omega_{z1}) + F_{\lambda y1} + R_{y1} + F_{y1} + G_{y1}\right]/(m+\lambda_{22}) \\ \dot{V}_{z1} = \left[m(V_{x1}\omega_{y1} - V_{y1}\omega_{x1}) + F_{\lambda z1} + R_{z1} + F_{z1} + G_{z1}\right]/(m+\lambda_{33}) \end{cases} \quad (1)$$

$$\begin{cases} \dot{\omega}_{x1} = \left[J_{y1} - J_{z1}\right)\omega_{y1}\omega_{z1} + M_{\lambda x1} + M_{x1} + M_{fx1}\right]/(J_{x1}+\lambda_{44}) \\ \dot{\omega}_{y1} = \left[J_{z1} - J_{x1}\right)\omega_{x1}\omega_{z1} + M_{\lambda y1} + M_{y1} + M_{fy1}\right]/(J_{y1}+\lambda_{55}) \\ \dot{\omega}_{z1} = \left[J_{x1} - J_{y1}\right)\omega_{x1}\omega_{y1} + M_{\lambda z1} + M_{z1} + M_{fz1}\right]/(J_{z1}+\lambda_{66}) \end{cases} \quad (2)$$

In Equation (1), $F_{\lambda x1}$, $F_{\lambda y1}$, and $F_{\lambda z1}$, represent the added forces; R_{x1}, R_{y1}, and R_{z1} represent the hydrodynamic forces; F_{x1}, F_{y1}, and F_{z1} represent the buoyancy forces; and G_{x1}, G_{y1}, and G_{z1} represent the gravitational forces. In Equation (2), $M_{\lambda x1}, M_{\lambda y1}$, and $M_{\lambda z1}$ represent the added moments; M_{x1}, M_{y1}, and M_{z1} represent the hydrodynamic moments; and M_{fx1}, M_{fy1}, and M_{fz1} represent the buoyancy moments.

The added forces and added moments can be expressed as follows:

$$\begin{cases} F_{\lambda x1} = -\dot{\lambda}_{11}V_{x1} - \lambda_{33}\omega_{y1}V_{z1} - \lambda_{35}\omega_{y1}^2 + \lambda_{22}\omega_{z1}V_{y1} + \lambda_{26}\omega_{z1}^2 \\ F_{\lambda y1} = -\dot{\lambda}_{22}V_{y1} - \dot{\lambda}_{26}\omega_{z1} - \lambda_{26}\dot{\omega}_{z1} + \lambda_{33}\omega_{x1}V_{z1} + \lambda_{35}\omega_{x1}\omega_{y1} \\ F_{\lambda z1} = -\dot{\lambda}_{33}V_{z1} - \dot{\lambda}_{35}\omega_{y1} - \lambda_{35}\dot{\omega}_{y1} - \lambda_{22}\omega_{x1}V_{y1} - \lambda_{26}\omega_{x1}\omega_{z1} \end{cases} \quad (3)$$

$$\begin{cases} M_{\lambda x1} = -\dot{\lambda}_{44}\omega_{x1} + (\lambda_{55} - \lambda_{66})\omega_{y1}\omega_{z1} + (\lambda_{26} + \lambda_{35})(\omega_{z1}V_{z1} - V_{y1}\omega_{y1}) \\ M_{\lambda y1} = -\dot{\lambda}_{55}\omega_{y1} - \dot{\lambda}_{35}V_{z1} - \lambda_{35}\dot{V}_{z1} + (\lambda_{66} - \lambda_{44})\omega_{x1}\omega_{z1} + \lambda_{26}\omega_{x1}V_{y1} \\ M_{\lambda z1} = -\dot{\lambda}_{66}\omega_{z1} - \dot{\lambda}_{26}V_{y1} - \lambda_{26}\dot{V}_{y1} + (\lambda_{44} - \lambda_{55})\omega_{x1}\omega_{y1} - \lambda_{35}\omega_{x1}V_{z1} \end{cases} \quad (4)$$

In the equations, $\lambda_{11}, \lambda_{22}$, and λ_{33} represent added mass; λ_{26} and λ_{35} represent added static moments; and $\lambda_{44}, \lambda_{55}$, and λ_{66} represent added moments of inertia. If the buoyancy carrier is a smooth axially symmetric body, the following relations hold: $\lambda_{22} = \lambda_{33}$, $\lambda_{55} = \lambda_{66}$, and $\lambda_{26} = -\lambda_{35}$. Consequently, Equations (3) and (4) can be simplified as follows:

$$\begin{cases} F_{\lambda x1} = -\dot{\lambda}_{11}V_{x1} + (\omega_{z1}V_{y1} - \omega_{y1}V_{z1})\lambda_{22} + \left(\omega_{y1}^2 + \omega_{z1}^2\right)\lambda_{26} \\ F_{\lambda y1} = -\dot{\lambda}_{22}V_{y1} - \dot{\lambda}_{26}\omega_{z1} - (\dot{\omega}_{z1} + \omega_{x1}\omega_{y1})\lambda_{26} + \lambda_{22}\omega_{x1}V_{z1} \\ F_{\lambda z1} = -\dot{\lambda}_{22}V_{z1} + \dot{\lambda}_{26}\omega_{y1} + (\dot{\omega}_{y1} - \omega_{x1}\omega_{z1})\lambda_{26} - \lambda_{22}\omega_{x1}V_{y1} \end{cases} \quad (5)$$

$$\begin{cases} M_{\lambda x1} = \dot{\lambda}_{44}\omega_{x1} \\ M_{\lambda y1} = -\dot{\lambda}_{55}\omega_{y1} + \dot{\lambda}_{26}V_{z1} + \left(\dot{V}_{z1} + \omega_{x1}V_{y1}\right)\lambda_{26} + (\lambda_{55} - \lambda_{44})\omega_{x1}\omega_{z1} \\ M_{\lambda z1} = -\dot{\lambda}_{55}\omega_{z1} - \dot{\lambda}_{26}V_{y1} - \left(\dot{V}_{y1} - \omega_{x1}V_{z1}\right)\lambda_{26} + (\lambda_{44} - \lambda_{55})\omega_{x1}\omega_{y1} \end{cases} \tag{6}$$

2.3. *Analysis of Buoyancy Vehicle's Attitude Stability*

By analyzing the various expressions of momentum and momentum moments for the underwater buoyancy carrier's motion in the water, it can be observed that each λ_{ij} represents an additional value, added to the buoyancy carrier's various inertia coefficients when it moves in an ideal fluid. The coefficients $\lambda_{ij}(i = 1,2,3; j = 1,2,3)$ have a mass dimension and are called added masses. The coefficients $\lambda_{ij}(i = 1,2,3;\ j = 4,5,6)$ have a static moment dimension and are referred to as added static moments. The coefficients λ_{ij} $(i = 4,\ 5,\ 6;\ j = 4\ ,5,\ 6)$ have a second-moment dimension. The coefficients with $i = k$ are added to the inertia moments, known as added inertia moments, while the coefficients with $i \neq k$ are added to the inertia products, termed added inertia products. In the computation of non-steady fluid forces and moments acting on the underwater buoyancy carrier, its axial asymmetry is usually neglected. Therefore, there are only eight coefficients that need to be determined: $\lambda_{11}, \lambda_{22}, \lambda_{33}, \lambda_{44}, \lambda_{55}, \lambda_{66}, \lambda_{26}$, and λ_{35}, with the following relationships:

$$\begin{array}{c} \lambda_{11} = \int_L \lambda_{11}(x)dx \\ \lambda_{22} = \int_L \lambda_{22}(x)dx \\ \lambda_{33} = \int_L \lambda_{33}(x)dx \\ \lambda_{44} = \int_L \lambda_{44}(x)dx \\ \lambda_{26} = \int_L x\lambda_{26}(x)dx \\ \lambda_{35} = -\int_L x\lambda_{35}(x)dx \\ \lambda_{55} = \int_L x^2\lambda_{55}(x)dx \\ \lambda_{66} = \int_L x^2\lambda_{66}(x)dx \end{array} \tag{7}$$

For a finite-length cylindrical body with a radius of 'a' in an unbounded, incompressible potential flow, the added mass per unit length is given by:

$$\lambda_{22}(x) = \lambda_{33}(x) = \rho\pi a^2 \tag{8}$$

Assuming that the diameter of the shell at the longitudinal coordinate x of the passive underwater buoyancy carrier is $D(x)$, and the volume of the shell for the passive underwater buoyancy carrier is V_b, we can calculate that:

$$\begin{array}{c} \lambda_{11} = \mu_x\rho V_b \\ \lambda_{22} = \lambda_{33} = \mu_y\rho V_b \\ \lambda_{55} = \lambda_{66} = \mu_y\rho\frac{\pi}{4}\int_L x^2D^2(x)dx \\ \lambda_{26} = -\lambda_{35} = \mu_y\rho\frac{\pi}{4}\int_L xD^2(x)dx = \mu_y\rho V_b x_b \end{array} \tag{9}$$

Here, μ_x represents the longitudinal added mass coefficient of the shell, typically ranging from 0.02 to 0.04, and μ_y represents the longitudinal flow correction coefficient, usually ranging from 0.95 to 0.98 [9].

In conclusion, for a spatially symmetrical six-degree-of-freedom motion model, it is essential to design an appropriate structure to maintain a relatively stable attitude motion under the influence of fluid forces. When the model tilts in a certain direction during motion, this structure can provide an additional inertia moment to counteract the tilting moment of the main structure and maintain its stable attitude [10]. Based on the above analysis, this paper employs the working principle of the PTMD to design a vibration reduction structure that reduces the unstable motion amplitude of the buoyancy carrier caused by fluid forces during its underwater operation.

3. PTMD System Design

3.1. The Working Principle of the PTMD System

The PTMD is a mechanical device used for structural vibration reduction. Its principle function involves utilizing the interaction between the mass and damping systems to achieve the effect of vibration reduction in a structure [11].

The mass component of the PTMD consists of one or multiple mass blocks, and the system's resonance frequency can be adjusted by changing the mass, quantity, and position of these blocks. When the system is subjected to external forces and undergoes vibration, the mass blocks generate inertia forces to counteract the external forces, thereby reducing the vibration amplitude of the system.

The conventional PTMD's damping system typically comprises multiple springs, which can be adjusted to control the system's stiffness. When the system experiences external forces and undergoes vibration, the springs generate reactive forces to counteract the external influences. They also serve to restrict the movement of the mass blocks, preventing them from rapidly destabilizing the entire damping system.

The working principle of the PTMD designed in this paper is similar. When the carrier experiences displacement in a certain direction due to fluid forces, the pendulum of the pendulum-tuned mass damper will swing in the opposite direction under the effect of inertia force, reducing the amplitude of the carrier's swing in that direction. Similarly, to achieve better vibration reduction results, the pendulum's swing must not be excessively free. Therefore, an electromagnetic damping system is adopted which allows the damping coefficient of the system to be adjusted through electromagnetic forces [12].

This system can be simply understood as follows: A magnetic force is applied to the pendulum, always directed towards the center of the permanent magnet. This force does not affect the direction of the pendulum's swing during actual operation but provides a constraint, allowing the pendulum to play its corresponding vibration reduction role. However, when the entire carrier tends to stabilize, the pendulum can quickly stabilize as well, limiting its free swinging and avoiding unnecessary damage.

3.2. Numerical Calculations for PTMD System Design

The SSXCTD buoyancy carrier can be regarded as a hollow cylindrical structure with uniform mass distribution. Therefore, the computational model of the PTMD-equipped SSXCTD buoyancy carrier can be simplified to a six-degrees-of-freedom cylindrical structure with uniform mass distribution [13]. The PTMD model consists of a rigid slender rod connected at one end to a metal mass block with a certain mass and at the other end to a three-degrees-of-freedom ball hinge, forming the PTMD model. The PTMD is fixed to the inner surface of the upper wall of the SSXCTD buoyancy carrier model using bolts, forming the computational model, as shown in Figure 4.

Figure 4. Computational model of a pontoon carrier equipped with PTMD.

The optimal parameters of the aforementioned PTMD computational model can be obtained using Den Hartog's formula [14,15]:

$$f = \frac{\sqrt{1+\frac{\gamma}{2}}}{1+\gamma} \tag{10}$$

$$\zeta_s = \sqrt{\frac{3\gamma}{8(1+\gamma)}} \tag{11}$$

$$\frac{|u_p|}{H} = \frac{g^2\sqrt{[f^2(1+\gamma)-g^2]^2+4g^2f^2\zeta_s^2(1+\gamma)^2}}{\sqrt{[\gamma f^2g^2-(g^2-1)(g^2-f^2)+4g^2\zeta_s\zeta_s f]^2+4g^2[\zeta_s f(g^2+\gamma g^2-1)+\zeta_s(g^2-f^2)]^2}} \tag{12}$$

For a given PTMD system with a structural damping ratio ζ_p, the optimal parameters γ, f, and ζ_s need to be determined through numerical iteration. Here, we consider the vibration response $\frac{|u_p|}{H}$ as a function of g. For a specific set of γ and f values, different ζ_s values (ζ_{s1}, $\zeta_{s2}, \zeta_{s3} \ldots \ldots \zeta_{si}$) are applied, and the maximum value of the function corresponding to each ζ_s value is found.

$$\{(\frac{|u_p|}{H})_{max1}, (\frac{|u_p|}{H})_{max2} \cdots (\frac{|u_p|}{H})_{maxi}\}$$

Next, the minimum value among the obtained function values is identified:

$$(\frac{|u_p|}{H})_{min1} = min\{(\frac{|u_p|}{H})_{max1}, (\frac{|u_p|}{H})_{max2} \cdots (\frac{|u_p|}{H})_{maxi}\}$$

For each specific set of γ and f values, the aforementioned data calculation process yields a value $(\frac{|u_p|}{H})_{mini}$. At this point, we can again identify the minimum value $\frac{|u_p|}{H}$ from the series of $(\frac{|u_p|}{H})_{mini}$ obtained through numerical iteration. The corresponding γ, f, and ζ_s values that lead to this minimum value $\frac{|u_p|}{H}$ will be the optimal parameters for the PTMD system.

When the excitation frequency remains constant, the system's natural frequency f and the phase of the amplitude response function ϕ_1 are related as follows:

$$\phi_1 = \tan^{-1}(\frac{2\xi_1 f}{1-f^2})$$

When $f = 1$, there is a 90° phase difference between the excitation frequency ξ_i and the phase of the amplitude response function ϕ_1, representing optimal system control. However, this is an idealized result that is challenging to achieve in practical applications. When $f \gg 1$, the phase difference between ξ_i and ϕ_1 is $-180°$, and a negative phase difference renders the computed system parameters invalid. Therefore, to obtain the best system parameters in real-world environments, it is essential to begin calculations with $f < 1$ and approach 1 gradually.

Therefore, in this study, the mass ratio γ was varied at 3%, 4%, 5%, 6%, 7%, and 8%, while the ratio of the PTMD's natural frequency to the main structure's natural frequency, denoted as f, was set to 0.98, 0.97, 0.96, 0.95, 0.94, and 0.93. These values were then input into Equation (3) and computed using MATLAB R2022a, with the results shown in Figure 5.

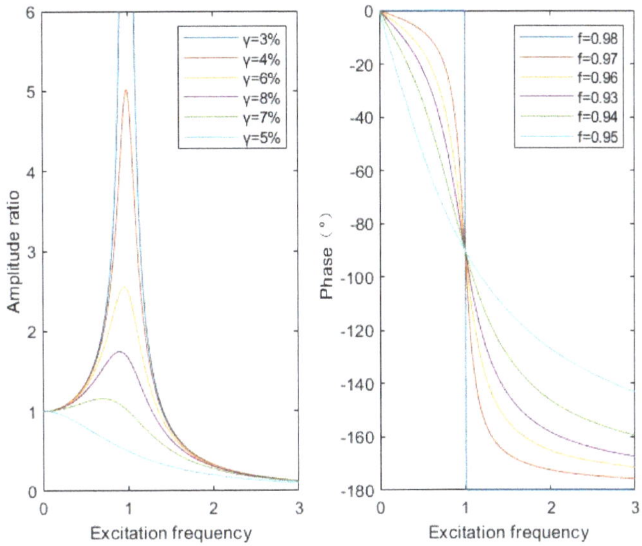

Figure 5. Amplitude and phase of response function versus excitation frequency for different mass ratios and their corresponding self-oscillation frequency ratios.

From the left side of the graph, showing the amplitude ratio variation with excitation frequency, it can be observed that, under the same external excitation frequency, when γ is 3%, the main structure exhibits a relatively large amplitude and significant structural vibration. However, when γ is 5%, the main structure's amplitude ratio is minimized, indicating the most ideal vibration reduction effect of the PTMD. From the right side of the graph, displaying the phase response variation with excitation frequency, it can be seen that when f is 0.96 (corresponding to γ being 5%), the phase response changes most rapidly, indicating that the system exhibits the most sensitive control performance [16].

In conclusion, for a structure with a determined structural damping, the optimal parameters for the PTMD system are as follows: $\gamma = 5\%, f = 0.96$, and $\zeta_s = 10\%$.

Because the natural period T of the PTMD system is given by $T = 2\pi\sqrt{\frac{L}{g}}$, the natural frequency ω_s of the PTMD system is given by $\omega_s = \frac{1}{2\pi}\sqrt{\frac{g}{L}}$. Therefore, the formula used to calculate the pendulum length (L) of the PTMD system is:

$$L = \frac{g}{4\pi^2\omega_s^2}$$

Furthermore, since $f = \frac{\omega_s}{\omega_p}$, we can express $\omega_s = f \cdot \omega_p$. By substituting the value of ω_s in the previous formula, we can calculate the optimal pendulum length (L) of the PTMD system, which is L = 92.3 mm.

3.3. Analysis of the Computational Results

The optimal mass ratio and pendulum length for the PTMD system have been determined through rigorous analysis and calculations. Subsequently, these optimal parameters were used to enhance and optimize the computational model. The enhanced model was then compared to the original model without the PTMD system (prior to improvement) through numerical simulations performed using MATLAB R2022a [17,18]. Figure 6 illustrates a comparison of the structural amplitude variations over time between the model equipped with the PTMD system before improvement and the model without the PTMD system.

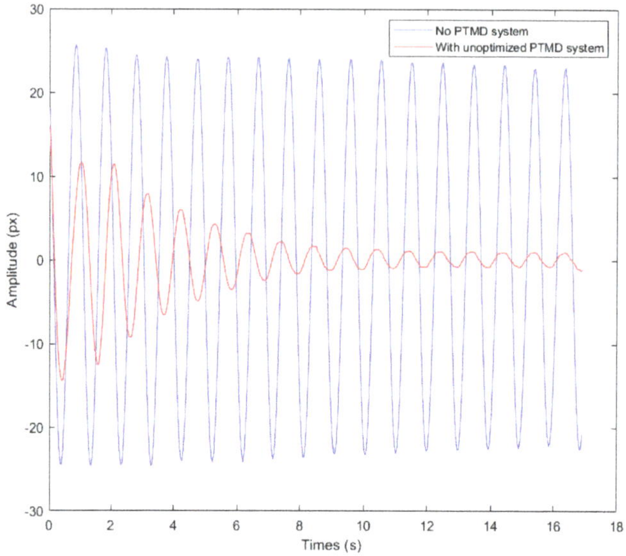

Figure 6. Comparison of the structural amplitude attenuation before optimization.

The amplitude curves reveal that the main structure experiences a slow decay in structural amplitude when subjected to external force without the PTMD system, necessitating several tens of seconds to reach a stable and vibration-free state. Conversely, the structure equipped with the PTMD system before optimization exhibits partial suppression of the main structure's amplitude. Within approximately a dozen seconds, the amplitude significantly decreases. However, despite the decay of the amplitude, a minor residual vibration persists in the main structure. Achieving complete elimination of the main structure's vibration at this stage demands a relatively extended duration.

Figure 7 illustrates a temporal comparison of structural amplitude variations between the model equipped with the enhanced PTMD system and the model lacking the PTMD system.

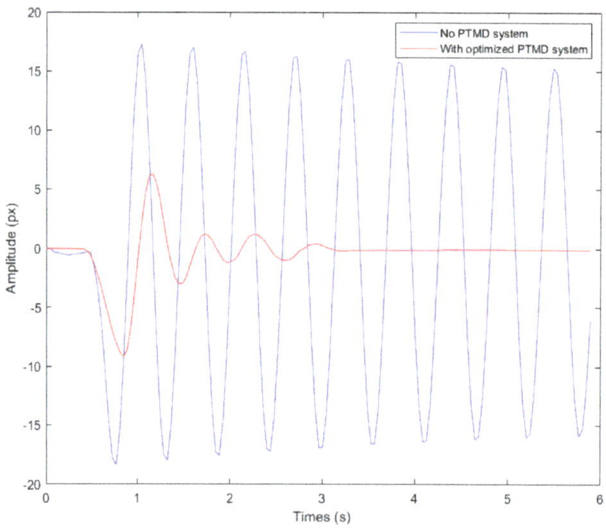

Figure 7. Comparison of amplitude attenuation of optimized structure.

The figure above clearly demonstrates the significant enhancement in the speed of the amplitude decay in the main structure achieved by the improved PTMD system. Within seconds, the main structure quickly returns to stability, effectively eliminating the small residual vibrations present in the pre-optimized scenario. The vibration reduction performance shows a substantial improvement.

4. Finite Volume Simulation of SSXCTD Floating Buoy's Motion and Attitude

4.1. Introduction to the SSXCTD Model

The SSXCTD floating buoy model, equipped with the pressure-tolerant motion device (PTMD) system, is illustrated in Figure 8. The buoy consists of four sequentially connected sections: the buoyancy-providing (Section 1), the attitude adjustment (Section 4), the near-sea surface attitude stabilization (Section 2), and the probe mounting (Section 5). The buoyancy-providing section and the attitude adjustment section are sealed, ensuring watertight integrity. In contrast, the near-sea surface attitude stabilization section and the probe mounting section are designed as open structures to maintain the internal and external pressure balance. The buoyancy-providing section is responsible for offering the necessary buoyancy for the buoy to ascend. The probe mounting section is specially designed to securely hold the SSXCTD probe. Inside the attitude adjustment section, the PTMD system is installed to control the buoy's motion and attitude during its underwater ascent. The near-sea surface attitude stabilization section is equipped with four evenly distributed attitude stabilizing fins (Section 3). These fins conform to the hull and remain locked in place under high-pressure conditions at great water depths. However, as the buoy ascends to shallower, low-pressure environments near the sea surface, these fins automatically open to achieve end-stage deceleration during the upward motion and stabilize the buoy's attitude at the sea surface.

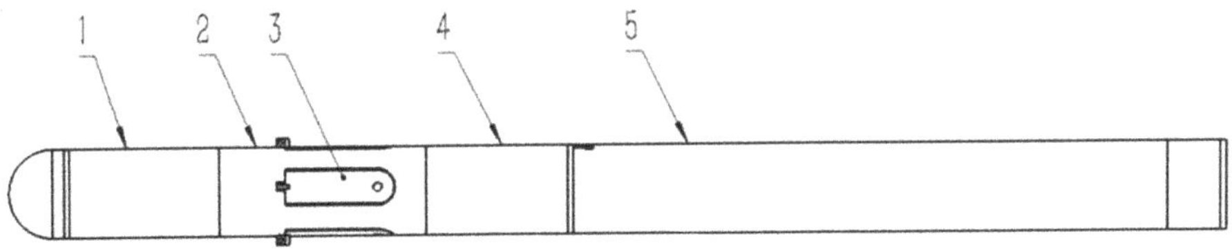

Figure 8. SSXCTD float carrier model.

4.2. Domain and Boundary Condition Settings for Calculation

To analyze the motion attitude of the SSXCTD floating buoy, we focus on two representative motion processes: the initial motion when the buoy leaves the mounting platform and the end-stage motion as it emerges from the water. These processes provide a comprehensive reflection of the buoy's motion state while optimizing the computational time and resources. In this study, we utilized ANSYS FLUENT 2021 R2's dynamic mesh technique to conduct dynamic simulations for these two motion processes. The dynamic mesh technique involves dividing the computational domain into background and moving regions, each with an independent mesh. These sub-regions may have overlapping, nesting, or covering relationships. Coupling between the overlapping regions is achieved by interpolating and matching the flow field information at their boundaries [19].

Figures 9 and 10 illustrate the background and moving regions for the two motion processes in this study. Both computational domains have rectangular shapes, measuring 20 m in length, 5 m in width, and 15 m in height. The left side of the background region functions as the velocity inlet, while the right side is designated as the pressure outlet. The velocity inlet is characterized by the field function of static water VOF wave velocity, whereas the pressure outlet is defined using the field function of static water VOF wave static pressure. The volume fractions for the velocity inlet and pressure outlet are determined using a composite field function that involves static water VOF light fluid volume fraction and static water VOF heavy fluid volume fraction.

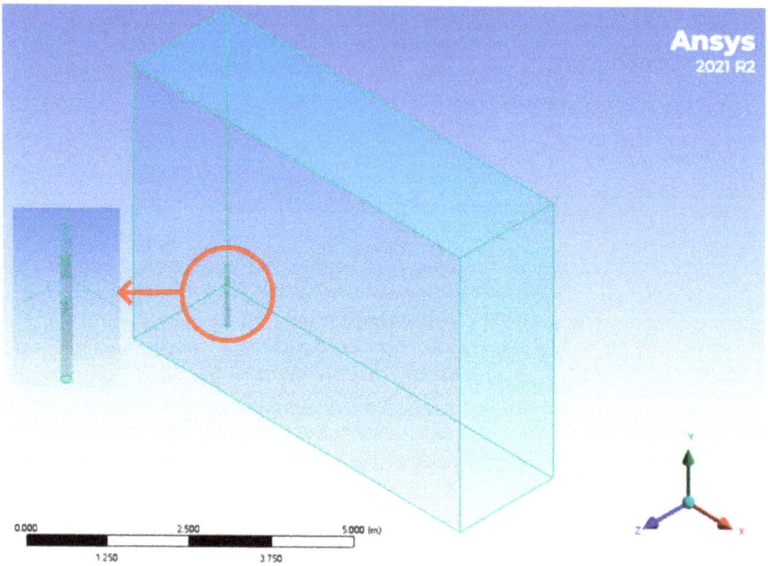

Figure 9. Calculation domain of the initial moment of the surfacing of SSXCTD float carriers.

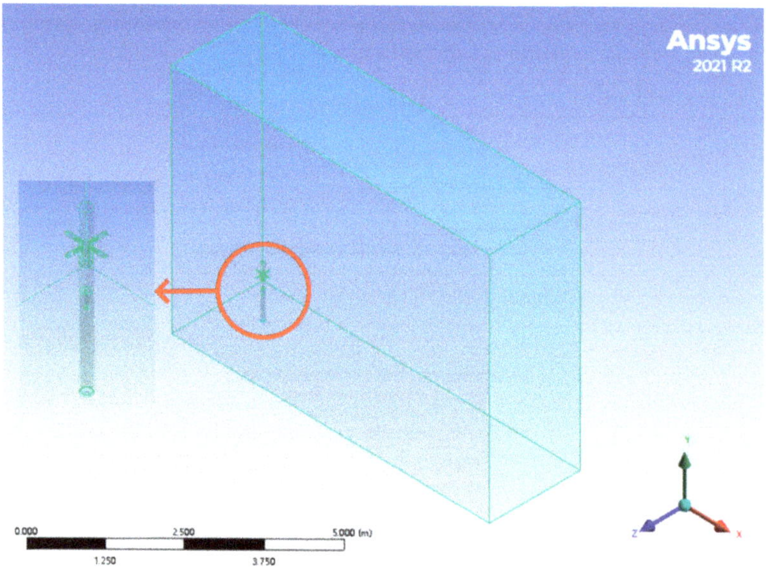

Figure 10. Calculation domain of the SSXCTD float carrier at the end moment of surfacing.

4.3. Computational Domain Meshing

The working depth of the SSXCTD platform is 1 km underwater. Therefore, for the scenario shown in Figure 7, the fluid medium when the SSXCTD buoyancy device leaves the mounting platform is 3 °C seawater, with a density (ρ) of 1030.5 kg/m^3, a flow velocity (v) of 2 m/s, and a dynamic viscosity (μ) of 1.01×10^{-3}Pa·s. For the scenario shown in Figure 8, the fluid medium when the SSXCTD buoyancy device is about to emerge from the water is 15 °C seawater, with a density (ρ) of 1025.91kg/m^3, a flow velocity (v) of 3 m/s, and a dynamic viscosity (μ) of 1.005×10^{-3}Pa·s [20]. The k-ε model is adopted for turbulence modeling, with the wall function set as Δy^+ = 20 [21]. The first layer's thickness of the prism layer on the surface of the buoyancy device is 4.344×10^{-4} mm, the total thickness of the prism layer is 9.467 mm, and the expansion factor is 2.8946. The surface mesh's maximum size is 0.1 mm, and a tetrahedral mesh is used to partition the background and moving regions, as shown in Figures 11 and 12, respectively.

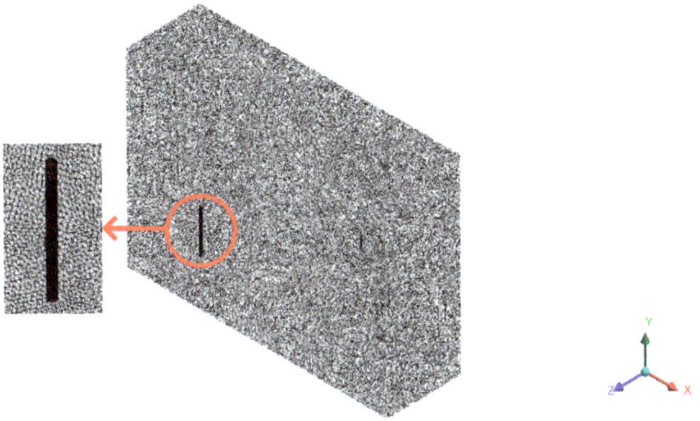

Figure 11. Gridding of the computational domain at the initial moment of the SSXCTD float carrier surfacing.

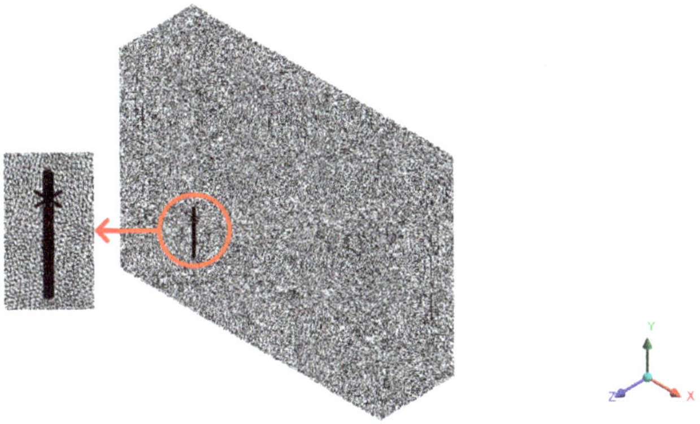

Figure 12. Gridding of the computational domain at the end moment of the SSXCTD float carrier surfacing.

4.4. Numerical Simulation Results

To enhance the accuracy of the simulation results for the two mentioned processes, this study utilizes a finite volume simulation time step of 0.0001 s and a total of 20,000 time steps. These settings allow for thorough finite volume simulation analyses and calculations, enabling a detailed study of the motion states of the floating platform at two different time points.

Figure 13 illustrates the instantaneous velocity variation cloud chart of the SSXCTD floating platform immediately after separating from the loading platform. The total duration of the floating platform's motion is 2 s, and there is a time interval of 0.5 s between two adjacent positions on the graph. From the figure, it is evident that the overall velocity of the floating platform is approximately 2.5 m/s at 2 s. Within 2 s of leaving the loading platform, the platform maintains a stable motion without any loss of stability, ascending with a relatively stable posture.

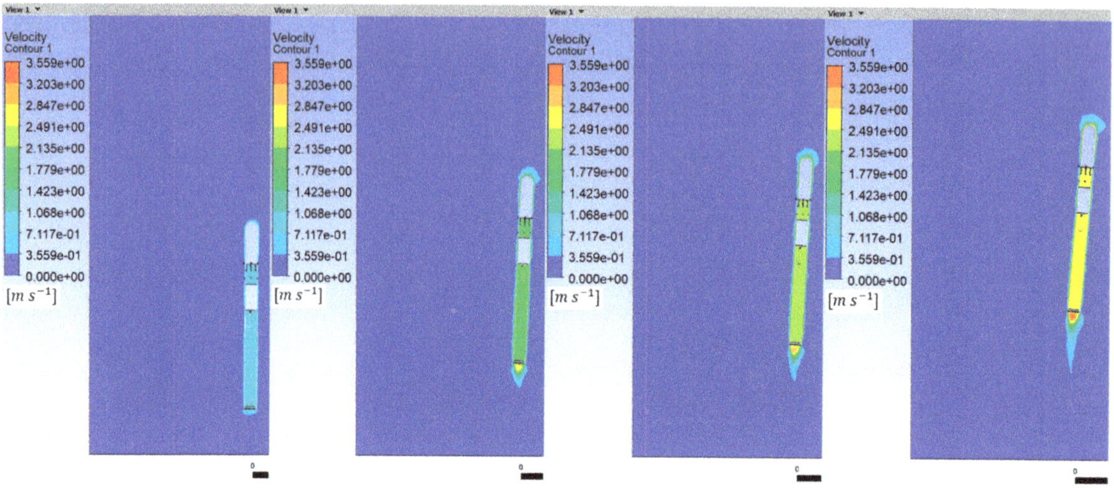

Figure 13. Velocity cloud at the initial moment of the surfacing of the SSXCTD float carrier.

Figure 14 illustrates the displacement–velocity curve of the floating platform at the initial moment of upwelling. The horizontal axis represents the displacement of the floating platform, while the vertical axis represents its velocity. From the graph, it is evident that after the floating platform separates from the loading platform, it starts to ascend due to its own buoyancy, and the upwelling velocity gradually increases. As the upwelling displacement reaches approximately 0.7 m, the velocity stabilizes at 2.51 m/s. The curve demonstrates that after attaining a stable velocity, the floating platform maintains a smooth upward posture without any vibration, aligning with the results shown in the velocity cloud chart at the initial moment of upwelling in Figure 13.

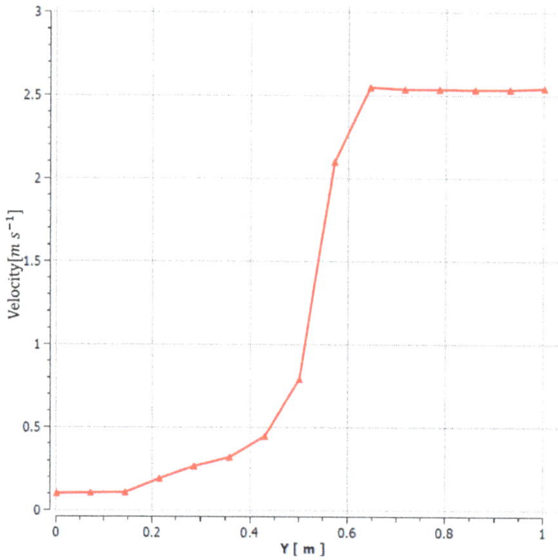

Figure 14. Displacement–velocity curve at the initial moment of float carrier surfacing.

Figure 15 displays the velocity cloud chart of the SSXCTD floating platform as it approaches the water surface, specifically during the 2 s interval after the previously mentioned attitude stabilizing fins have opened. The graph clearly indicates that, at the moment the fins open, the velocity of the floating platform notably decreases. However, it continues to ascend under the influence of its own buoyancy. Furthermore, during the final moments of the upwelling process, the floating platform maintains its stable motion and continues to rise steadily until it reaches the water surface.

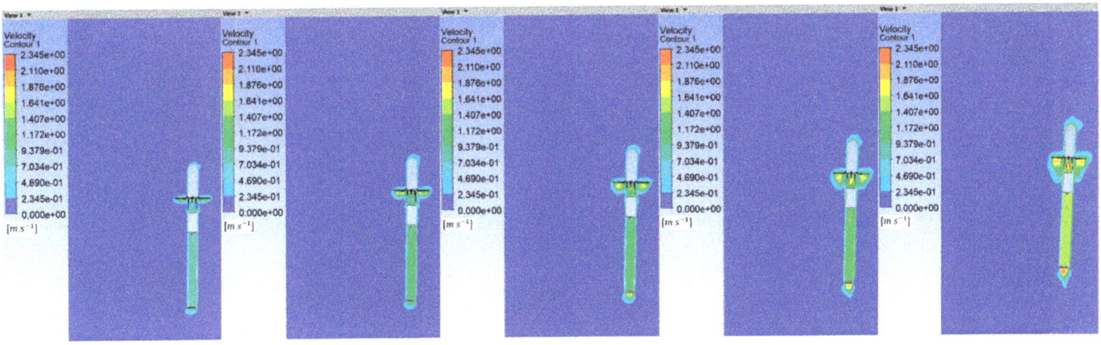

Figure 15. Velocity cloud at the end moment of float carrier surfacing.

Figure 16 illustrates the displacement–velocity curve of the floating platform at the moment when the stabilizing fins open during the final phase of upwelling. From the graph, it is evident that at the instant when the stabilizing fins open, the upwelling velocity of the floating platform rapidly decreases. Subsequently, it continues to ascend at a stable speed of 1.48 m/s. Once the velocity stabilizes, the floating platform maintains a steady upward posture without any signs of instability, and the upwelling speed remains relatively constant until it reaches the water surface. The results depicted in this curve are consistent with the velocity cloud chart shown in Figure 15.

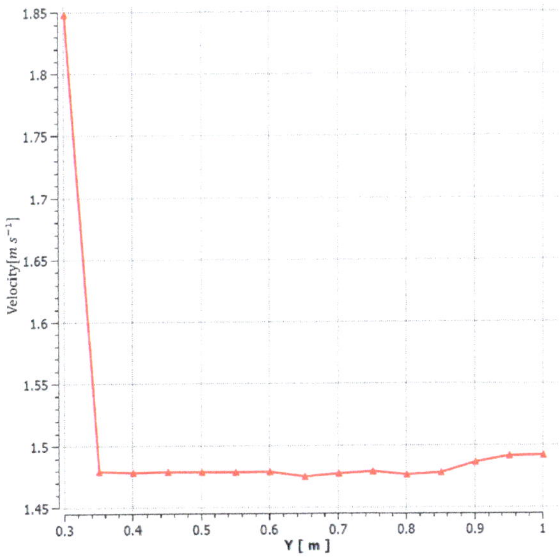

Figure 16. Displacement–velocity curve after the opening of stabilizing flaps at the final moment of floatation of the float carrier.

From the above simulation results, it can be seen that during both the initial and the end moment of the floating of the float carrier, the floating motion attitude of the float carrier is kept stable and there is no loss of stability. It can also be seen from Figures 15 and 16 that the attitude stabilizing airfoils have an obvious effect in slowing down the deceleration of the float carrier at the end of the floating moment and stabilizing the attitude. Therefore, from the results of the simulation calculations, it can be concluded that the PTMD has a remarkable effect on the stabilization of the motion attitude in the floating process of the float carrier.

4.5. Comparison of Numerical Simulation Results

The obtained numerical simulation results in this paper are compared with the underwater non-propulsive carrier design proposed by Zhang Haiyang et al. [22] from the National Key Laboratory of Robotics at the Shenyang Institute of Automation, Chinese Academy of Sciences, which was not equipped with a PTMD and attitude-stabilizing fins. Figure 17 depicts the structural rendering of the underwater non-propulsive carrier designed by Zhang Haiyang et al. [22].

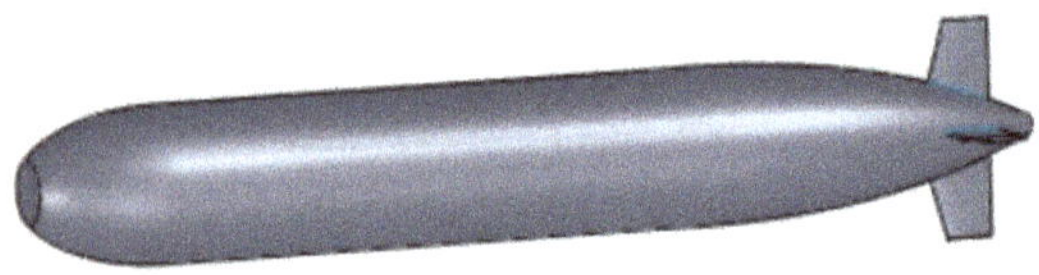

Figure 17. Structural rendering of the underwater non-propulsive carrier without a PTMD.

Utilizing the overlapping grid technique in STAR-CCM+, dynamic simulations were conducted for the buoyant ascent process of the carrier. The simulated fluid medium maintained a temperature (T) of 15 °C, a density (ρ) of 1025.91 kg/m3, a velocity (v) of 3 m/s, and a dynamic viscosity (μ) of 1.005 × 10^{-3} Pa·s, representing seawater. The turbulence model employed for simulation was the k-ε model with a wall spacing parameter Δy^+ of 20. Figure 18 illustrates the initial motion attitude of the non-propulsive underwater carrier without a PTMD, while Figure 19 portrays the motion attitude of the non-propulsive underwater carrier without a PTMD at a specific moment during its ascent process.

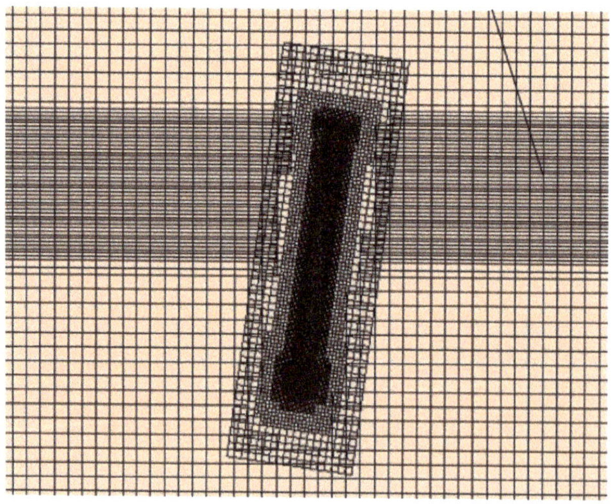

Figure 18. A motion attitude diagram of the underwater non-propulsive carrier without a PTMD at the initial moment of its buoyant ascent.

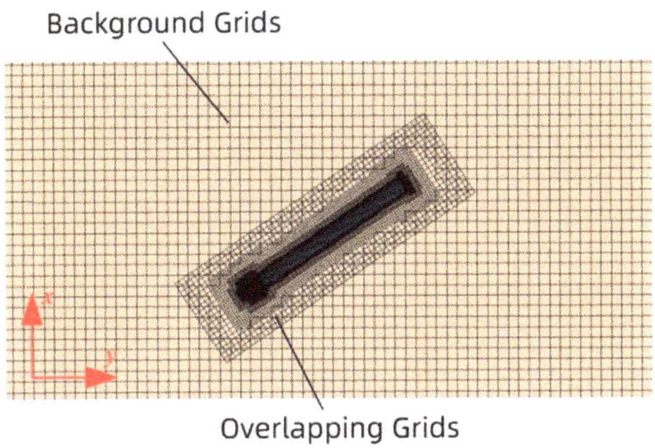

Figure 19. A motion attitude diagram of the underwater non-propulsive carrier without a PTMD at a specific moment during its buoyant ascent process.

From Figures 18 and 19, it is evident that the carrier without a PTMD exhibited significant angular deviations in its motion attitude during the buoyant ascent process, leading to severe instability. Conversely, as indicated in Figures 13–16, the carrier equipped with a PTMD demonstrated a considerable improvement in attitude stability during the ascent process.

5. Conclusions

At present, the research on and applications of pressure-tuned mass damper (PTMD) systems in the field of marine detection instruments and equipment are limited. Therefore, exploring the application and research of PTMD systems in non-traditional areas presents significant potential for development. This study primarily aimed to establish computational models, conduct theoretical analyses, and perform numerical simulations to investigate the application of PTMD systems in the design and study of attitude stability control for a SSXCTD floating platform.

Firstly, we established a kinematic model for the SSXCTD float carrier, enabling an analysis to identify the factors influencing the carrier's motion and attitude stability. Subsequently, we proposed a novel approach and design concept for a miniaturized pendulum-tuned mass damper which demonstrated excellent performance and applicability in mitigating flow-induced vibrations in the SSXCTD float carrier.

Secondly, we adopted a combined approach, utilizing MATLAB and ANSYS FLUENT, to conduct simulations and comparative analyses of the investigated pendulum-tuned mass damper's control effectiveness concerning float carrier flow-induced vibrations. This approach surpassed traditional finite element analysis methods, offering heightened computational accuracy and efficiency. Moreover, the visual representation of the results enabled a more precise simulation of the system's effectiveness.

Through the theoretical analysis, calculations, and numerical simulations conducted in this study, the following conclusions can be drawn:

(1) For an axially symmetric cylindrical object subjected to a unidirectional force in a flow field, its motion can be described as six-degrees-of-freedom motion. When it moves in directions other than the direction of the applied force, applying an additional moment in that direction can counteract the motion deviation and enhance its motion stability.

(2) The PTMD system can provide additional mass and an additional moment to the main structure during its motion. Based on the structural damping ratio of the SSXCTD floating device, the optimal parameters of the PTMD system were calculated using the Den Hartog formula as follows: the mass ratio between the mass block and the main structure,

γ, is 5%; the ratio of natural frequencies, f, is 0.97; the damping ratio, ζ_s, is 10%; and the optimal pendulum length, L, is 92.3 mm.

(3) The numerical simulation and finite volume analysis results demonstrate that the PTMD system exhibits excellent performance in controlling the flow-induced vibrations of the SSXCTD floating device, which is of significant importance in improving the accuracy and stability of measurements during operation.

In conclusion, this study provides an effective solution by applying the PTMD and other structural vibration reduction systems in marine instruments such as SSXCTD, effectively addressing the issue of flow-induced vibrations in ocean measurement equipment. The research presented in this paper has valuable implications for the design and optimization of other marine measurement instruments that may encounter similar vibration challenges.

6. Patents

The content studied in this paper has been awarded a certificate of invention patent by the People's Republic of China.

Name of the invention: A self-correcting attitude-correcting underwater unpowered float carrier, method of use.

Inventors: Jie Liu, Yongchao Cui, Libin Du, Zezheng Liu, and Haijing He.

Patent number: ZL 2022 1 1401584.4

Author Contributions: Conceptualization, L.D.; methodology, J.L.; software, Y.C.; validation, J.L. and Y.M.; formal analysis, Z.L.; investigation, Y.M.; resources, Y.C. and Y.M.; data curation, Y.C.; writing—original draft preparation, Y.C.; writing—review and editing, J.L. and Y.M.; visualization, Y.C.; supervision, L.D.; project administration, L.D. All authors have read and agreed to the published version of the manuscript.

Funding: This research received no external funding.

Institutional Review Board Statement: Not applicable.

Informed Consent Statement: Not applicable.

Data Availability Statement: The data that support the findings of this study are openly available in Baidu Netdisk at https://pan.baidu.com/s/1_wZZj1JlBOSZPQWLlK7qqw?pwd=7777 (29 August 2023).

Conflicts of Interest: The authors declare no conflict of interest. The funders had no role in the design of the study; in the collection, analyses, or interpretation of data; in the writing of the manuscript; or in the decision to publish the results.

References

1. Shi, X.; Su, Q. Development status of expendable ocean instruments and equipment. *Acoust. Electron. Eng.* **2015**, *4*, 46–48.
2. Chen, W.; Zhang, L.; Sun, X.; Duan, J. Overview of the application and development of expendable ocean temperature-depth profiler. *J. Oceanogr.* **2020**, *44*, 114–122.
3. Hu, Y.; Xu, J.; Cheng, C.; Zhou, H.; Cai, X. Passive vibration control of floating offshore wind turbine based on electromechanical inertial capacitance. *J. Mech. Eng.* **2023**, *59*, 202–211.
4. Tan, J.; Ho, S.C.M.; Zhang, P.; Jiang, J. Experimental Study on Vibration Control of Suspended Piping System by SingleSided Pounding Tuned Mass Damper. *Appl. Sci.* **2019**, *9*, 285. [CrossRef]
5. Guan, X.; Guo, H.; Wang, H.; Wang, Z. Parameter optimization of TMD based on genetic algorithm. *J. Xi'an Univ. Archit. Technol. (Nat. Sci. Ed.)* **2023**, *55*, 211–216. [CrossRef]
6. Xia, H. Collision Damping and Vibration Reduction Control of Offshore Wind Turbine under Combined Wind and Wave Action. Master's Thesis, Wuhan University of Technology, Wuhan, China, 2020. [CrossRef]
7. Lockheed Martin. *Submarine-Launched Expendable Profiling System*; Lockheed Martin: Washington, DC, USA, 2003.
8. Geng, B.; Xu, Z.; Pei, Y. Underwater motion modeling and simulation of unmanned carriers. *Comput. Simul.* **2017**, *34*, 6–9+67.
9. Wang, C.; Shi, W. Optimal Design and Application of a Multiple Tuned Mass Damper System for an In-Service Footbridge. *Sustainability* **2019**, *11*, 2801. [CrossRef]
10. Rong, J. Study on the attitude motion stability of unmanned carriers. *Adv. Hydrodyn. (Ser. A)* **1998**, *1*, 88–94.

11. Ouyang, L. Design and Analysis of Tuned Mass Damper. Ph.D. Thesis, Huazhong University of Science and Technology, Wuhan, China, 2017.
12. Li, H. Structural Vibration Control Research Based on Strong Magnetic Array Tuned Mass Damper. Ph.D. Thesis, Dalian University of Technology, Dalian, China, 2022. [CrossRef]
13. Cai, Q.; Zhang, Y.; Wang, Y.; Zhang, P. Ballistic characterisation of uplift of large depth unpowered vehicle. *Comput. Simul.* **2012**, *29*, 38–41+50.
14. Den Hartog, J.P. *Mechanical Vibrations*, 4th ed.; McGraw Hill: New York, NY, USA, 1956.
15. Liu, X.; Shi, W.; Chen, X. Introduction to pendulum-type TMD and analysis of its vibration reduction performance. *Struct. Eng.* **2012**, *28*, 66–71. [CrossRef]
16. Liu, L.; Tan, P.; Li, X.; Zhang, Y.; Zhou, F. Analysis of phase and control effect of TMD control system. *J. Vib. Shock* **2015**, *34*, 160–165. [CrossRef]
17. Shi, W.; Wang, L.; Yan, J.; Shan, J. Research on the vibration reduction performance of an improved self-tuning pendulum-type TMD. *J. Huazhong Univ. Sci. Technol. (Nat. Sci. Ed.)* **2018**, *46*, 40–45. [CrossRef]
18. Zhang, Z.; Zhai, J.; Zhang, X. Water trajectory simulation of unmanned carriers based on Simulink. *Silicon Val.* **2011**, *22*, 20–21+50.
19. Zhang, H.; Gu, H.; Lin, Y.; Gao, H.; Feng, M. Analysis of inclined climbing up-floating characteristics of unmanned carriers. *Chin. Ship Res.* **2020**, *15*, 38–47. [CrossRef]
20. Yuan, H.; Guan, Q.; Zhang, T. South Pacific ocean current speed prediction based on LSTM-BP combined model. *Bull. Oceanogr. Limnol.* **2022**, *44*, 10–16. [CrossRef]
21. Wang, K.; Li, F.; Yang, X. Estimation of sea current velocity using ocean environmental noise. *Appl. Acoust.* **2022**, *41*, 263–269.
22. Zhang, H. Research on Hydrodynamic Characteristics of Deep-Sea Vehicle without Power Uplift. Master's Thesis, Northeastern University, Xi'an, China, 2020. [CrossRef]

Journal of
Marine Science and Engineering

MDPI

Article

A Strain Fitting Strategy to Eliminate the Impact of Measuring Points Failure in Longitudinal Bending Moment Identification

Gengdu Xu [1,2,3], Jin Gan [1,2,*], Jun Li [4], Huabing Liu [3] and Weiguo Wu [1,3]

1 Key Laboratory of High Performance Ship Technology, Wuhan University of Technology, Ministry of Education, Wuhan 430063, China; 248833@whut.edu.cn (G.X.); mailjt@163.com (W.W.)
2 Departments of Naval Architecture, Ocean and Structural Engineering, School of Transportation, Wuhan University of Technology, Wuhan 430063, China
3 Green & Smart River-Sea-Going Ship, Cruise and Yacht Research Centre, Wuhan University of Technology, Wuhan 430063, China; huabingliu@whut.edu.cn
4 China Ship Development and Design Center, Wuhan 430064, China; junleec@126.com
* Correspondence: ganjinwut@163.com

Abstract: The identification of longitudinal bending moments is a critical component in the health monitoring of ship structures. This study examines the effect of the failure of measurement points on the accuracy of bending moment identification and presents a solution using an XGboost fitting method. The impact of failure point position and quantity on strain fitting accuracy and bending moment identification was investigated by performing a four-point bending experiment in typical failure scenarios. Further numerical analysis was conducted to identify potential sources of errors in the measurement process. Additionally, several XGBoost-based fitting schemes were tested under practical conditions to provide reliable fitting suggestions. The results indicated that the XGboost strain fitting method outperforms conventional methods for removing failed measurement points, resulting in improved accuracy of identification. When the most critical failure condition occurs (i.e., the deck plate measurement points and deck stiffener measurement points fail), the XGboost method can still estimate the strain at the failure points with acceptable accuracy. These results also hold in complex load scenarios. Moreover, in the practical measurement conditions, the arrangement of measuring points includes two sections that are sufficient to support the fitting of failed measurement points by using the XGboost method. The XGboost strain fitting method exhibits promising potential in strain fitting applications.

Keywords: ship structure health monitoring; longitudinal bending moment identification; XGboost fitting; four-point bending experiment; influence coefficient matrix

Citation: Xu, G.; Gan, J.; Li, J.; Liu, H.; Wu, W. A Strain Fitting Strategy to Eliminate the Impact of Measuring Points Failure in Longitudinal Bending Moment Identification. *J. Mar. Sci. Eng.* **2023**, *11*, 2282. https://doi.org/10.3390/jmse11122282

Academic Editor: Vincenzo Crupi

Received: 25 October 2023
Revised: 16 November 2023
Accepted: 21 November 2023
Published: 30 November 2023

1. Introduction

To this day, safety and lightweight hull structures remain the primary objectives pursued by ship designers. In order to accurately adjust hull design parameters, a high level of proficiency in assessing the navigation environment is essential. With this in mind, the structural health monitoring (SHM) system was introduced. In recent years, interest in SHM has increased with the development of large-scale and intelligent ships [1,2]. Generally, the hull can be regarded as an ultra-long, hollow, and thin-walled beam with distinctive structural characteristics that challenge its ability to withstand longitudinal bending. The issue of the longitudinal bending of the hull has been a subject of great concern among researchers for many years [3,4]. Therefore, the monitoring of longitudinal bending is one of the most important modules of hull structural health monitoring [5–7].

At present, the methods to monitor longitudinal bending fall into two common categories, namely, longitudinal stress monitoring and longitudinal bending moment identification. Compared with the method of stress monitoring, longitudinal bending moment identification can be applied to not only provide design loads but also learn lessons in order to enhance requirements for next-generation vessels, exhibiting a brighter outlook [8–11]. During navigation, irregular wave loads would produce vertical, lateral, and torsional moments on the hull girder [12]. These moments can be determined via the relationship between the theoretical influence coefficient function and the strain value measured on a section. For years, researchers have verified this principle and adopted it in practical applications. Liu [13] proposed several methods for establishing the influence coefficient matrix based on this principle. The method based on FEM analysis is highly efficient in constructing the influence coefficient matrix of complex ship hulls. In view of this principle, Li et al. [14,15] referenced beam theory and proposed a construction method of the influence coefficient matrix for longitudinal bending moment identification. They conducted hull girder experiments and verified that this method is effective. By using this principle, Yu et al. [12] established a longitudinal bending identification method based on LBSG (long base strain gauges) and monitored the longitudinal bending moment time history of a large container ship during navigation. However, in practical cases, strain signals display faults due to the failure of measuring points. This limitation would reduce the identification accuracy and significantly mislead the safety assessment of the hull structure.

Up to now, the strategy to address the issue of measuring point failure in practical structures did not gained much attention [16,17]. In general, removing failed points is a widely used strategy to reduce the identification error to a certain extent in practical engineering due to its simplicity [16]. Nevertheless, the information on stress near the failed point is still inaccessible. To this end, estimating the strain of failed measure points was found to be more beneficial to ice load identification than the method of removing failed points on a typical polar ship's side grillage [17]. Aided by FEM analysis, the researchers adopted cubic function to estimate the strain value at any position by investigating the potential relationships between the coordinates of measuring points. This scheme is similar to the method of fitting missing values. Nowadays, in some datasets where the correlation between data points is low, it is typical to establish direct mappings between values in order to estimate missing values. Previous research studies have concluded that mean imputation and K-nearest neighbor (KNN) methods [18,19] are effective for missing value fitting. However, due to the lack of consideration of the correlation between different variables, the fitting performance of these methods has potential for further improvement.

In order to enhance the fitting accuracy of missing values, ensemble learning methods such as MissForest and XGboost are introduced to fit missing values [20]. Compared with the XGboost method, the solution efficiency of MissForest is not satisfactory due to the requirement of establishing multiple random forest models. It is worth mentioning that the XGBoost method has become widely adopted in recent years due to its efficient and flexible characteristics, which enable it to capture complex data dependencies with high precision [21]. Thanks to its robustness to outliers, XGboost has demonstrated the applicability of large-scale data and performs well in predicting static travel time [22]. In addition, XGboost shows a better fitting ability of missed values compared to the D-GEX algorithm and linear regression method [23]. In summary, the methods to process the situation with failed measuring points deserve intensive exploration.

This work aims to utilize a strain fitting method based on XGboost to improve the identification accuracy of longitudinal bending moment. Different from missing value fitting, strain fitting presents unique challenges due to the limited number of measurement points, resulting in a small-scale dataset that makes it difficult to capture dependencies. In addition, the involvement of measurement errors and the intricate stress distribution under longitudinal bending exacerbates the level of difficulty even further. Nevertheless, by exploring the relationship between bending moment identification and the positions of measurement points, along with a comprehensive evaluation of XGboost, it becomes possible to establish a simplified training set comprising highly correlated measurement points. Through this approach, the fitting method can be brought closer to practical requirements while effectively mitigating failure issues.

As shown in Figure 1, to investigate the issue of failed measurement points in the context of longitudinal bending moment identification under a practical scenario, we conducted some experiments which are organized as follows:

(1) A four-point bending experiment based on a box girder was conducted to realize the loading identification based on the measured signals.
(2) The correlation between the measurement point position and identification accuracy is explored in the experiment scenario to investigate the impact of measuring point failure on longitudinal bending moment identification.
(3) Based on the investigation of correlation, 17 failure conditions were designed to assess the effectiveness of the fitting method based on XGboost, and the method's fitting capabilities were thoroughly evaluated.
(4) To clarify the sources of fitting errors and further explore the applicability of fitting method, the connections between the fitting values and the training set were comprehensively studied under typical load cases using finite element analysis. Then, the minimum training set required for fitting was determined for the simplification of the training set.
(5) Recommendations were provided for the implementation of the strain fitting method for longitudinal bending moment identification.

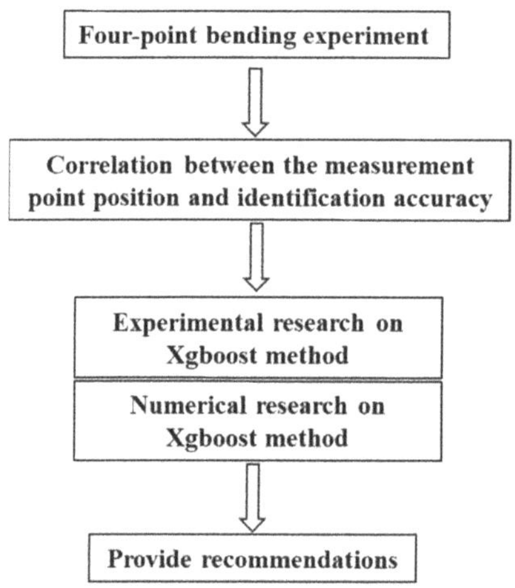

Figure 1. Workflow.

2. Experiment on Longitudinal Bending Moment Identification

2.1. Experimental Details

At present, researchers primarily investigate longitudinal bending through designing four-point bending box girder experiments [24–27]. Therefore, we conduct a four-point bending experiment utilizing a stiffened box girder model for longitudinal bending moment identification research. By subjecting the middle section of the model to a four-point bending configuration, the longitudinal bending of a cabin in ship hull can be simulated.

2.1.1. Design of the Box Girder

To ensure that the model's structure closely resembles that of a ship, the design of the stiffened plate on the box girder is based on the relevant literature [28]. The geometric dimensions of the model are illustrated in Figure 2 and Table 1. The model consists of two extension parts and a middle part, with bulkheads installed to separate them. To simulate the practical ship structure, except bulkheads, longitudinal and transverse stiffeners are arranged on the box girder. The stiffeners are arranged on the outer surface of the model to facilitate the installation of measuring points. Throughout the entire length of the model, the longitudinal stiffeners remain continuous, and the structures of sections are strengthened by transverse stiffeners. The elastic modulus of steel of the model is 206 GPa.

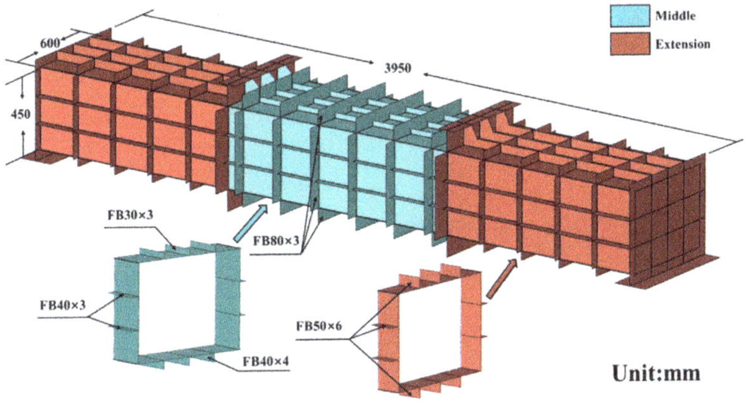

Figure 2. Design of the stiffened box girder.

Table 1. Dimensions of the components.

Component (mm)		Dimension (Extension Part)	Dimension (Middle Part)
Entire model	Length	1250	1450
	Breadth	600	600
	Depth	450	450
Deck	Plate thickness	6	3
	Longitudinal stiffener	FB50 × 6	FB30 × 3
	Transverse stiffener	FB50 × 6	FB80 × 3
Bottom	Plate thickness	6	4
	Longitudinal stiffener	FB50 × 6	FB40 × 4
	Transverse stiffener	FB80 × 3	FB80 × 3
Side	Plate thickness	6	3
	Longitudinal stiffener	FB50 × 6	FB40 × 3
	Transverse stiffener	FB50 × 6	FB80 × 3
Bulkhead	Both ends	6	6
	Middle	6	6

2.1.2. Experiment Setup

As shown in Figure 3, the box girder is placed on two rigid support seats with a round bar at the bottom. The whole model is loaded using the MTS loading system, and the vertical force is equally divided by a square beam to the two bulkheads of the box girder. During the experiment, loads on bulkheads are kept at the same value and increased by the same step to ensure that the middle part is under bending conditions. The bending moment is calculated as follows:

$$M_{mid} = F_t \times L \tag{1}$$

where F_t is the bending load on a bulkhead and L is the length of an extension part. The load conditions of this experiment are 5t ($F_t = 2.5t$), 10t, 15t, 20t, 25t, and 30t, which are the safety load levels for this box girder (after applying the load, the box girder did not fail).

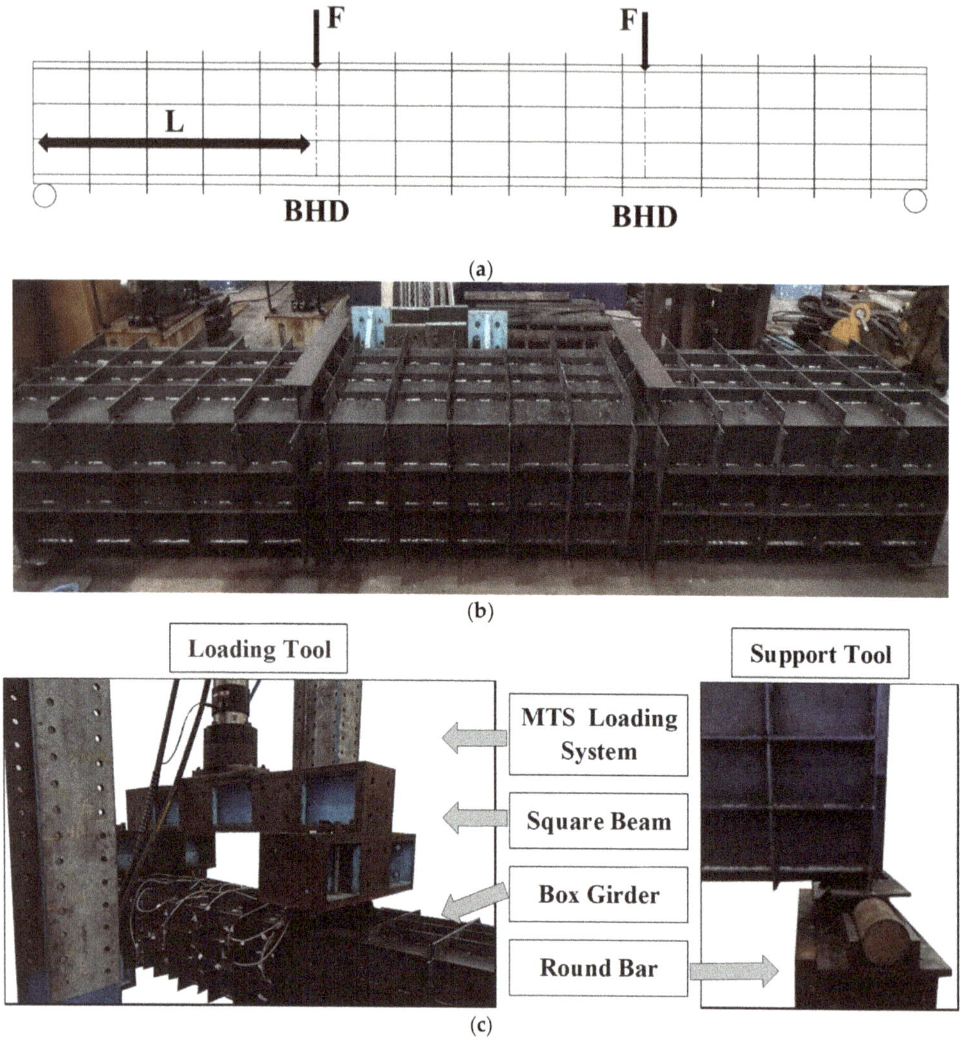

Figure 3. Experimental setup. (**a**) Four-point bending. (**b**) Test model. (**c**) Loading tool.

The layout of measuring points is shown in Figure 4, where S (. . .) is the number of points on the stiffener and P (. . .) is the number of points on the deck. All measuring points are in the center of the panel. In our quest to explore the correlation between measuring point position and the accuracy of longitudinal bending moment identification, it is necessary to monitor the longitudinal strain across all panels and stiffeners. For this purpose, a substantial number of measuring points were arranged on the left side and deck. This extensive placement of measuring points ensures comprehensive data collection for a comprehensive analysis of the longitudinal strain distribution throughout the structure. The scale of arrangement on the right side and the bottom is small, as they are used to verify the symmetry of the model. The strain gauge we adopted is a unidirectional resistance strain gauge with a resistance of 120 Ω and a sensitivity of 2%.

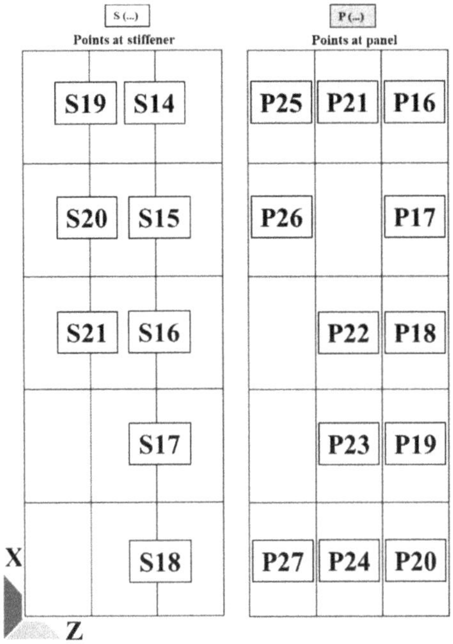

(a)

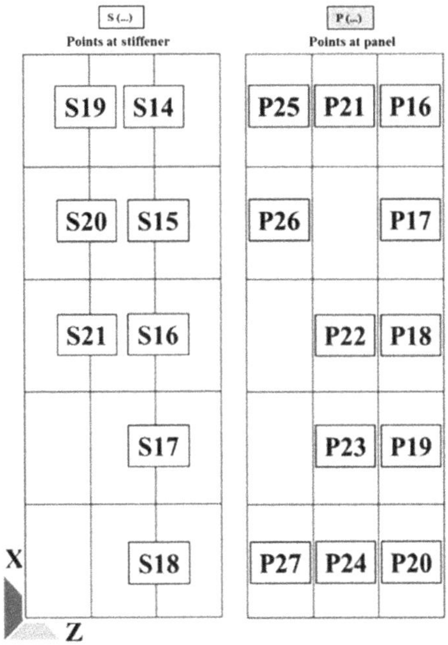

(b)

Figure 4. *Cont.*

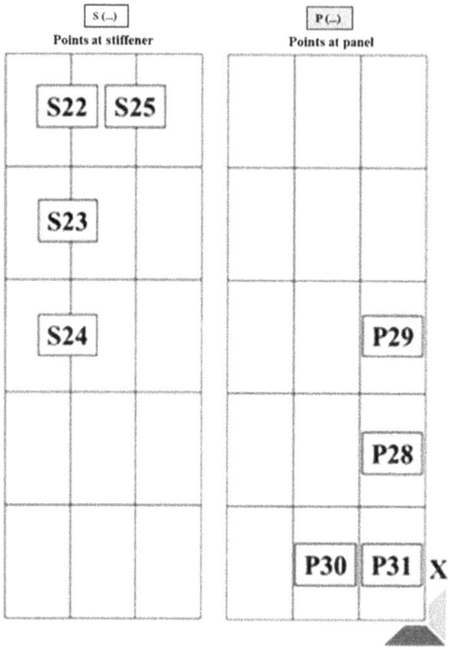

(c)

Figure 4. Layout of strain gauges on the box girder. (**a**) Layout of strain gauges on the deck. (**b**) Layout of strain gauges on the left side. (**c**) Layout of strain gauges on the right side.

2.2. Longitudinal Bending Moment Identification Method

During navigation, the normal stress on a section of the hull girder is balanced. The relationship between strain and bending moment is as follows:

$$\sigma = Mz(EI)^{-1} \tag{2}$$

where E is the elastic modulus; σ is the longitudinal strain of a measuring point on the section; I is the section moment of inertia; and z is the vertical distance between the measuring point and the neutral axis. Therefore, the load–strain transfer matrix can be estimated as follows [14,15]:

$$S = A_V M_V \tag{3}$$

where S is the vector which is made up of the measured longitudinal strain on a section, M_V is the longitudinal bending moment, and A_V is the longitudinal moment–strain transfer vector composed of the ideal longitudinal strain.

2.3. Gradient Descent Method for Load Identification

According to Equation (3), the key to moment identification is to obtain the least squares solution of the bending moment. To this end, we adopt the adaptive gradient descent method to identify the bending moment in this study, which is as follows [29]:

First, operate the initial parameters: $F_1 = 0, d_1 = 0, \alpha_1 = 0, f_1 = 0$.

When $n > 1$,

$$f_{n-1} = (S - AF_{n-1})^2 \tag{4}$$

$$d_{n-1} = \left[(S - AF_{n-1})^2\right]' \tag{5}$$

$$\alpha_{n-1} = \arg\min_{\alpha>0}\{[S - A(F_{n-1} + \alpha d_{n-1})]^2\} \tag{6}$$

$$F_n = F_{n-1} + \alpha_{n-1} d_{n-1} \tag{7}$$

where F_n is the load to be identified; and d_n, α_n, f_n are the intermediate variables. The relative error of bending moment identification of each section is as follows:

$$R_E = \| M_A - M \| / M \tag{8}$$

where R_E is the relative error, M_A is the identified bending moment, and M is the test bending moment which can be calculated as follows:

$$M = F \times 1250 \text{ mm} \tag{9}$$

Figure 5a–e present the results of the longitudinal bending moment identification, along with the corresponding relative errors, for various load conditions ranging from 5t to 30t. The sectional interval between each measurement is set at 200 mm. The maximum relative error of the identification result is less than 4%, while the average relative error is calculated to be 2.3%. These identification errors stem from the deviation between the strain measuring value and the simulated value. During the experiment, due to the existence of unavoidable geometric defects in the structure and interference from the practical environment, the deviation of the strain value is hard to be eliminated. In this study, the identification error is very low. This indicates that the accuracy of strain measurement is within an acceptable range, and the longitudinal moments can be accurately identified in the practical scenario.

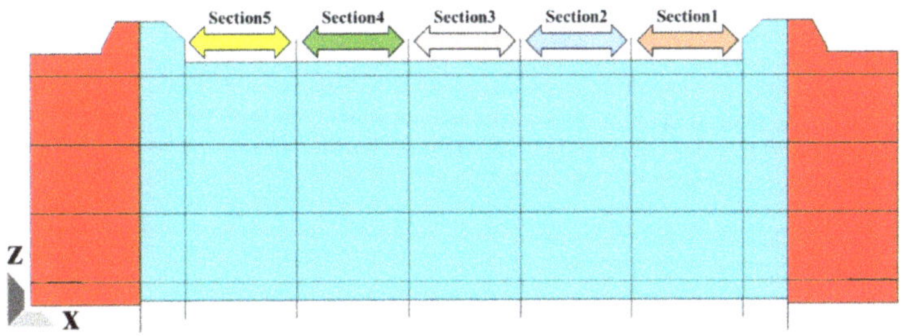

(a)

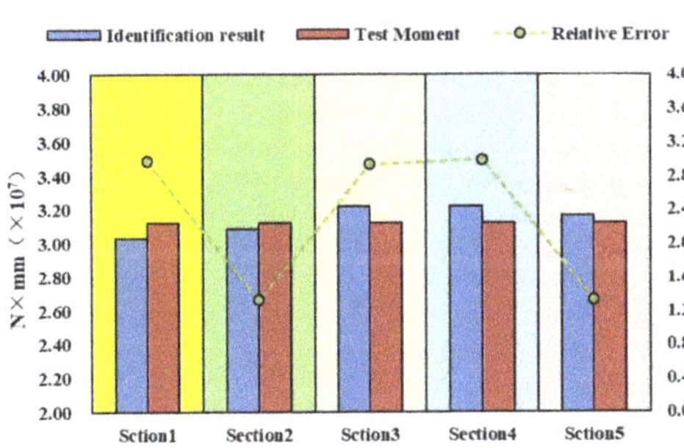

(b)

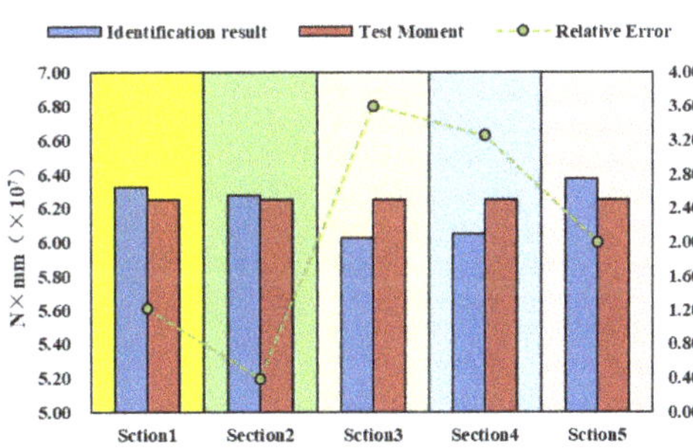

(c)

Figure 5. *Cont.*

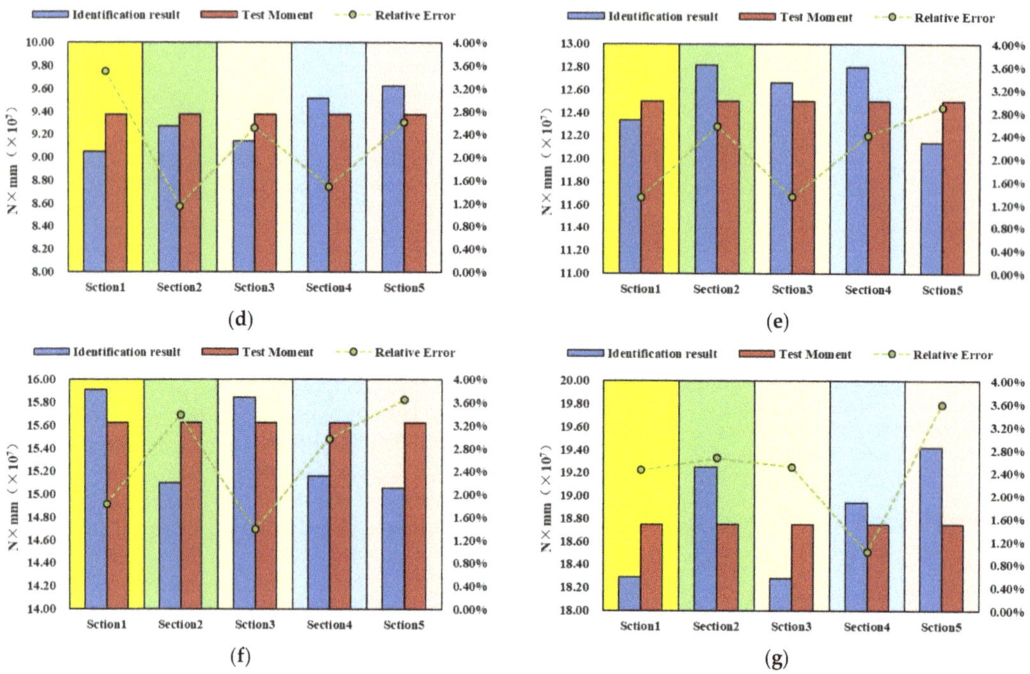

Figure 5. Bending moment identification of different sections. (**a**) Distribution of identified sections. (**b**) 5t. (**c**) 10t. (**d**) 15t. (**e**) 20t. (**f**) 25t. (**g**) 30t.

3. XGboost Fitting of the Strains at Measuring Points

XGboost integrates the cart tree model, which is a weak regression model to form a strong regression model. A cart tree model can be built as follows [21]:

$$\min_{j,s}[\min_{c_1} \sum_{x_i \in R_1(j,s)} (y_i - c_1)^2 + \min_{c_2} \sum_{x_i \in R_2(j,s)} (y_i - c_2)^2] \quad (10)$$

$$R_1(j,s) = \left\{x \mid x^{(j)} \leq s\right\}, R_2(j,s) = \left\{x \mid x^{(j)} > s\right\} \quad (11)$$

$$\hat{c}_m = \frac{1}{N_m} \sum_{x_i \in R_m(j,s)} y_i \quad (12)$$

where R_1, R_2 is the divided input space and c_1, c_2 is the corresponding output value. By repeating Equation (12) for R_1, R_2 until the stopping criterion is met, the input space is divided into m areas. $R_1, R_2 \ldots R_m$ generates a decision tree. In this study, the stopping criterion is restricted by setting the maximum depth of the cart tree.

As shown in Figure 6, the principle of the XGboost method is building a new cart tree model which is learning a new function to fit the residual predicted last iteration. The final regression result can be obtained by adding the regression values of all cart tree models.

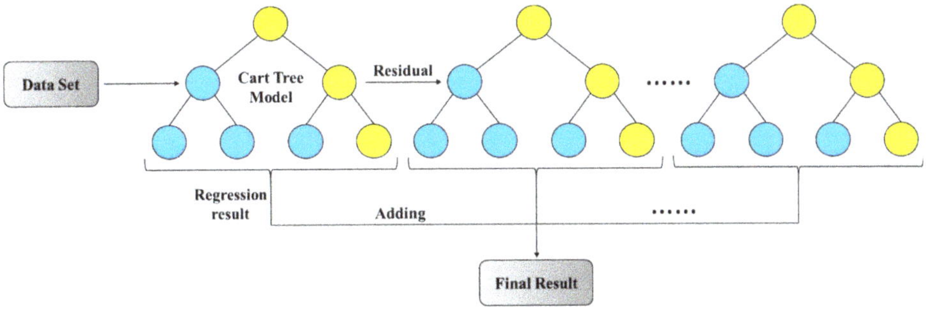

Figure 6. XGboost regression method.

The objective function of XGboost to predict residual consists of the loss function and the regularization term, as follows [21]:

$$L = \sum_i l(y_i, \hat{y}_i) + \sum_k \Omega(f_k) = \sum_{i=1}^{N} l\left(y_i, \hat{y}_i^{(t-1)} + f_t(\mathbf{x_i})\right) + \Omega(f_t) \quad (13)$$

where $l(y_i, \hat{y}_i)$ is the loss function; $\Omega(f_k)$ is the regularization term; y_i is the sample; and $\hat{y}_i$ is the predicted value of y_i. In this study, the loss function is a linear function which is as follows:

$$l(y_i, \hat{y}_i) = MSE(y_i, \hat{y}_i) \quad (14)$$

For each iteration, the objective function is optimized as follows:

(1) By expending the second-order Taylor of loss function term and removing the constant term $l\left(y_i, \hat{y}_i^{(t-1)}\right)$, the loss function term can be optimized as follows:

$$L = \sum_{i=1}^{N} (l(y_i, \hat{y}_i^{(t-1)}) + g_i f_t(\mathbf{x_i}) + \frac{1}{2} h_i f_t^2(\mathbf{x_i})) + \Omega(f_t) \quad (15)$$

$$L = \sum_{i=1}^{N} (g_i f_t(\mathbf{x_i}) + \frac{1}{2} h_i f_t^2(\mathbf{x_i})) + \Omega(f_t) \quad (16)$$

where $g_i = \frac{\partial l\left(y_i, \hat{y}_i^{(t-1)}\right)}{\partial \hat{y}_i^{(t-1)}}, h_i = \frac{\partial^2 l\left(y_i, \hat{y}_i^{(t-1)}\right)}{\partial^2 \hat{y}_i^{(t-1)}}$.

(2) By expanding the regularization term and removing the constant term, the regularization term can be optimized as follows:

$$L = \sum_{i=1}^{N} (g_i f_t(\mathbf{x_i}) + \frac{1}{2} h_i f_t^2(\mathbf{x_i})) + \gamma T + \frac{1}{2} \lambda \sum_{j=1}^{T} w_j^2 \quad (17)$$

$$\Omega(f_t) = \gamma T + \frac{1}{2} \lambda \sum_{j=1}^{T} w_j^2 \quad (18)$$

where T is the number of leaf nodes and w is the score of leaf nodes. γ, λ are regularization parameters. As $f_t(\mathbf{x_i})$ is a function about w, the objective functions can be converted to functions about w, that is, $f_t(\mathbf{x_i}) = w_{q(x_i)}$, as follows:

$$L = \sum_{j=1}^{T} [(\sum_{i \in I_j} g_i) w_j + \frac{1}{2} (\sum_{i \in I_j} h_i + \lambda) w_j^2] + \gamma T \quad (19)$$

(3) By solving the minimum value of the objective function, the optimal tree structure can be obtained as follows:

$$w_j = -\frac{\sum_{i \in I_j} g_i}{\sum_{i \in I_j} h_i + \lambda} \tag{20}$$

$$L = -\frac{1}{2}\sum_{j=1}^{T} \frac{\left(\sum_{i \in I_j} g_i\right)^2}{\sum_{i \in I_j} h_i + \lambda} + \gamma T \tag{21}$$

By utilizing the unfailed points as the training set, it is possible to perform regression analysis to estimate the strain values at the failed points. For the parameter options of the XGboost method, when the fitting residual is less than 10^4 or the number of iterations exceeds 1000, the iteration stops. And the maximum depth of the tree is 10.

4. Investigations into Strain Fitting Method Based on the XGboost Method

4.1. Correlation between Measuring Point Position and Identification Accuracy

In practical load monitoring, positioning measurement points at locations exhibiting strong correlation is more effective in accurately identifying bending moments. This viewpoint is similar to the results of related research studies focused on identifying ice loads [30] and impulse loads [14]. Moreover, measurement points with strong correlation can be used to build training sets for strain fitting with strong features. Therefore, it becomes imperative to investigate the correlation between the positions of measurement points and the identification accuracy of the bending moment.

Based on the above moment identification method, the accuracy of moment identification is related to the strain on a section of the hull. Consequently, this study specifically examines the impact of measuring point position on a single section. By substituting the functional measuring points with failed points (the strains at failed points are replaced with zero), several failure conditions of measuring points are introduced. The degree of correlation between the failed measuring point position and moment identification can be regarded as increasing as the error in moment identification becomes greater. Through this approach, we investigate the effect of measuring point failures on the accuracy of longitudinal bending moment identification, thereby obtaining insights into the correlation between measuring point position and identification.

4.1.1. Correlation between Measuring Points at Different Locations and Identification Accuracy

As shown in Figure 7, under the condition of longitudinal bending, the longitudinal strain distribution on the hull's side exhibits a linear function distribution, assuming ideal conditions, while the deck displays a uniform distribution. This indicates differences in the strain distribution characteristics between the side and the deck of the hull. Hence, this study categorizes the locations of failed measuring points into two types: deck and side.

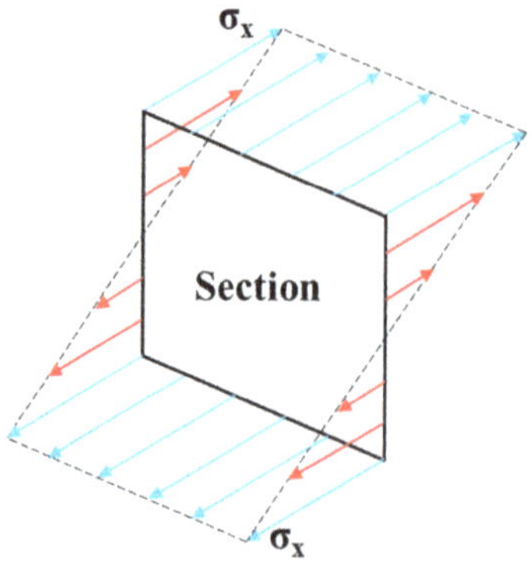

Figure 7. Strain distribution on the hull under longitudinal bending.

In order to investigate the underlying correlations, section 1 was regarded as the failure section due to its arrangement of a relatively higher number of measuring points compared to other sections. Seven failed conditions are designed, as shown in Figure 8.

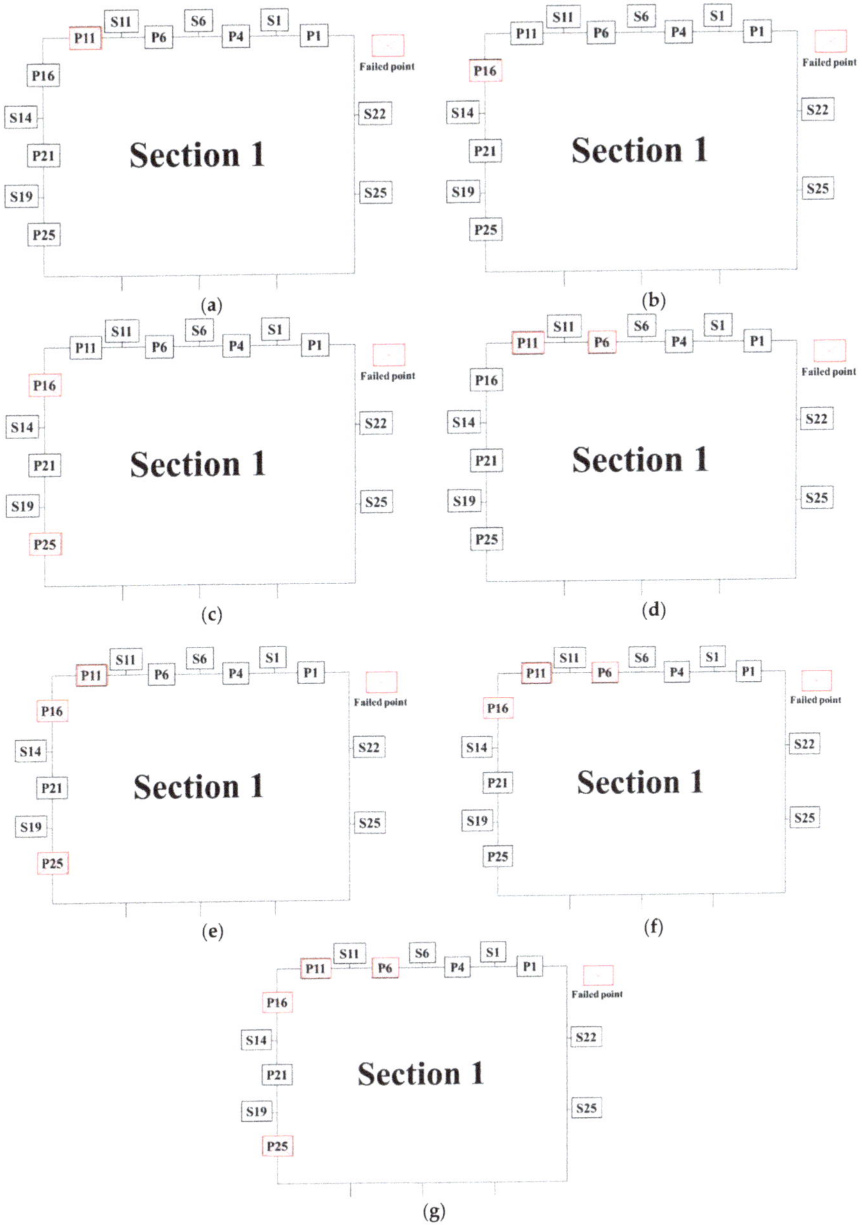

Figure 8. Failure conditions at different locations. (**a**) One failed point (deck). (**b**) One failed point (side). (**c**) Two failed points (side). (**d**) Two failed points (deck). (**e**) Three failed points (deck: one, side: two). (**f**) Three failed points (deck: two, side: one). (**g**) Four failed points (deck: two, side: two).

The failed points are considered for longitudinal bending moment identification. The identification results and relative error of different conditions are shown in Figure 9a–f; the variables are the conditions a–g shown in Figure 7. As the number of failed points increases, the relative error also rises. Once the failed test points exceed two, the identification relative error surpasses 7%. Among these seven conditions, the maximum identification relative error is more than 30%, which emphasizes that the failed points significantly affect the identification of the bending moment. By exploring the trend of the error curve, when there are more failed points on the deck, the relative error of identification is greater (the relative error of condition a is greater than that of condition b; the relative error of condition d is greater than that of condition c; the relative error of condition f is greater than that of condition e; and the relative error of condition e is smaller than that of condition d). This suggests a stronger correlation between failed points on the deck plate and bending moment identification, as opposed to failed points on the side.

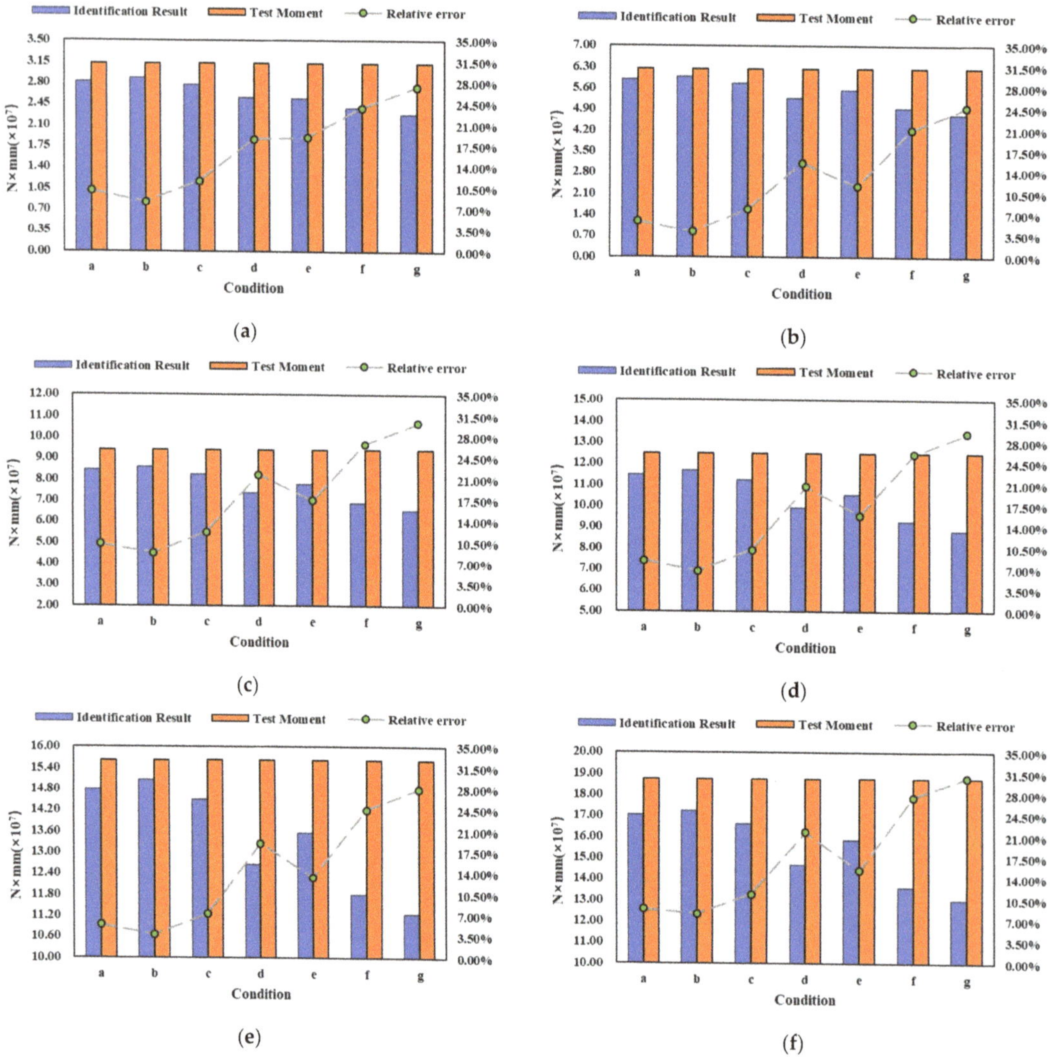

Figure 9. Bending moment identification of point failures at different locations. (**a**) 5t. (**b**) 10t. (**c**) 15t. (**d**) 20t. (**e**) 25t. (**f**) 30t.

4.1.2. Correlation between Measuring Points at Different Structures and Identification Accuracy

In general, the bearing capacity of the plate and the stiffener on a stiffened plate is different [31]. The failed points on the stiffener and the failed points on the plate have different correlations with moment identification. To this end, in this study, the located structures of failed points are divided into two types: plate and stiffener. As shown in Figure 10, six failure conditions were designed to investigate the correlation between measuring points at different structures and identification accuracy.

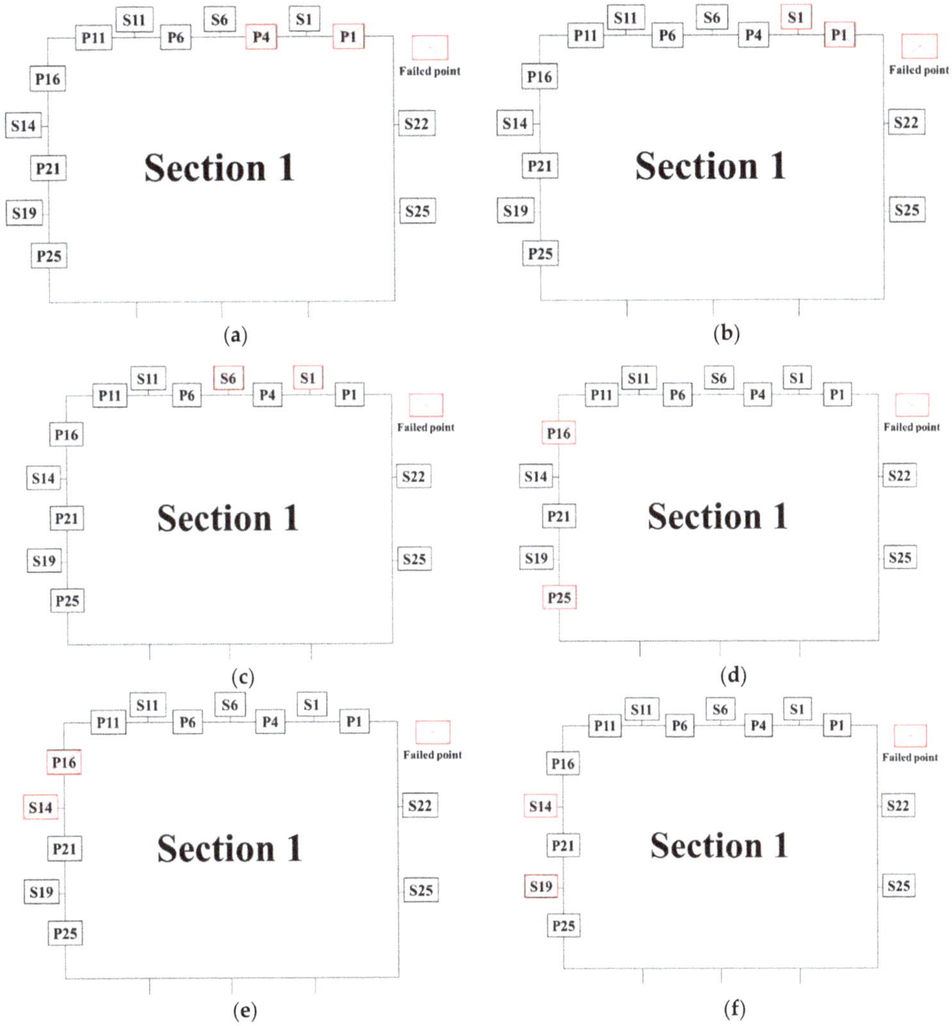

Figure 10. Failure conditions at different structures. (**a**) Two failed points on the deck (plate). (**b**) Two failed points on the deck (plate, stiffener). (**c**) Two failed points on the deck (stiffener). (**d**) Two failed points on the side (plate). (**e**) Two failed points on the side (plate, stiffener). (**f**) Two failed points on the side (stiffener).

The identification results and relative error of different conditions are shown in Figure 11. Among the identification results of conditions a to c, the relative error in condition a is the lowest, whereas condition c exhibits the highest relative error. This indicates that the measuring points located on the deck stiffener have a more pronounced influence on the accuracy of moment identification compared to those on the deck plate. Similarly, when examining the identification results of conditions d to f, it becomes evident that condition d displays the highest relative error, whereas condition f demonstrates the lowest relative error. This suggests that the measuring point on the side plate exhibits a stronger correlation with the identification of the bending moment as compared to the measuring point on the side stiffener. Furthermore, the relative error observed in conditions a to c is greater than that observed in conditions d to f, which further substantiates that the deck measuring points are more strongly correlated with the accurate identification of the bending moment.

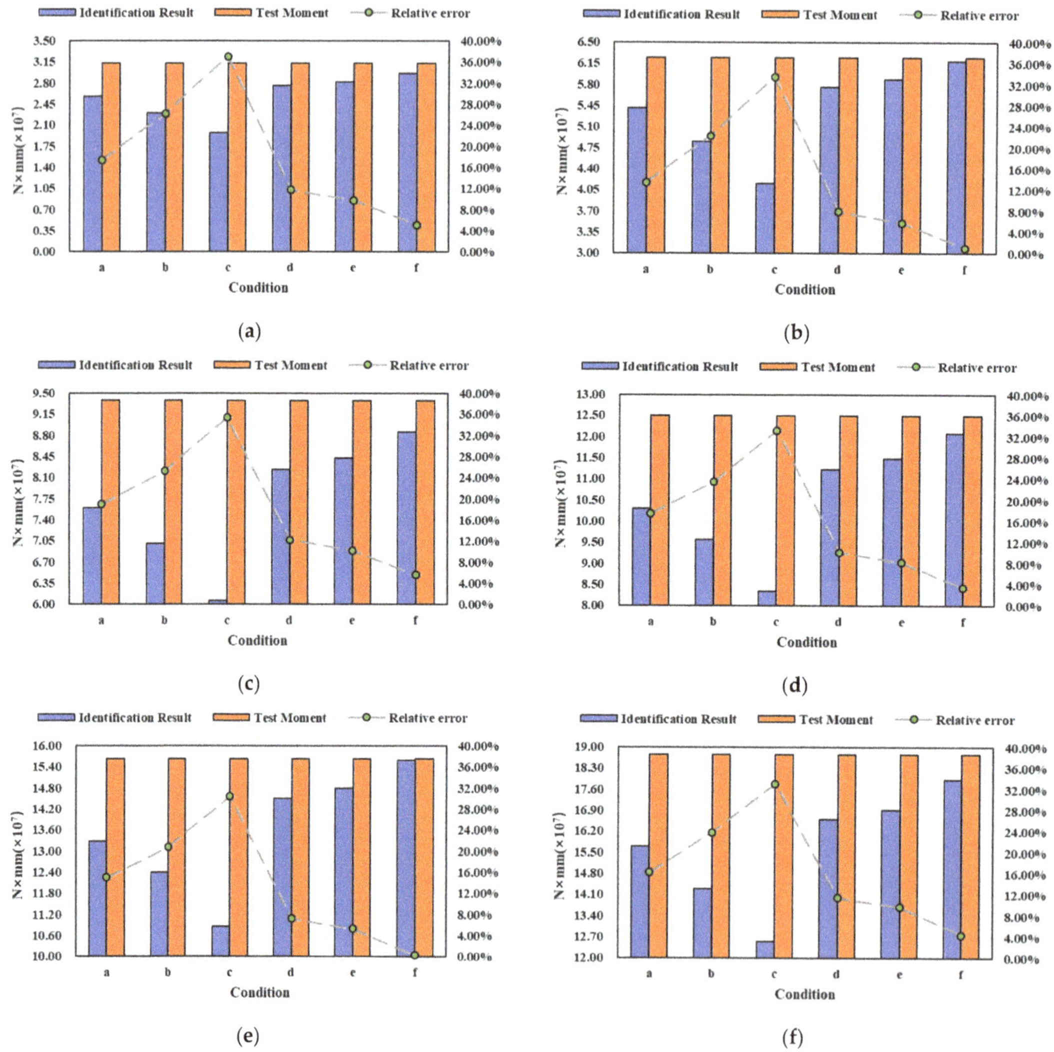

Figure 11. Bending moment identification of point failure at different structures. (**a**) 5t. (**b**) 10t. (**c**) 15t. (**d**) 20t. (**e**) 25t. (**f**) 30t.

The identification error as shown in cases a–c indicates that the failure of highly correlated measuring points can significantly impact the accuracy of bending moment identification which requires further investigation. On the other hand, it should be noted that the aforementioned conclusion may not be universally applicable across all ships. Since the identification results are obtained by inverting the influence coefficient matrix, this investigation is synonymous with the correlation between the measurement point position and the influence coefficient matrix.

According to beam theory, the strain on a section increases in proportion to its distance from the neutral axis. Consequently, influence coefficients within the matrix are also proportionately greater when measurement points are positioned further from the neutral axis. In situations where a portion of the input vector signal is lost, the related influence coefficient can have a notable impact on the accuracy of the identification results. This impact is particularly significant when the influence coefficient is higher. In the experimental model tested in this study, the neutral axis is in close proximity to P21. Consequently, the highest influence coefficient within the matrix falls on the deck stiffener (S1, S6, and S11), followed by the deck plate (P1, P4, P6, and P11), with the side plate (P16 and P25) ranking third, and the side stiffener (S14 and S19) exhibiting the lowest coefficient of influence. This order is in complete agreement with the previously established correlation conclusions.

4.2. Strain Fitting Based on Experiment Results

Removing failed measuring points is a common method for processing failed signals. To explore the improvement of the XGboost method on moment identification, this study adopts two methods to process failure signals as follows:

(a) Remove failed points.
(b) Replace the failed strain with a fitting strain. The XGboost fitting method is used to fit the failed points shown in Figures 7 and 9. The unfailed points on the whole model are used as training sets, and the failed points are used as test sets.

The effectiveness of strain fitting is examined by comparing the improvement of moment identification accuracy under various failure conditions. Strain fitting is performed on the above failure conditions.

4.2.1. Strain Fitting of Failed Points on Different Positions

Strain Fitting at Different Locations

The bending moment identification results after fitting the failed points at different locations are shown in Figure 12. After implementing strain fitting, the range of relative error decreases significantly from 4–31% to 0–5%. This enhancement in accuracy underscores the effectiveness of strain fitting in improving the identification of the bending moment. In parallel, the method of removing failed points also demonstrates an improvement in identification accuracy. Following the removal of failed points, the range of relative error decreases to 0–8%. In conditions a, b, and c, it becomes challenging to determine a superior method solely by comparing the identification errors of experimental results. However, in conditions d to g, as the number of failed points increases, the method of removing failed points yields higher identification errors. Furthermore, compared to the method of removing failed points, the relative error trend after strain fitting displays a smoother pattern. This indicates that the strain fitting method not only enhances identification accuracy but also renders the bending moment identification results more stable and reliable, albeit with a slight sacrifice in identification accuracy.

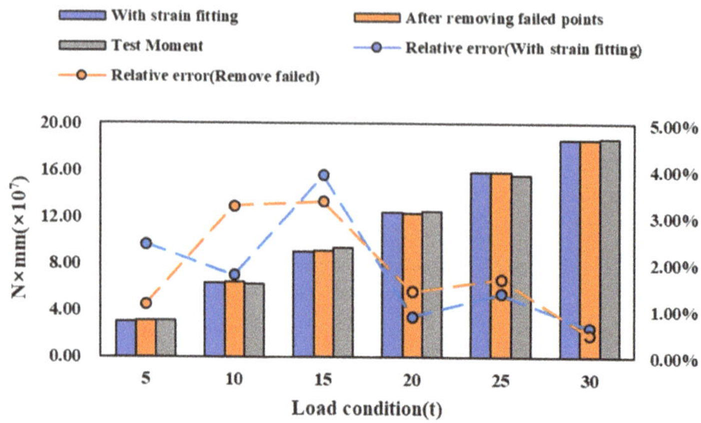

(a)

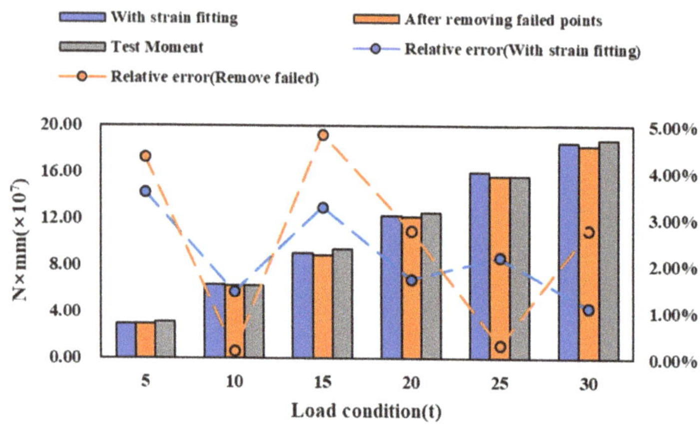

(b)

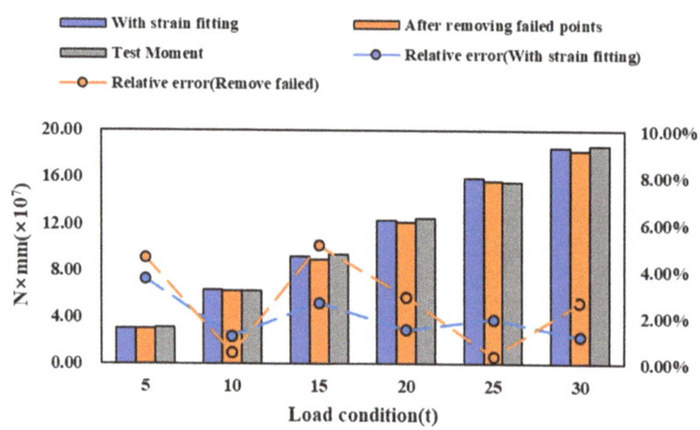

(c)

Figure 12. *Cont.*

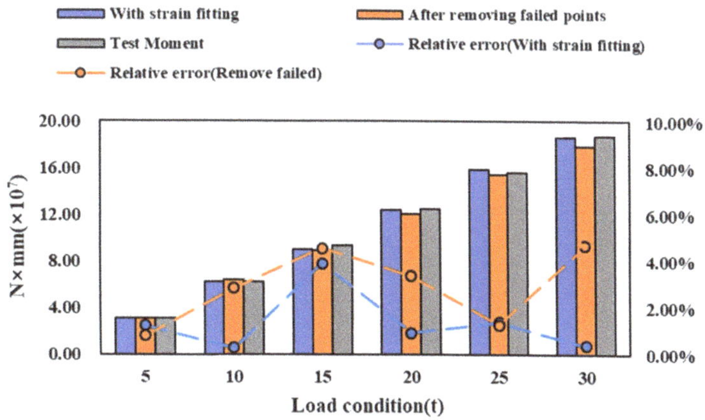

(d)

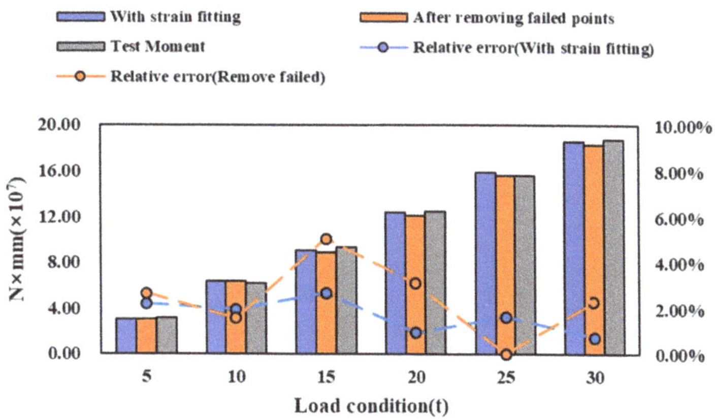

(e)

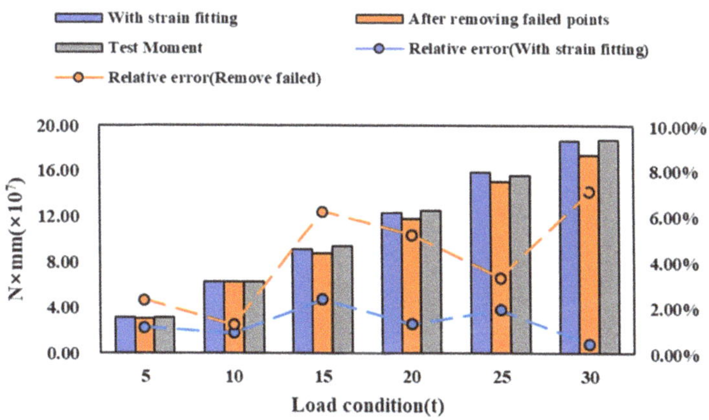

(f)

Figure 12. *Cont.*

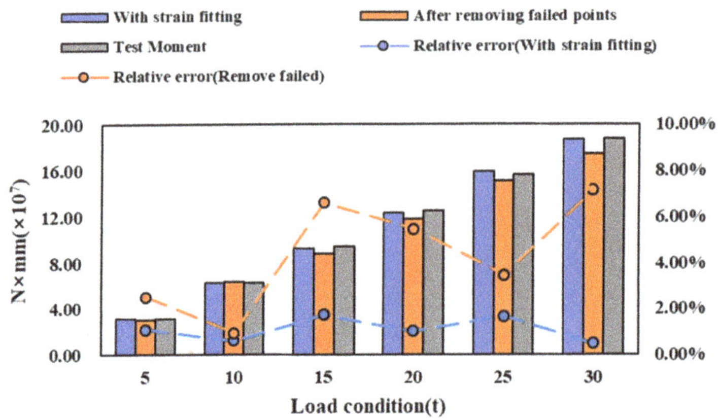

(g)

Figure 12. Bending moment identification of strain fitting at different locations. (**a**) Condition a. (**b**) Condition b. (**c**) Condition c. (**d**) Condition d. (**e**) Condition e. (**f**) Condition f. (**g**) Condition g.

The strain values of each failed point at different locations before and after XGboost fitting are shown in Figure 13. Despite a certain degree of error between the fitted strain values and the actual values, the fitted strain can still provide a reasonable estimation of the strain values at each measuring point. In conditions where failures occur on a small scale (e.g., conditions a to d), the fitted strain values for points P16 and P11 remain unchanged. However, as the scale of failure at the measuring points increases (e.g., conditions e to g), particularly when adjacent measuring points such as P11, P16, and P6 are affected, the strain fitting values differ from those observed under small-scale failure conditions due to the changes in the training set.

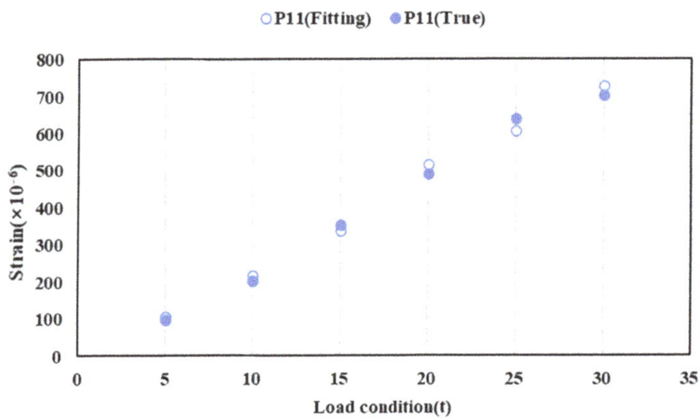

(a)

Figure 13. *Cont.*

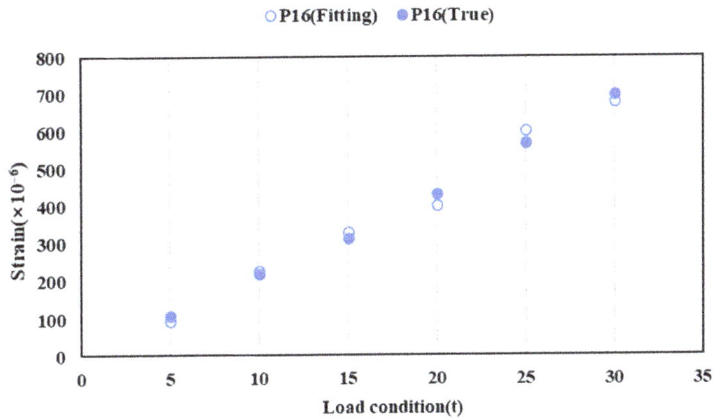

(b)

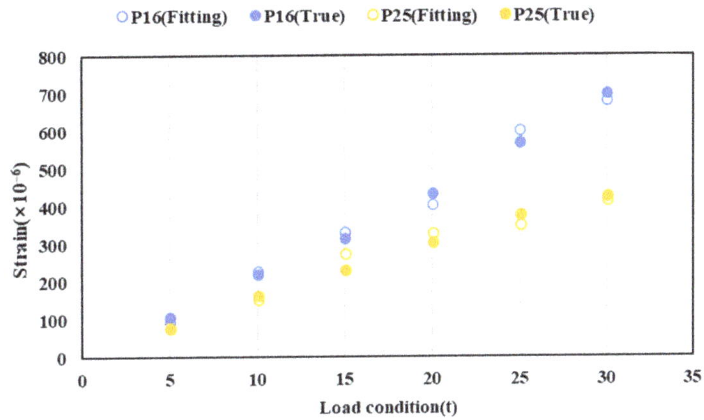

(c)

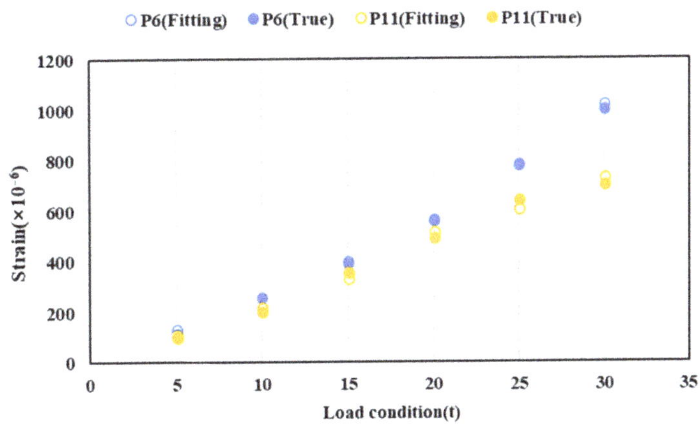

(d)

Figure 13. *Cont.*

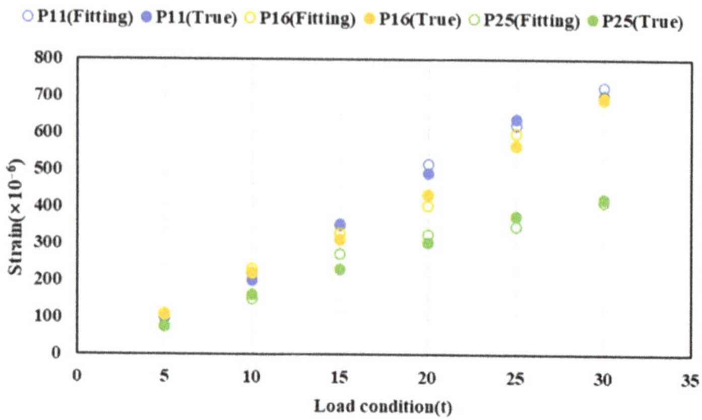

(e)

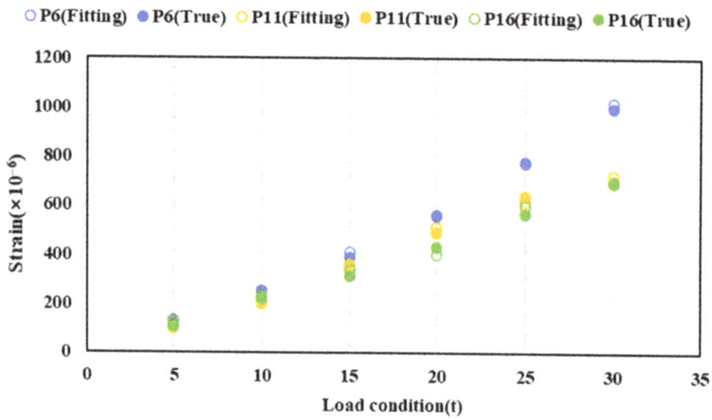

(f)

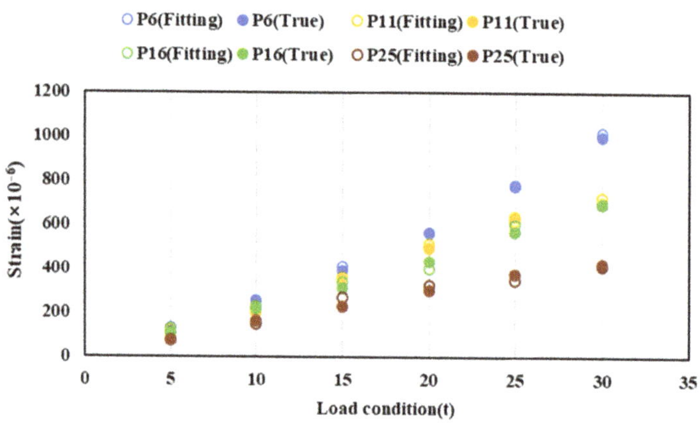

(g)

Figure 13. Strain fitting at different locations. (**a**) Condition a. (**b**) Condition b. (**c**) Condition c. (**d**) Condition d. (**e**) Condition e. (**f**) Condition f. (**g**) Condition g.

Strain Fitting at Different Structures

The bending moment identification results after fitting the failed points at different structures are shown in Figure 14. The range of relative error in the identification results decreases from 0–40% to 0–4% through the implementation of strain fitting. Removing the failed points also improves bending moment identification, and the range of relative error reduces to 0–10%. When the failed points are located on the deck stiffener (conditions b and c), the strain fitting method outperforms the method of removing the points. However, both methods prove effective in improving identification accuracy in the other conditions (conditions d to f). Taking overall performance into consideration, the strain fitting method is deemed more suitable for enhancing identification accuracy.

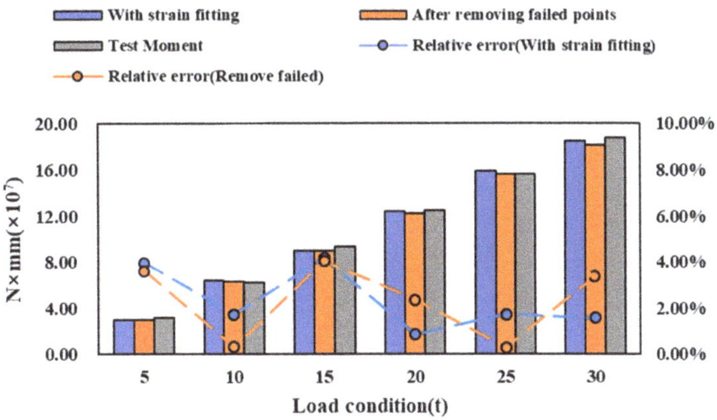

(a)

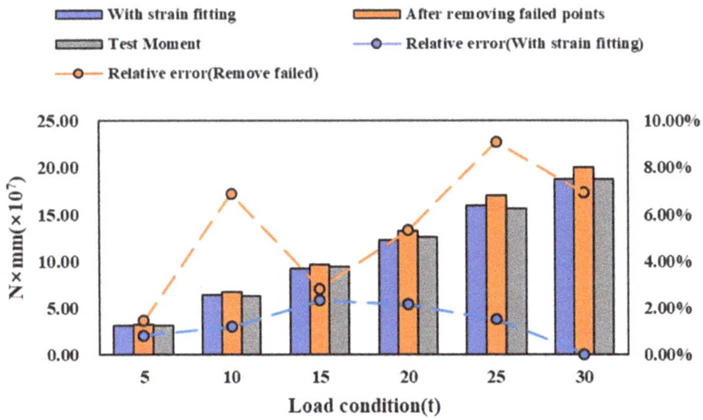

(b)

Figure 14. *Cont.*

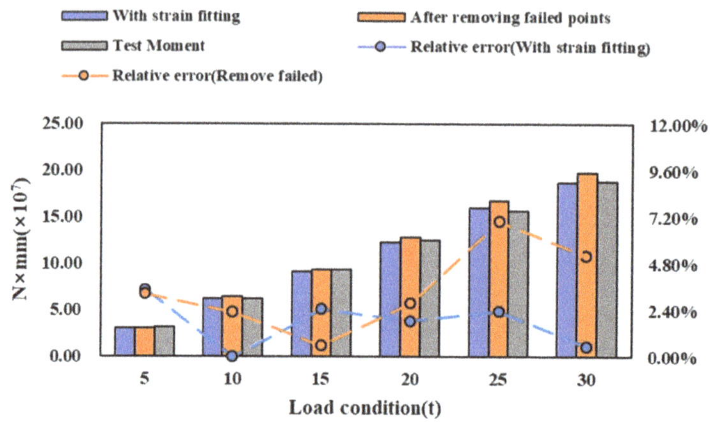

(c)

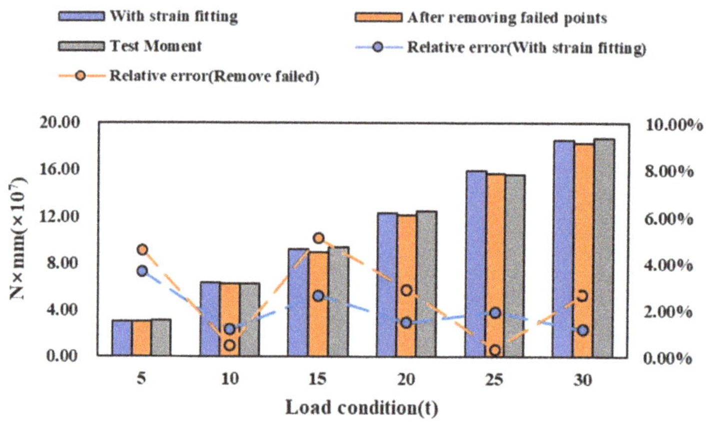

(d)

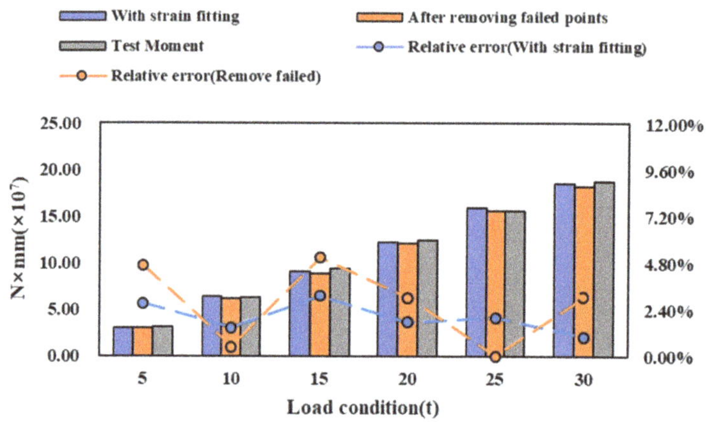

(e)

Figure 14. *Cont.*

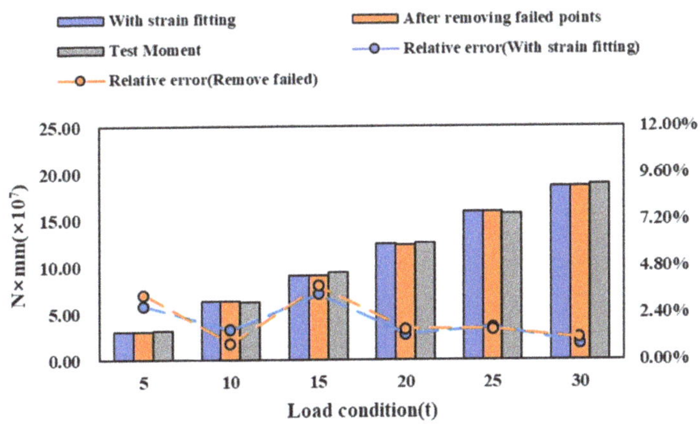

(f)

Figure 14. Bending moment identification of strain fitting at different structures. (**a**) Condition a. (**b**) Condition b. (**c**) Condition c. (**d**) Condition d. (**e**) Condition e. (**f**) Condition f.

The strain values of each failed point at different structures before and after XGboost fitting are shown in Figure 15. The estimated value is close to the valid and actual value. The fitting accuracy of condition f is worse than the others under low load conditions (condition e and f, 5t).

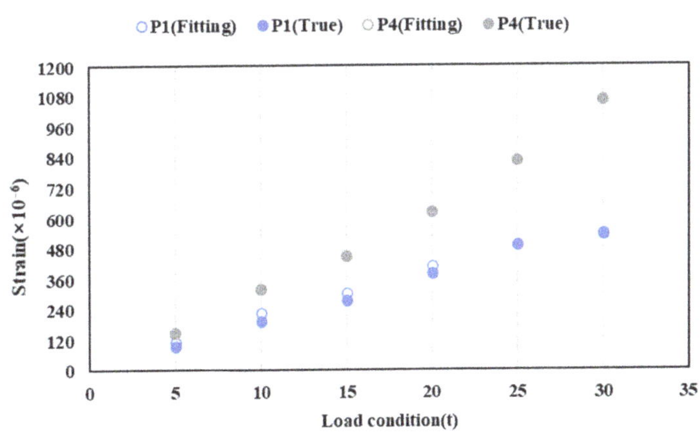

(a)

Figure 15. *Cont.*

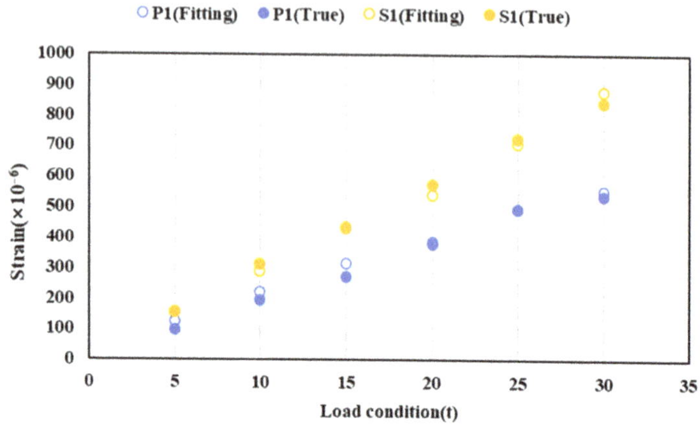

(b)

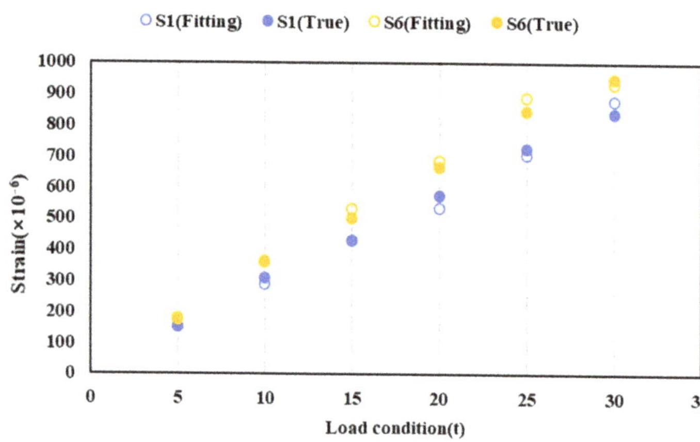

(c)

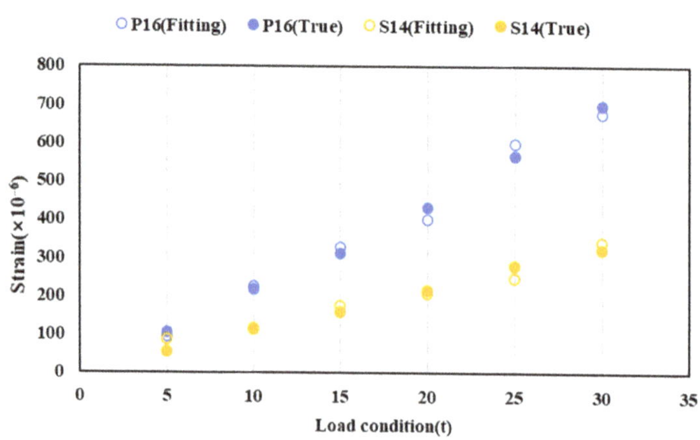

(d)

Figure 15. *Cont.*

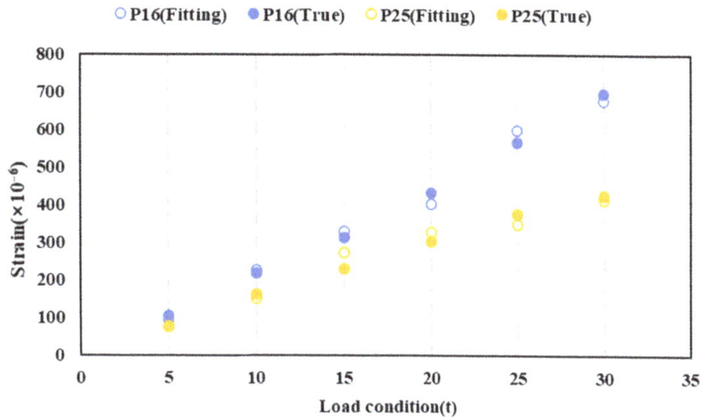

(e)

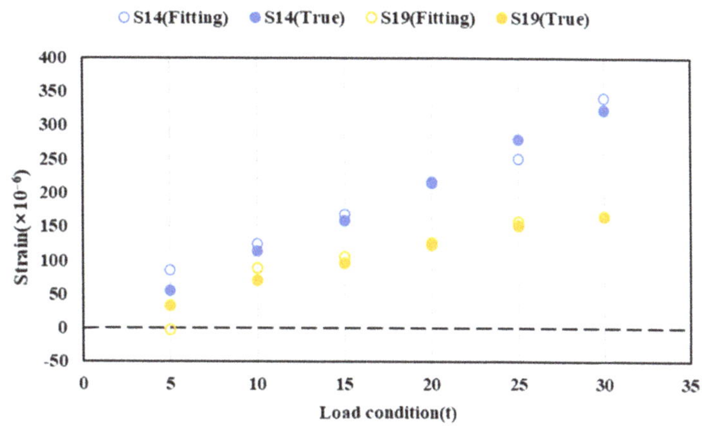

(f)

Figure 15. Strain fitting at different structures. (**a**) Condition a. (**b**) Condition b. (**c**) Condition c. (**d**) Condition d. (**e**) Condition e. (**f**) Condition f.

4.2.2. Strain Fitting on Different Sections

Based on the above exploration, the XGboost method is effective in improving bending moment identification under the failure of section 1. To further verify the effectiveness of the XGboost method, random failure conditions are designed on other sections which are shown in Figure 16.

The longitudinal bending moment identification results are shown in Figure 17. By employing two different methods to handle failure measurement points, the relative error in bending moment identification is reduced to less than 10%. In conditions characterized by a small-scale layout of measurement points (conditions a to c), the strain fitting method outperforms the method of removing measurement points when the load is low (0–15t). Removing measurement points diminishes the number of signals involved in the identification process. When the load is low and the true strain is small, the measured values become more susceptible to interference from other factors. Under such circumstances, a reduced number of measurement points leads to lower identification accuracy. In contrast, regardless of the layout of measuring points, the strain fitting method consistently exhibits a more stable trend in identification error.

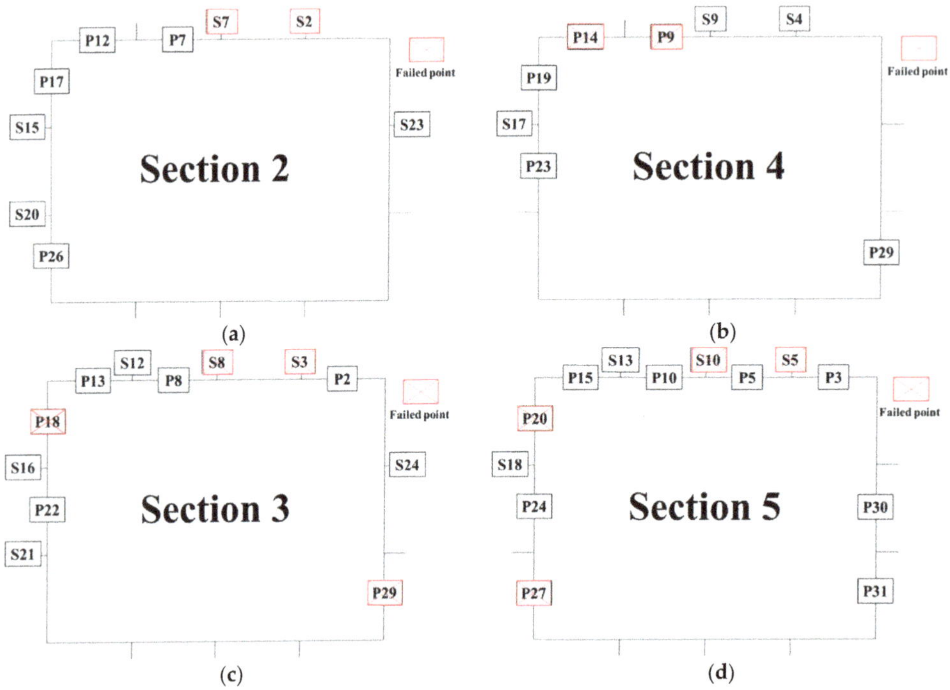

Figure 16. Conditions of lack of points. (**a**) Two points deleted on the deck stiffener. (**b**) Two points deleted on the side plate. (**c**) Four points deleted (deck: two; side: two). (**d**) Four points deleted (deck: two; side: two).

The fitting strain of measuring points is shown in Figure 18. The strain of other conditions is almost consistent with the actual values. Based on the above analysis, the XGboost method can effectively improve the identification of the longitudinal bending moment in the case of failure of the measuring points. Compared to common methods of removing failed points, the advantages of this method are as follows:

(a) It makes the identification results more stable.
(b) It shows a better improvement capability in cases of high-correlation measurement point failure and small-scale measurement point layouts.
(c) Using strain fitting can estimate the approximate strain of failed points, which is more conducive to structural safety assessment.

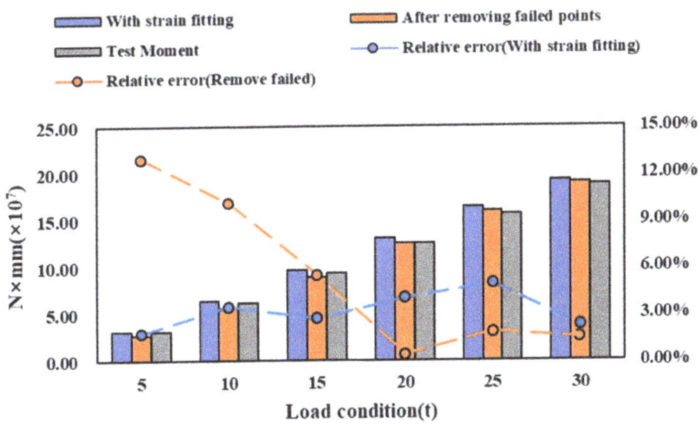

(a)

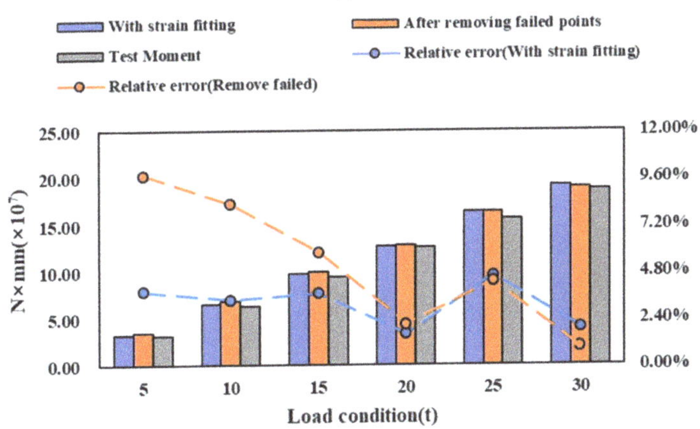

(b)

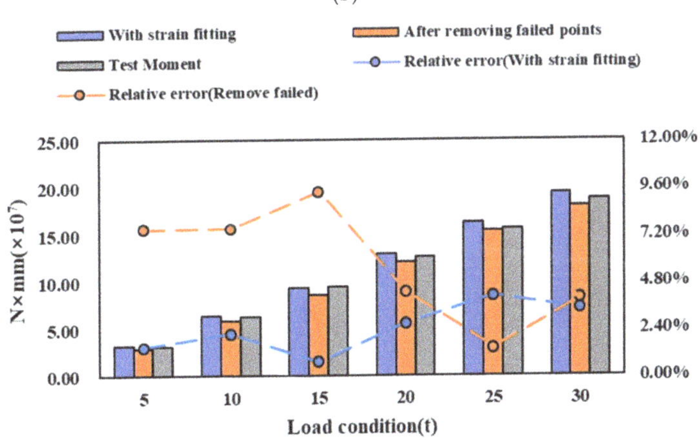

(c)

Figure 17. *Cont.*

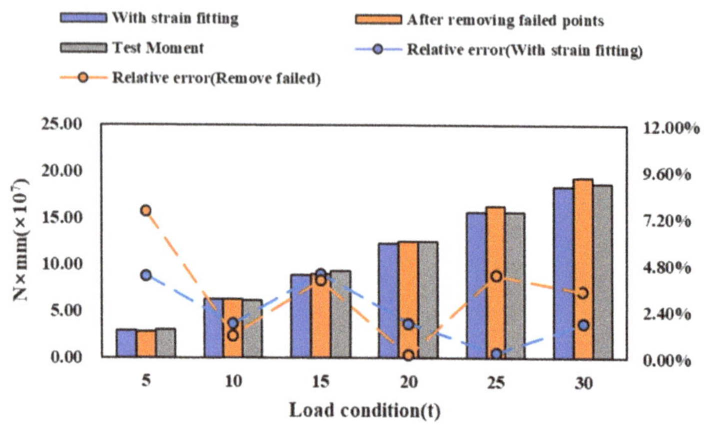

(**d**)

Figure 17. Longitudinal bending moment of strain fitting under different layouts of points. (**a**) Condition a. (**b**) Condition b. (**c**) Condition c. (**d**) Condition d.

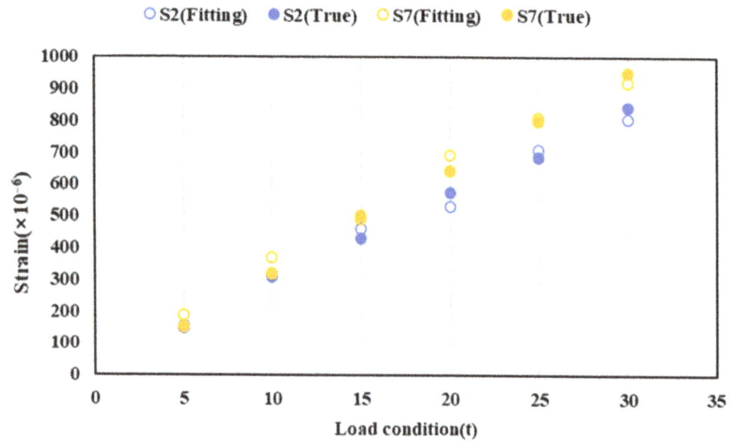

(**a**)

Figure 18. *Cont.*

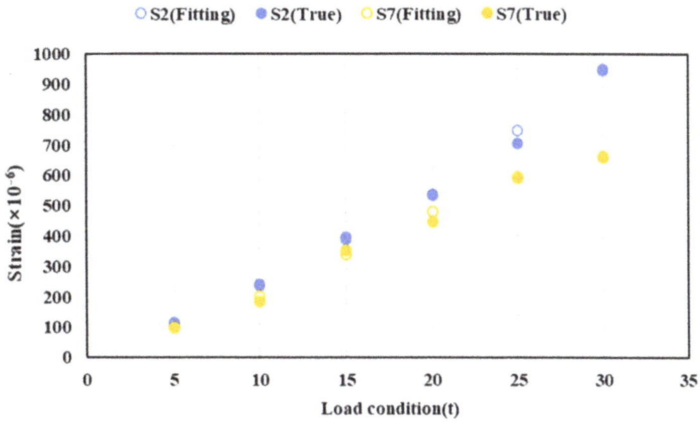

(**b**)

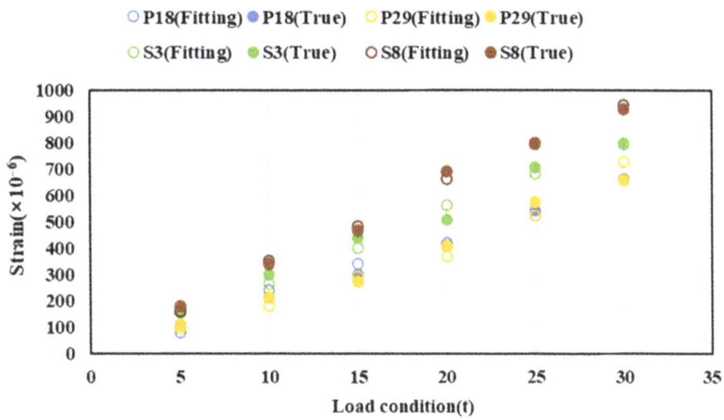

(**c**)

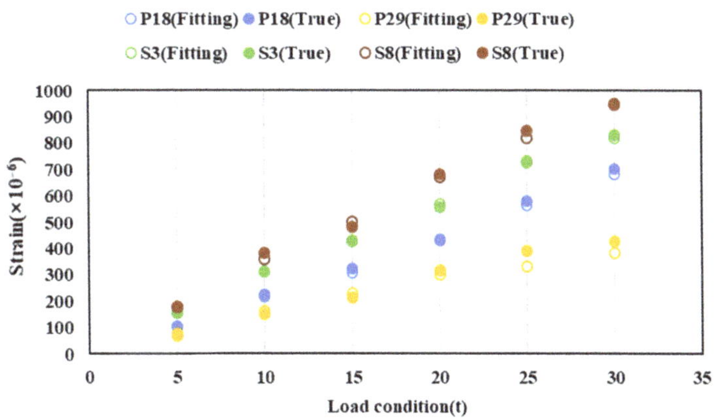

(**d**)

Figure 18. Strain fitting under the layout of points. (**a**) Condition a. (**b**) Condition b. (**c**) Condition c. (**d**) Condition d.

4.3. Strain Fitting Based on Numerical Analysis

Based on the above experimental results, discrepancies persist between the fitted strain values and the actual values, and it is hard to determine whether these arise from measurement errors or the algorithm itself, based solely on the experimental results. Moreover, in addition to the longitudinal bending moment, the ship is affected by other types of load such as the lateral bending moment and torsion during navigation [12]. This makes the strain distribution on the different sections more complex, and the difficulty of fitting increases. Therefore, further numerical research is necessary, utilizing finite element analysis to provide additional insights.

4.3.1. Finite Element Model

The finite element model consists of shell elements. As the preloading cycles were used to eliminate the residual stresses, the numerical analysis did not consider the influence caused by residual stress. For the numerical analysis, the support boundary of the finite element model is free only with the x-displacement at one end and the y-rotation at both ends. This condition is to generate the actual boundary of the box girder in the experiment. The bending loads are applied at the bulkheads of the model. In this study, the numerical results of longitudinal strain on the sections are extracted to establish the moment–strain transfer matrix.

4.3.2. Convergence Study on the Mesh Size of Finite Elements

This study adopted ABAQUS 14.0 for numerical analysis. To study the convergence of the mesh size of finite element model, nine sizes (i.e., 10, 15, 20, 25, 30, 35, 40, 50, and 80 mm) are set to analyze the mean square error (*MSE*) between the solutions of the z-displacement in the middle of the box girder under two adjacent sizes. In this study, the *MSE* can be calculated as follows [17]:

$$MSE = \frac{1}{m}\sum_{i=0}^{m}\left(u_i^{\mathrm{f}} - u_i^{\mathrm{l}}\right)^2 \tag{22}$$

where u_i^{f} is the z-displacement of the former size and u_i^{l} is the z-displacement of the latter size. As shown in Figure 19, when a bending load of 1×10^5 N is applied to the box girder, the *MSE* increases with the increase of the mesh size. When the mesh size is less than 40 mm, the *MSE* grows slowly. Therefore, the mesh size of 20 mm is adopted for numerical analysis. The number of elements for the model is 11,511.

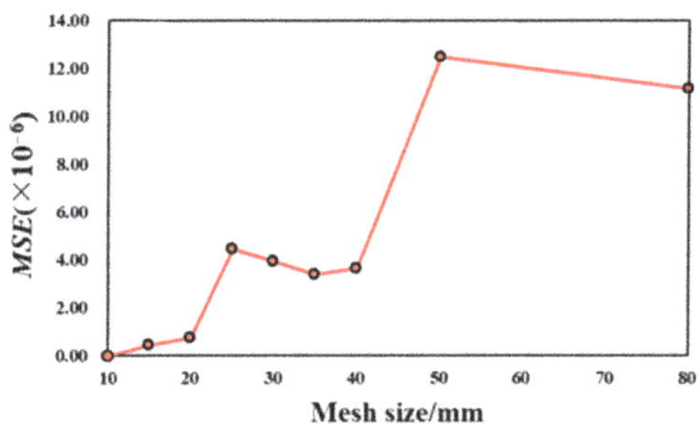

Figure 19. MSE under different mesh sizes.

4.3.3. Analysis of the Sources of Fitting Errors

A finite element model was adopted to explore the influence of lateral bending moment and torsion on the XGboost fitting method. Based on the above research, under the longitudinal bending condition, the effect of the failed measuring points on the deck stiffener is greater than that on the deck plate, and the opposite is true on the side. Thus, two failed points are designed on the deck stiffener and the others are on the side plate, as shown in Figure 20.

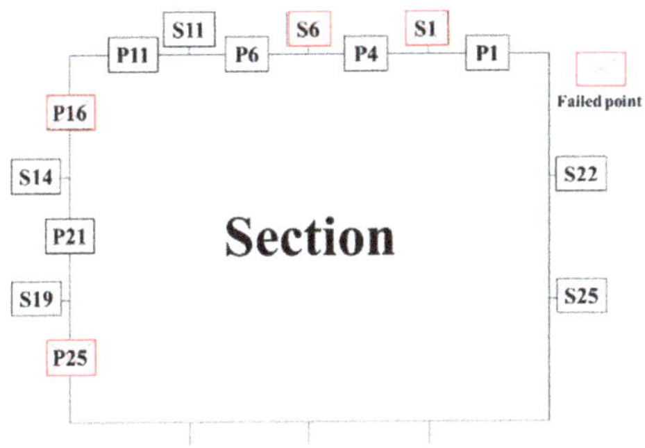

Figure 20. Measuring point failure under different loading types.

In order to identify the longitudinal bending moment, the influence of other loads is required to be restrained as follows [14,15]:

$$S = \begin{bmatrix} A_V & A_H & A_T \end{bmatrix} \begin{bmatrix} M_V \\ M_H \\ M_T \end{bmatrix} \quad (23)$$

where M_V is the longitudinal bending moment of the section, M_H is the lateral bending moment of the section, and M_T is the torque of the section. $\begin{bmatrix} A_V & A_H & A_T \end{bmatrix}$ is the longitudinal moment–strain transfer matrix composed of longitudinal bending moment–strain transfer vectors (A_V), lateral bending moment–strain vectors (A_H), and torque–strain vectors (A_T). To investigate the reason of the experimental errors, the conditions of loading under longitudinal bending are designed as shown in Table 2.

Table 2. Conditions of loading without interfering.

Condition No.	Longitudinal Bending Moment (N × mm)
1	1.000×10^8
2	5.000×10^7
3	2.500×10^7

The results of longitudinal bending moment identification after strain fitting are shown in Table 3. The relative error is below 0.1% which is not shown in the table. The strain fitting results, which are almost consistent with the actual values, are shown in Figure 21. This result suggests the reliability of the finite element model, with most of the fitting errors in the experimental results originating from measurement errors.

Table 3. Identification results without interfering.

Condition No.	Identified Moment (N × mm)	True Moment (N × mm)
1	9.994×10^8	1.000×10^8
2	4.997×10^7	5.000×10^7
3	2.499×10^7	2.500×10^7

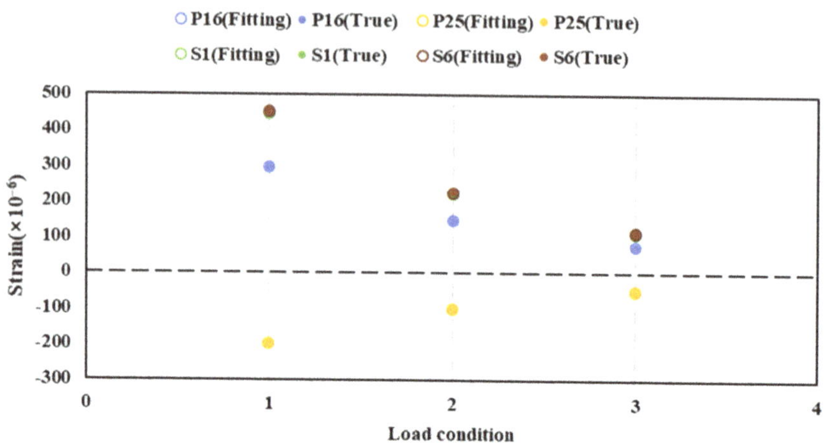

Figure 21. Strain fitting without interfering.

4.3.4. Impact of Lateral Bending

To explore the impact of lateral bending on the XGboost fitting method, lateral bending is applied to the model and it is enhanced. The loading conditions are shown in Table 4. We designed some cases where the lateral bending moment is greater than the longitudinal bending moment. The finite element analysis results of condition 1 are shown in Figure 22a. The maximum longitudinal strain is situated in the corner of the box girder, whereas the strain in the middle of the deck and sides is relatively lower. The lateral strain and shear strain experienced by the model are significantly smaller in magnitude compared to the longitudinal strain. The longitudinal strains on the model deck display a linear distribution, indicating that the influence of lateral bending cannot be disregarded. To account for the impact of lateral bending moment, the above mentioned influence coefficient matrix (Equation (23)) can be utilized to decrease its effects.

Table 4. Conditions of loading under lateral bending.

Condition No.	Lateral Bending Moment (N × mm)	Longitudinal Bending Moment (N × mm)
1	2.500×10^7	1.000×10^8
2	5.000×10^7	5.000×10^7
3	1.000×10^8	2.500×10^7

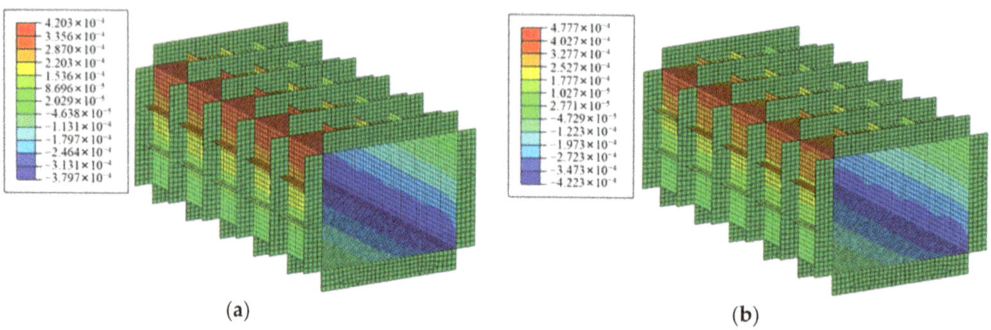

Figure 22. Longitudinal strain under different conditions. (**a**) Lateral bending. (**b**) Torsion-lateral bending.

The longitudinal bending moment identification results under different conditions are shown in Table 5. After strain fitting, the relative error of bending moment identification results is very small. Figure 23 shows the strain fitting results under different conditions. The fitted value is almost consistent with the actual value. Therefore, the XGboost method is effective under the condition of lateral-longitudinal bending.

Table 5. Identification results under lateral bending.

Condition No.	Identified Moment (N × mm)	True Moment (N × mm)
1	1.000×10^8	1.000×10^8
2	5.000×10^7	5.000×10^7
3	2.498×10^7	2.500×10^7

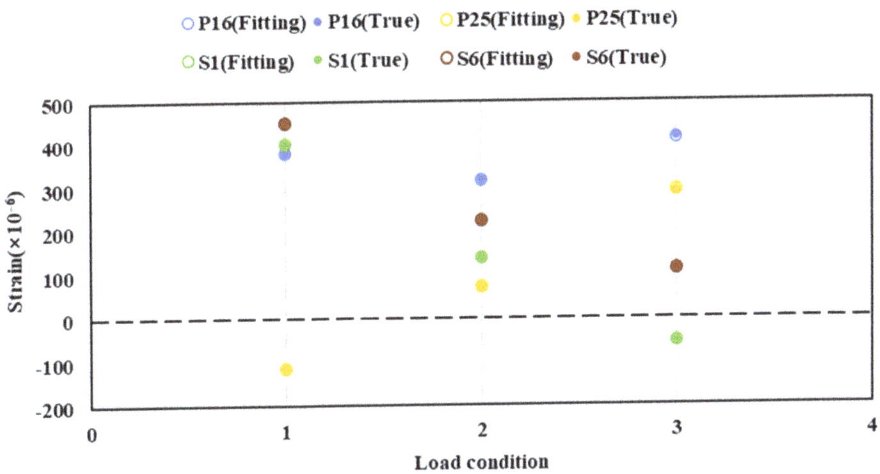

Figure 23. Strain fitting under lateral bending.

4.3.5. Impact of Torsion-Lateral Bending

To explore the influence of torsion on the identification of the longitudinal bending moment based on the XGboost fitting method, torsion is applied to the model as shown in Table 6.

Table 6. Conditions of loading under torsion-lateral bending.

Condition No.	Lateral Bending Moment (N × mm)	Longitudinal Bending Moment (N × mm)	Torque (N × mm)
1	2.5×10^7	1×10^8	5×10^7
2	5×10^7	5×10^7	5×10^7
3	1×10^8	2.5×10^7	5×10^7

Figure 24 shows the relationship between the longitudinal strain distribution of the deck under torsion-lateral bending and that under lateral bending on a section. The relationship between cases is linear. The slope of the liner function in Figure 24 is 1.002, and the intercept is −0.3921. The slope close to 1 and the low intercept indicate that the strain in the two cases is close to being the same. Therefore, the influence of torsion on the longitudinal strain of the model is small.

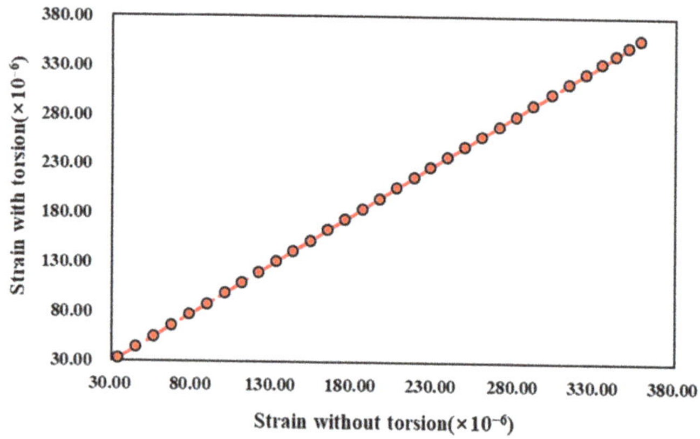

Figure 24. Relationship between strain distribution of two cases.

The longitudinal bending moment identification results under different conditions are shown in Table 7 and the strain fitting results are shown in Figure 25. The identification accuracy and fitting accuracy are good. For this model, torsion has little effect on the identification of the longitudinal bending moment and strain fitting.

Table 7. Identification results under torsion-lateral bending.

Condition No.	Identified Moment (N × mm)	True Moment (N × mm)
1	1.000×10^8	1.000×10^8
2	5.000×10^7	5.000×10^7
3	2.499×10^7	2.500×10^7

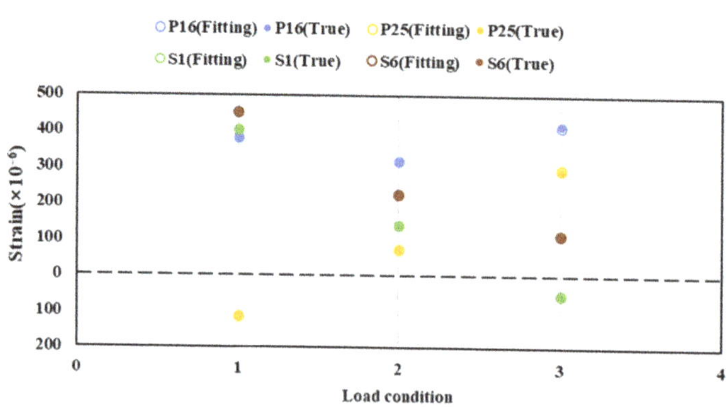

Figure 25. Strain fitting under torsion-lateral bending.

4.4. Investigation of Fitting Schemes

Based on the above exploration, the XGboost fitting method has excellent performance in improving the accuracy of moment identification. However, in practical moment monitoring, parts of the monitored sections are far away. Using strain signals from all sections as the training set might reduce the accuracy of longitudinal bending moment identification. Therefore, the training set requires to be streamlined, and its connection with fitting value should be further investigated. To this end, based on the FEM analysis under longitudinal bending-lateral bending-torsion (conditions in Section 4.3.5), training conditions are proposed as shown in Table 8 to investigate the fitting schemes. The section number in the table represents the section included in the training set. The layout of measuring points

the deck stiffener or deck plate. These suggestions ensure the establishment of a training set with strong features to effectively fit failed points, which align with the guidelines set by classification societies [6,7]. On the other hand, for strain fitting based on XGboost, arranging the measurement points into two sections, which is a common method employed in practical longitudinal bending moment monitoring [12,13,34–36], is deemed sufficient. Consequently, the strain fitting method does not impose any special requirements on the layout of measurement points, making it widely applicable for longitudinal bending moment monitoring systems on ships. Additionally, apart from improving the identification of the longitudinal bending moment, the XGboost fitting method extends its potential applications in the supplement strain values to help load identification (such as ice load identification [37,38]) or deformation monitoring (such as iFEM [39,40]).

6. Conclusions

In this study, a novel method, the XGboost fitting method, is proposed to reduce the impact of the failure of measuring points on the identification of the longitudinal bending moment. Suggestions for point arrangement and the scheme for strain fitting were deeply investigated by experimental and numerical analysis. The main conclusions are as follows:

(1) The failure of highly correlated measuring points has a significant impact on the accuracy of identifying the total longitudinal bending moment. This result indicates that the points on the deck stiffener have the highest correlation, followed by the deck plate, the side plate, and the side stiffener.
(2) Based on the experimental results, the XGboost fitting method can effectively improve the identification accuracy of the longitudinal bending moment in the case of the failure of measuring points. Compared to common methods of removing failed points, the XGboost method makes the identification results more stable and shows a better improvement capability in the case of high-correlation measurement point failure. In this study, after using the strain fitting method, the relative error decreased to less than 6%, and that was less than 5% in most conditions. By contrast, the relative error of some conditions still reaches 10% by removing failed points.
(3) There is a fitting error between the fitted strain and the measured strain due to the measurement error. However, the trend of the fitted values is almost consistent with the actual values. Therefore, this method is able to estimate strain to an approximate value.
(4) Based on the FEM analysis, the XGboost fitting method is effective for complex load conditions. After further numerical investigation, it is suggested as the optimal fitting scheme which adopts the failed section and its adjacent section as the training set.

Author Contributions: Conceptualization, G.X. and J.G.; Methodology, G.X. and J.G.; Software, G.X.; Formal analysis, G.X.; Data curation, J.L.; Writing—original draft, G.X.; Writing—review & editing, H.L.; Supervision, W.W. All authors have read and agreed to the published version of the manuscript.

Funding: This research received no external funding.

Data Availability Statement: Data are contained within the article.

Conflicts of Interest: The authors declare no conflict of interest.

References

1. Katsoudas, A.S.; Silionis, N.E.; Anyfantis, K.N. Structural Health Monitoring for Corrosion Induced Thickness Loss in Marine Plates Subjected to Random Loads. *Ocean Eng.* **2023**, *273*, 114037. [CrossRef]
2. Rahgozar, R.; Bitaraf, M. A Summary Evaluation of Output-Only Damage-Sensitive Features for Structural Health Monitoring of Offshore Platforms Subjected to Ambient Loads. *Ocean Eng.* **2022**, *266*, 112892. [CrossRef]
3. Gordo, J.; Guedes Soares, C.; Faulkner, D. Approximate Assessment of the Ultimate Longitudinal Strength of the Hull Girder. *J. Ship Res.* **1996**, *40*, 60–69. [CrossRef]
4. Tanaka, Y.; Ando, T.; Anai, Y.; Yao, T.; Fujikubo, M.; Iijima, K. Longitudinal Strength of Container Ships Under Combined Torsional and Bending Moments. In Proceedings of the Nineteenth International Offshore and Polar Engineering Conference, Osaka, Japan, 21–26 July 2009; OnePetro: Richardson, TX, USA, 2009.

5. American Bureau of Shipping. Guide for Hull Condition Monitoring System 2020. Available online: https://ww2.eagle.org/content/dam/eagle/rules-and-guides/current/conventional_ocean_service/73_Hull_Condition_Monitoring_2016/hull-condition-monitoring-guide-july20.pdf (accessed on 23 February 2023).
6. China Classification Society. Rules for Intelligent Ship. Available online: https://www.ccs.org.cn/ccswz/specialDetail?id=201900001000009739 (accessed on 23 February 2023).
7. Nippon Kaiji Kyokai. Rules for Hull Monitoring Systems. Available online: https://www.classnk.or.jp/account/zh/Rules_Guidance/ssl/tech_rules.aspx (accessed on 24 February 2023).
8. Li, M.; Boulougouris, E.; Lazakis, I.; Theotokatos, G. Wave-Induced Vertical Bending Moment Estimation by Onboard Tiltmeters Units on Container Ship. In Proceedings of the International Conference on Maritime Safety and Operations, Glasgow, UK, 13–14 October 2016; Available online: https://strathprints.strath.ac.uk/58370/1/Li_etal_MSO2016_Wave_induced_vertical_bending_moment_estimation_by_onboard_tiltmeters.pdf (accessed on 24 February 2023).
9. Tatsumi, A.; Iijima, K.; Fujikubo, M. Estimation of Still-Water Bending Moment of Ship Hull Girder Using Beam Finite Element Model and Ensemble Kalman Filter. In Proceedings of the ASME 2022 41st International Conference on Ocean, Offshore and Arctic Engineering, Hamburg, Germany, 5–10 June 2022; American Society of Mechanical Engineers Digital Collection: Washington, DC, USA, 2022.
10. Cusano, G.; La Marca, L.S. Evaluation and Forecasting of Elapsed Fatigue Life of Ship Structures by Analyzing Data from Full Scale Ship Structural Monitoring. *J. Shipp. Ocean Eng.* **2015**, *5*, 59–74. [CrossRef]
11. Moreira, L.; Soares, C.G. Neural Network Model for Estimation of Hull Bending Moment and Shear Force of Ships in Waves. *Ocean Eng.* **2020**, *206*, 107347. [CrossRef]
12. Yu, H.; Ha, M.; Choi, J.; Tai, J.S. Design and Implementation of a Comprehensive Full-Scale Measurement System for a Large Container Carrier. In Proceedings of the Design & Operation of Container Ships conference, London, UK, 22–23 November 2006; RINA: Genoa, Italy, 2006; pp. 51–60.
13. Liu, D. Fatigue Assessment Methodfor Hull Structure Based on Strength Monitoring System. Master's Thesis, Harbin Engineering University, Harbin, China, 2018.
14. Li, J. Research on Load Inversion Method and Model Experiment of Ultra Large Ships. Master's Thesis, Harbin Engineering University, Harbin, China, 2021.
15. Li, J.; Dong, W.; Liu, N.; Feng, G.; Ren, H. Theoretical and Experimental Study on Load Inversion of Ship Hull Girder. *Shipengineering* **2022**, *44*, 41–46. [CrossRef]
16. Liu, Y.; Xing, Y.; Shen, Z. Identification of Cross-section Loads Based on Measured Strain of Missile Body. *Acta Amamentaii* **2016**, *37*, 332–337.
17. Wang, J.; Chen, X.; Duan, Q.; Ji, S. Eliminating the Influence of Measuring Point Failure in Ice Load Identification of Polar Ship Structures. *Ocean Eng.* **2022**, *261*, 112082. [CrossRef]
18. Shrive, F.M.; Stuart, H.; Quan, H.; Ghali, W.A. Dealing with Missing Data in a Multi-Question Depression Scale: A Comparison of Imputation Methods. *BMC Med. Res. Methodol.* **2006**, *6*, 57. [CrossRef] [PubMed]
19. Troyanskaya, O.; Cantor, M.; Sherlock, G.; Brown, P.; Hastie, T.; Tibshirani, R.; Botstein, D.; Altman, R.B. Missing Value Estimation Methods for DNA Microarrays. *Bioinformatics* **2001**, *17*, 520–525. [CrossRef] [PubMed]
20. Stekhoven, D.J.; Buhlmann, P. MissForest-Non-Parametric Missing Value Imputation for Mixed-Type Data. *Bioinformatics* **2012**, *28*, 112–118. [CrossRef]
21. Chen, T.; Guestrin, C. XGBoost: A Scalable Tree Boosting System. In Proceedings of the 22nd ACM SIGKDD International Conference on Knowledge Discovery and Data Mining, San Francisco, CA, USA, 13–17 August 2016; pp. 785–794.
22. Kankanamge, K.D.; Witharanage, Y.R.; Withanage, C.S.; Hansini, M.; Lakmal, D.; Thayasivam, U. Taxi Trip Travel Time Prediction with Isolated XGBoost Regression. In Proceedings of the 2019 Moratuwa Engineering Research Conference (MERCon), Moratuwa, Sri Lanka, 3–5 July 2019; pp. 54–59.
23. Li, W.; Yin, Y.; Quan, X.; Zhang, H. Gene Expression Value Prediction Based on XGBoost Algorithm. *Front. Genet.* **2019**, *10*, 1077. [CrossRef] [PubMed]
24. Saad-Eldeen, S.; Garbatov, Y.; Soares, C.G. Experimental Assessment of the Ultimate Strength of a Box Girder Subjected to Severe Corrosion. *Mar. Struct.* **2011**, *24*, 338–357. [CrossRef]
25. Xu, W.; Iijima, K.; Wada, R.; Fujikubo, M. Experimental Evaluation of the Post-Ultimate Strength Behavior of a Ship's Hull Girder in Waves. *J. Mar. Sci. Appl.* **2012**, *11*, 34–43. [CrossRef]
26. Saad-Eldeen, S.; Garbatov, Y.; Guedes Soares, C. Effect of Corrosion Severity on the Ultimate Strength of a Steel Box Girder. *Eng. Struct.* **2013**, *49*, 560–571. [CrossRef]
27. Akhras, G.; Gibson, S.; Yang, S.; Morchat, R. Ultimate Strength of a Box Girder Simulating the Hull of a Ship. *Can. J. Civ. Eng.* **1998**, *25*, 829–843. [CrossRef]
28. Hughes, O. *Ship Structural Design: A Rationally-Based, Computer-Aided Optimization Approach*; Wiley-Interscience: New York, NY, USA, 1983.
29. Hauser, R. *Line Search Methods for Unconstrained Optimisation, Lecture 8*; Oxford University Computing Laboratory: Oxford, UK, 2007.
30. Kong, S.; Cui, H.; Tian, Y.; Ji, S. Identification of Ice Loads on Shell Structure of Ice-Going Vessel with Green Kernel and Regularization Method. *Mar. Struct.* **2020**, *74*, 102820. [CrossRef]

31. Ma, H.; Xiong, Q.; Wang, D. Experimental and Numerical Study on the Ultimate Strength of Stiffened Plates Subjected to Combined Biaxial Compression and Lateral Loads. *Ocean Eng.* **2021**, *228*, 108928. [CrossRef]
32. Santhanam, R.; Uzir, N.; Raman, S.; Banerjee, S. Experimenting XGBoost Algorithm for Prediction and Classification of Different Datasets. In Proceedings of the National Conference on Recent Innovations in Software Engineering and Computer Technologies (NCRISECT), Pittsburgh, PA, USA, 23–24 March 2017.
33. Zhang, X.; Yan, C.; Gao, C.; Malin, B.A.; Chen, Y. Predicting Missing Values in Medical Data Via XGBoost Regression. *J. Healthc. Inform. Res.* **2020**, *4*, 383–394. [CrossRef]
34. Choi, J.; Kang, Y. Two-Plane Hull Girder Stress Monitoring System for Container Ship. *J. Ship Ocean Technol.* **2004**, *8*, 17–25.
35. Park, J.; Kang, Y.-T. Strain Decomposition Method in Hull Stress Monitoring System for Container Ship. *J. Ship Ocean Technol.* **2003**, *7*, 56–65.
36. Yu, H.; Choi, J.; Park, G.; Han, S.; Tai, S.; Ha, M. Full Scale Measurement of a Large Container Carrier on the Far East—Europe Route. In Proceedings of the SNAME Maritime Convention, Houston, TX, USA, 15–17 October 2008; OnePetro: Richardson, TX, USA, 2008.
37. Suominen, M.; Kujala, P.; Romanoff, J.; Remes, H. Influence of Load Length on Short-Term Ice Load Statistics in Full-Scale. *Mar. Struct.* **2017**, *52*, 153–172. [CrossRef]
38. Lee, T.; Lee, J.; Kim, H.; Rim, C. Field Measurement of Local Ice Pressures on the ARAON in the Beaufort Sea. *Int. J. Nav. Archit. Ocean Eng.* **2014**, *6*, 788–799. [CrossRef]
39. Kefal, A.; Oterkus, E. Displacement and Stress Monitoring of a Chemical Tanker Based on Inverse Finite Element Method. *Ocean Eng.* **2016**, *112*, 33–46. [CrossRef]
40. Kefal, A.; Oterkus, E. Displacement and Stress Monitoring of a Panamax Containership Using Inverse Finite Element Method. *Ocean Eng.* **2016**, *119*, 16–29. [CrossRef]

Journal of
Marine Science and Engineering

MDPI

Article

Displacement Values Calculation Method for Ship Multi-Support Shafting Based on Transfer Learning

Yibin Deng [1,*], Yuefan Li [2], Hanhua Zhu [1] and Shidong Fan [2]

[1] School of Naval Architecture, Ocean and Energy Power Engineering, Wuhan University of Technology, Wuhan 430063, China; hh.zhu@163.com
[2] School of Transportation and Logistics Engineering, Wuhan University of Technology, Wuhan 430063, China; liyuefan@whut.edu.cn (Y.L.); sdfan@whut.edu.cn (S.F.)
* Correspondence: dengyb@whut.edu.cn

Abstract: Deviations between the design and actual shafting occur due to limitations in ship construction accuracy. Consequently, accurately obtaining the relationship between the actual shafting load and displacement relationship based on the design shafting becomes challenging, leading to inaccurate solutions for bearing displacement values and low alignment efficiency. In this research article, to address the issue of incomplete actual shafting data, a transfer learning-based method is proposed for accurate calculation of bearing displacement values. By combining simulated data from the design shafting with measured data generated during the adjustment process of the actual shafting, higher accuracy can be achieved in calculating bearing displacement values. This research utilizes a certain shafting as an example to carry out the application of the bearing displacement value calculation method. The results show that even under the action of shafting deviation, the actual shafting load and displacement relationship model can become more and more accurate with the shafting adjustment process, and the accuracy of bearing displacement values calculation becomes higher and higher. This method contributes to obtaining precise shafting adjustment schemes, thereby enhancing alignment quality and efficiency of ship shafting.

Keywords: multi-support shafting; shafting alignment; transfer learning; bearing displacement value calculation

Citation: Deng, Y.; Li, Y.; Zhu, H.; Fan, S. Displacement Values Calculation Method for Ship Multi-Support Shafting Based on Transfer Learning. *J. Mar. Sci. Eng.* **2024**, *12*, 36. https://doi.org/10.3390/jmse12010036

Academic Editor: Marco Cococcioni

Received: 1 November 2023
Revised: 11 December 2023
Accepted: 19 December 2023
Published: 22 December 2023

1. Introduction

The alignment quality of the propulsion shafting is a critical factor influencing ship propulsion performance [1]. High-quality alignment of shafting is an important guarantee for safe navigation [2,3]. Poor alignment quality can result in excessive bearing force, abnormal wear, or even shafting failure, leading to power loss and hindered sailing capability [4–7]. For example, elementary mistakes have been discovered in the shaft alignment of the HMS Prince of Wales: both the GBP 3 billion warship's starboard and portside shafts are misaligned, causing the shafts to be offset. This situation has seriously affected its normal navigation [8]. The alignment quality of the propulsion shafting relies on meticulous design and installation processes. Therefore, strict specifications must be followed during the inspection of its design, manufacturing, and installation procedures to ensure alignment quality [9,10].

Shafting alignment involves precise installation according to design shafting specifications—a meticulous process within shipbuilding procedures. The bearing height adjustments typically require accuracy to 0.01 mm. When a ship under construction moves from the shipyard to the wharf, the hull is deformed due to the change of support forms and ambient temperature. In addition, the main engine hoisting will also cause local deformation of the hull and the shaft line will be deformed accordingly. Therefore, the shafting alignment and adjustment work is usually arranged after the ship is launched and the main engine is hoisted. However, this causes another problem: since it becomes difficult to establish a precise bearing height reference standard after launching the ship, evaluating the installation quality solely based on direct measurement of bearing height is not feasible. Consequently, the shafting alignment quality is inspected directly by measuring the bearing load.

Shafting alignment is conducted through an iterative process involving bearing load checking, bearing displacement values calculation, and implementation of bearing displacements until all the bearing loads pass the check. Bearing load checking entails measuring the bearing load to determine if it falls within the allowable deviation range. If not, a scheme for adjusting the shaft needs to be developed by calculating and implementing appropriate displacement values for the bearings. Among them, calculating accurate bearing displacement values is crucial in achieving proper shafting alignment, as it reduces adjustment times and improves efficiency.

Currently, the trial-and-error method is commonly employed to determine bearing displacement values by repeatedly comparing deviation direction and magnitude between measured loads and design loads until they converge towards each other. This method works well for short shafts with a single support where only one variable needs to be adjusted iteratively. However, for a multi-support long shaft system with the engine room arranged in the front or middle of the hull, it becomes a multivariate solution problem with increased constraints that limit effective adjustment of bearings. Additionally, there exists a mutual coupling relationship between bearing loads which makes it challenging to find combinations of height adjustments that satisfy all specifications solely through the trial-and-error method. Therefore, achieving efficient shafting alignment primarily relies on experienced technicians.

To overcome the limitations of the trial-and-error method, a more scientific and rational approach involves utilizing the design shafting bearing load and displacement relationship as a basis for determining the bearing displacement values. This approach considers the interdependent relationship between bearing loads and employs numerical calculations using methods such as the three moment method [11,12], transfer matrix method [13], singular function method [14], or finite element method [15] to obtain the influence number matrix representing changes in bearing load caused by unit changes in bearing height. Based on this, the minimum bearing load of the stern tube [16] or the minimum adjustment displacement [17] is taken as the optimization objective function, the allowable deviation and rotation angle of the bearing load are taken as the constraint conditions, and optimization algorithms such as linear programming [18] and quadratic programming [19] are employed to calculate the bearing displacement values. In order to reduce the computational complexity of the shafting alignment caused by the optimization algorithm, Deng et al. [20] take the simulated data of the design shafting numerical model as a training sample to represent the relationship between bearing load and height. It then becomes possible to input measured loads from the current shafting state with designed loads into this neural network model during shafting alignment site operations, thereby obtaining current heights and design heights for each bearing which can then be compared to derive their corresponding displacement values. It is worth noting that the aforementioned method is based on the design shafting model, while the actual shafting during alignment is formed through processing, manufacturing, and installation processes as depicted in Figure 1. However, deviations exist between them due to manufacturing errors, positioning errors, and assembly errors [21,22], leading to incomplete agreement between the actual and design shafting bearing load and displacement relationship. Therefore, the effectiveness of this method for alignment depends on the magnitude of these deviations. Given the current limitations in shipbuilding accuracy, further improvements are required to enhance its practical applicability.

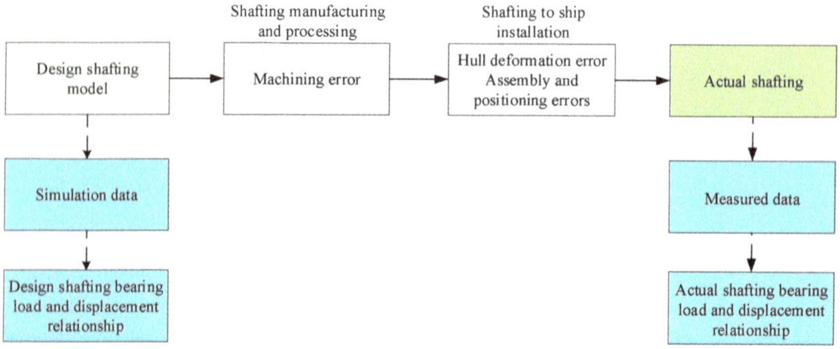

Figure 1. Design shafting to actual shafting evolution.

Due to the limitations in shipbuilding accuracy and the complex coupling characteristics between shafting bearing loads, multi-support shafting alignment becomes a more challenging task. The personnel responsible for shafting assembly often encounter numerous repeated adjustment attempts, which can lead to difficulties in identifying the adjustment rules, loss of direction, and even greater deviations in bearing loads with each subsequent adjustment. Consequently, this results in prolonged alignment periods for the shafting and low efficiency in adjustments. For instance, one study [23] mentioned a delay of more than 4 months in dock assembly after ship launching, with a shafting adjustment period lasting as long as 31.5 weeks. Another study [24] mentioned that the intermediate shafting adjustment of a ship took as long as 35 days, accounting for 72% of the total shafting installation period and significantly impeding ship construction progress.

Building upon previous study [20], Deng et al. considered the deviation between the design shafting and the actual shafting, respectively using the model simulation data representing the design shafting and the measured data generated in the shafting alignment process to represent the actual shafting, and trained the neural network model with different confidence levels, so as to improve the accuracy of determining bearing displacement values [25]. However, since the bearing height datum cannot be found after the ship is launched, the bearing height defined by this method is not easy to obtain in engineering applications.

The key factor for accurately calculating bearing displacement values lies in establishing an accurate actual shafting load and displacement relationship. This relationship evolves from the design shafting, undergoing various error condition. The design shafting is completely known, and the measured data of the actual shafting can be obtained gradually during the alignment process. Therefore, establishing the relationship between the actual shafting bearing load and displacement relationship becomes a modeling problem under incomplete information conditions. It is essential to construct the actual shafting bearing load and displacement relationship model based on the design shafting model and limited measured data.

Transfer learning, as an emerging machine learning algorithm, has been widely employed to address incomplete information problems by overcoming isolated learning methods [26]. By transferring relevant knowledge from known models, transfer learning reduces the reliance of target models on extensive target data and enables construction with only a small amount of target data available. Shahin et al. [27] utilized a transfer learning method based on regularization to eliminate the negative impact caused by missing information in incomplete target domain datasets and developed a prediction model suitable for the target domain. Zheng et al. [28] applied fuzzy sets for data processing and conducted co-evolutionary transfer learning using different types of training data to overcome limited training samples when constructing practical models. Xiao et al. [29] considering transfer learning as their primary solution, reduced weightage assigned to negative samples through loss functions to tackle the problem of inaccurate and incomplete vehicle trajectory collection data, and constructing a trajectory prediction framework aligned with actual environments.

This research article proposes a calculation method for determining the bearing displacement values of multi-support shafting based on transfer learning. By utilizing the transfer learning method, the actual shafting bearing load and displacement relationship model is constructed using both the design shafting model and measured data from the adjustment process. Then, bearing displacement values are further obtained. This research mainly comprised (1) establishing a neural network model to characterize the functional relationship between the load values before displacement, the load values after displacement, and displacement values of the intermediate bearing, and (2) utilizing the transfer learning approach, deriving the actual shafting bearing load and displacement relationship based on the design shafting model and measured data from adjustment process. As more measured data accumulates during the adjustment process, this model progressively improved in accuracy. (3) Considering the mounting positioning deviation of the stern bearing and main engine in different degrees, the application effect of this method was evaluated by the simulated process of adjusting a particular ship's shafting.

2. Calculation Method of Bearing Displacement Values Based on Transfer Learning

2.1. General Idea of Calculating the Bearing Displacement Value

The overall concept is illustrated in Figure 2. The left side of the figure depicts the elements associated with design shafting, including the establishment of a numerical model based on design parameters and the utilization of simulation data to derive a bearing load and displacement relationship model for the design shafting. On the right side, elements related to actual shafting are presented, primarily encompassing installation and adjustment processes. Two main connections exist between the designed and actual shafting: firstly, the design shafting is subject to various error factors that contribute to the formation of the actual shafting; secondly, by leveraging the shafting bearing load and displacement relationship of the designed shafting, transfer learning is combined with measured data from the actual shafting to establish the bearing load and displacement relationship for the actual shafting. This model is then used to acquire bearing displacement values and implement bearing displacement, conduct load and displacement tests, and verify bearing loads. In the iterative process of shafting alignment, the number of cycles is also the number of shafting adjustments. Transfer learning can be performed during each cycle, so the accuracy of the neural network model characterizing the actual bearing load and displacement relationship will improve with the increase of adjustment times. Consequently, more accurate bearing displacement values are obtained. Finally, evaluating the effectiveness of calculating bearing displacement values relies on the number of adjustments and assessing errors between measured loads and design loads.

2.2. Modeling the Design Shafting Bearing Load and Displacement Relationship

A study [20] constructed a neural network model of design shafting bearing load and height relationship, as depicted in Figure 3. However, measuring the bearing height in engineering is challenging and its practical application is difficult. In this research, the transfer learning method approach was applied by considering the design shafting bearing load and displacement relationship model as the fundamental model and combining it with the measured data from the actual shafting alignment process. For a shafting with 'n' intermediate bearings, Figure 4 illustrates the neural network topology constructed. The input of the model consists of each intermediate bearing load value before and after displacement, while the output represents the corresponding displacement values. The number of intermediate hidden layers was set to 2. The training sample of the model was calculated using the design shafting numerical model, which also held the necessary data for the actual shafting alignment process.

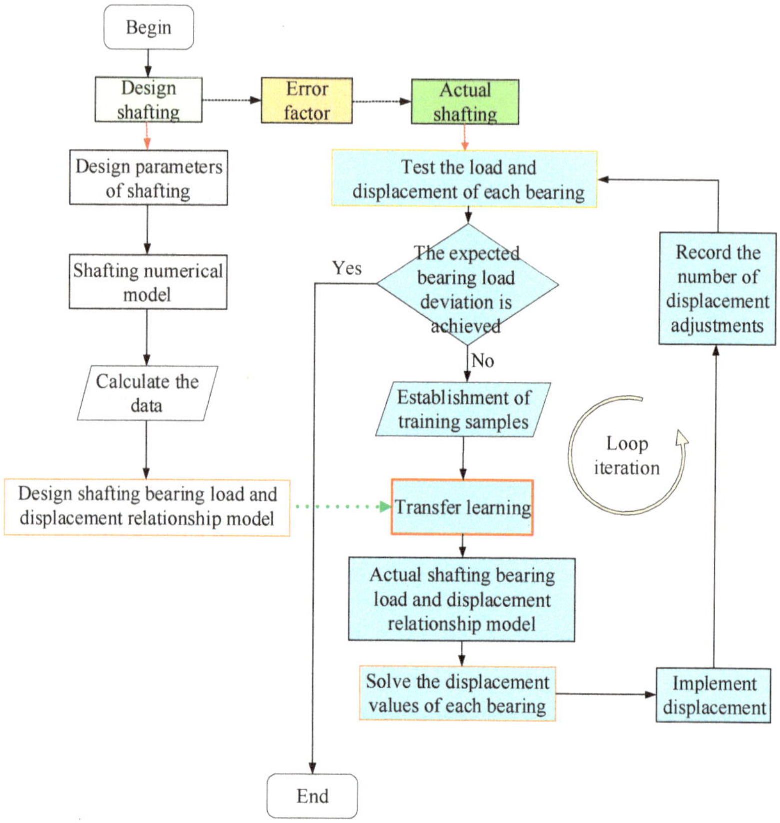

Figure 2. Flowchart of the method.

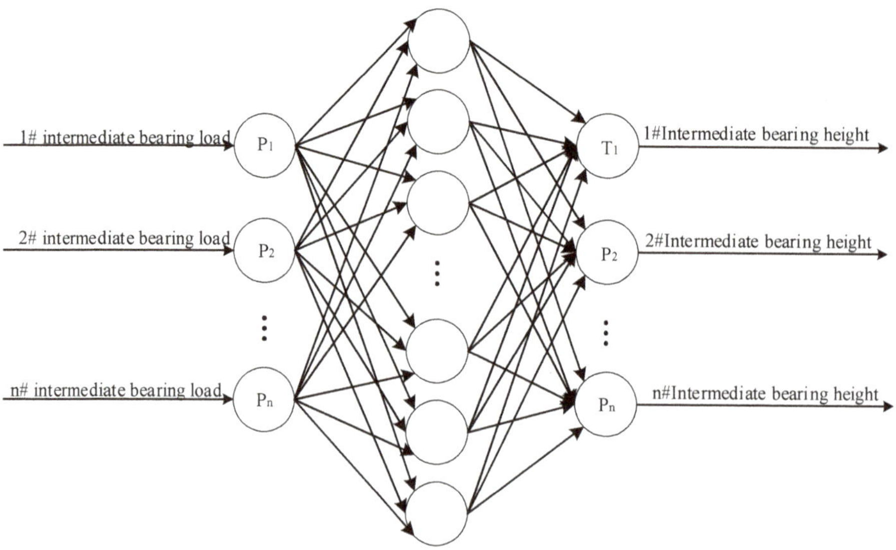

Figure 3. Neural network topology based on bearing load and bearing height.

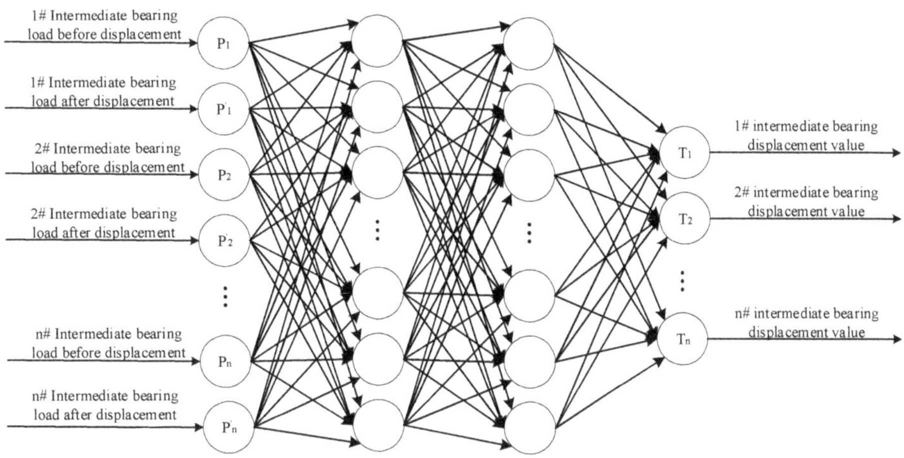

Figure 4. Neural network topology based on bearing load and bearing displacement.

2.3. *The Measurement of Shafting Alignment Parameters and the Construction of Training Samples*

Displacement sensors were placed on each intermediate bearing to accurately measure their displacements, while the dynamometer method, jacking method or strain gauge method were employed to measure loads on these bearings. During the shafting adjustment process, initial loads on all bearings [$f_1 1$, $f_2 1$, ..., $f_n 1$] were initially measured before any displacements were implemented. Subsequently, adjustments were made based on calculated displacement values, followed by measurements of individual bearing displacements [$d_1 1$, $d_2 1$, ..., $d_n 1$], and loads after displacement adjustments [$f_1 2$, $f_2 2$, ..., $f_n 2$]. The loads before and after displacement for each bearing, along with their corresponding displacement values obtained during the adjustment process, constituted a set of measured samples representing the actual shafting [$f_1 1$, $f_2 1$, ..., $f_n 1$; $f_1 2$, $f_2 2$, ..., $f_n 2$; $d_1 1$, $d_2 1$, ..., $d_n 1$].

Accumulating a larger volume of measured samples enhances the effectiveness of transfer learning. Existing measured data can be combined to expand the sample set. For instance, considering two adjacent sample groups [$f_1 1$, $f_2 1$, ..., $f_n 1$; $f_1 2$, $f_2 2$, ..., $f_n 2$; $d_1 1$, $d_2 1$, ..., $d_n 1$] and [$f_1 2$, $f_2 2$, ..., $f_n 2$; $f_1 3$, $f_2 3$, ..., $f_n 3$; $d_1 2$, $d_2 2$, ..., $d_n 2$], a new sample group can be generated by merging bearing displacement [$f_1 1$, $f_2 1$, ..., $f_n 1$; $f_1 3$, $f_2 3$, ..., $f_n 3$; $d_1 1 + d_1 2$, $d_2 1 + d_2 2$, ..., $d_n 1 + d_n 2$]. 'N' groups of adjacent samples can create ($\frac{n^2}{2} - \frac{3}{2}n + 1$) groups of new samples, thereby expanding the number of measured samples to a certain extent.

2.4. *Transfer Learning Method*

There are four main transfer learning methods: instance-based transfer learning, feature-based transfer learning, parameter-based transfer learning, and relation-based transfer learning. Given the resemblance between the design shafting model and the actual shafting model, the samples obtained from the two models had no missing items and shared a certain common connection. Consequently, this research employed two transfer learning methods: sample-based and model-based.

2.4.1. Sample-Based Transfer Learning

The sample-based transfer learning method reuses data samples according to certain weight generation rules, so as to achieve the purpose of improving results. The simulation sample is used as the source data, the measured sample is used as the target data, and the source data and the target data are weighed by weight generation rules. The weight is calculated according to Equations (1) and (2):

$$y_i = N_i / \sum_i^n N_i \tag{1}$$

In Equation (1), y_i is the proportion of samples of a certain type in the total samples; N_i is the number of samples of a certain type.

$$W(y_i) = 1/\log(C + y_i) \tag{2}$$

In Equation (2), $W(y_i)$ is the weight of a certain type of sample; C is a constant and $C > 0$.

The sample weight is added to the final loss function, and the influence of the measured samples on the loss function is adjusted to improve the attention of the neural network model to a small amount of measured data [30], and finally build an accurate actual shafting bearing load and displacement relationship. With the shafting alignment process, new measured data can be obtained and the measured sample set can be expanded, further improving the accuracy of the neural network model.

The loss function is calculated according to Formula (3), and the loss function adding sample weight is calculated according to Formula (4):

$$E = \frac{1}{2N}(T - Y)^2 = \frac{1}{2N}\sum_{i=1}^{N}(t_i - y_i)^2 \tag{3}$$

In Equation (3), T is the real result; Y is the network output result; i is the i-th data; t_i is the real result corresponding to the i-th data; y_i is the network output result corresponding to the i-th data; N is the number of training samples; $T - Y$ is the error between each training sample and the real result.

$$E' = \frac{1}{2N}(T - Y)^2 * W(y) = \frac{1}{2N}\sum_{i=1}^{N}(t_i - y_i)^2 * W(y_i) \tag{4}$$

In Equation (4), E' is the weight-adjusted loss function.

2.4.2. Model-Based Transfer Learning

The primary aim of this method is to identify shared parameter information between the source and target domains, facilitating migration between the two. The model training process is divided into two parts: universal learning and feature learning. Fundamentally, this method involves initializing model parameters based on acquired knowledge rather than relying solely on random initialization [31]. It effectively combines static experimental simulation data with real-time information, enabling model adaptation to dynamic environments and the construction of accurate models [32,33].

In the context of shafting alignment, certain model parameters can be shared between the design shafting bearing load-displacement relationship and the actual shafting bearing load-displacement relationship due to their correlation. These model parameters encompass knowledge acquired from the simulation sample set, some of which is also applicable to the measured sample set. Consequently, parameter sharing becomes feasible through this approach, allowing for the sharing of specific parameters.

Typically, the first layer is not particularly related to the specific dataset, while the last layer of the network is closely related to the selected dataset and its task goals. The features extracted by the first layer are referred to as general features, whereas those obtained by the last layer are termed specific features [34]. Given discrepancies between design shafting and actual shafting, the output layer of the design shafting neural network model is closely related to the simulation sample set. Therefore, utilizing output layers from the design shafting neural network models for actual shafting models is not suitable. Consequently, adjustments focus solely on specific feature layers, while other layers remain unchanged during this method. In instances of suboptimal performance, further fine-tuning of the remaining layers can be undertaken by reducing the initial learning rate to one-tenth of the value used during initial training.

The model-based transfer learning process is depicted in Figure 5, and involves copying some parameters from the design shafting load and displacement relationship neural network model as parameters of the actual shafting load and displacement relationship model. Subsequently, the network parameters of the output layer are randomly initialized. By training using measured sample sets, an accurate bearing load and displacement relationship model for actual shafting can be obtained through this approach.

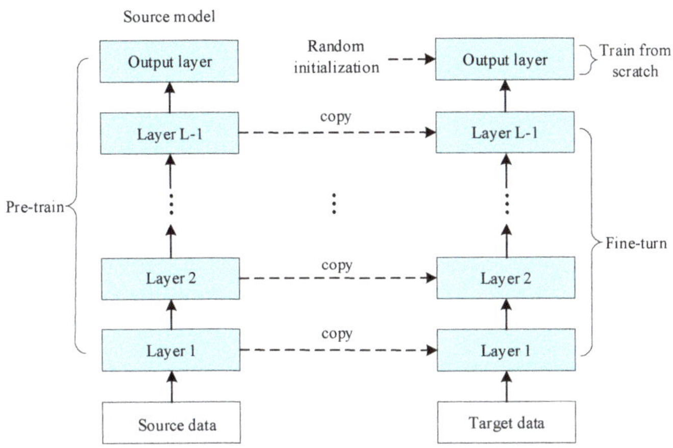

Figure 5. Schematic diagram of model-based transfer learning.

2.5. Calculating the Bearing Displacement Values

Based on the neural network topology constructed in Section 2.2, both measured loads and design loads for each bearing were input into the neural network model, and the model output was a set of bearing displacement values.

3. Application Examples

3.1. Research Object

The propulsion shafting of a large cruise rescue ship was taken as the research subject. The total length of the shafting was 43.8 m, including three intermediate shafts with lengths of 8 m, 8 m, and 7.461 m. The main shaft section of each intermediate shaft had an outer diameter of 305 mm and an inner diameter of 120 mm. The stern shaft had a length of 18.146 m, with its main shaft segment having an outer diameter of 350 mm and an inner diameter of 120 mm. Positioned in front of the stern shaft was the propeller, while behind the intermediate shaft lay the main engine crankshaft. This multi-support shafting consisted of four intermediate bearings. The structural diagram can be seen in Figure 6.

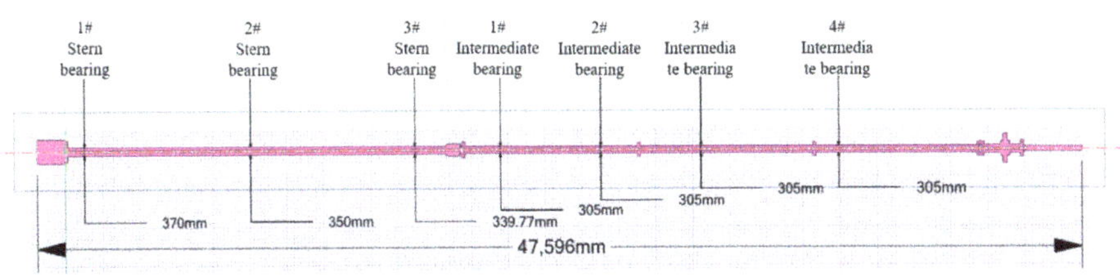

Figure 6. Schematic diagram of propulsion shafting structure.

All design parameters were meticulously known, enabling the creation of a comprehensive numerical model using finite element software. As per the specifications delineated in the shafting design calculation sheet, constraint conditions, load, and displacements were applied to each part of the shafting model. Displacement constraints were specifically applied to the three stern bearings in accordance with their designated heights. Moreover, the section of the shafting tail extending beyond the hull was treated as a free end. A concentrated force equivalent to the weight of the propeller was applied at the propeller installation position and the buoyancy of the propeller was applied, accounting for gravity across the entire shafting. To accurately replicate bearing support characteristics, axial and radial springs were incorporated at the contact surface of each intermediate bearing with the shafting. The stiffness of the these springs replaced that of the entire bearing support system. The radial springs simulated actual bearing support behavior. When determining adjustments for bearing displacement, only radial displacement at intermediate bearings was considered by modifying the length of corresponding radial springs. Additionally, an axial spring, possessing sufficient elastic modulus and approximate rigidity, was introduced to confine the overall axial displacement of the entire shafting [20].

By substituting the bearing design height into the model for computation, the finite element analysis revealed a maximum error of 4.74% between the intermediate bearing load and the design load. The minimum discrepancy was 0.77%, with an average of 2.73%. These values aligned with the specification requirements, affirming the model's adherence to the design criteria and its reliability for both calculation and simulation samples.

Critical factors influencing shafting alignment were the stern bearing and the main engine, where positioning errors and installation deviations significantly impact the alignment process. Table 1 compares various deviation scenarios between the designed shafting and the actual shafting. Notably, positive positioning error refers to a deviation in the vertical upward direction, while positive axial deviation denotes a stretching direction.

Table 1. Different error scenarios.

Serial Number	Deviation Factors
Case 1	Main engine positioning deviation + 2 mm
Case 2	Main engine axial mounting deviation + 5 mm
Case 3	Main engine positioning deviation + 2 mm and stern shaft positioning deviation − 0.5 mm
Case 4	Main engine positioning deviation + 5 mm, stern shaft positioning deviation − 2 mm and main engine unit axial installation deviation + 10 mm

The actual shafting in this research was simulated by modifying the model parameters according to the deviation in Table 1 on the basis of the design shafting numerical model. This actual shafting numerical model was employed to simulate displacement adjustment processes, with its calculated data serving as measured data for the actual shafting.

3.2. *Analysis Ideas*

The calculation methods of displacement values investigated in this article are shown in Table 2. The primary focus lies in investigating the efficacy of two transfer learning methods across different shafting error scenarios. For comparative purposes, Method 1, utilizing the design shafting bearing load and height relationship model, is introduced. This method employs simulated samples to train a neural network, generating the design shafting bearing load and displacement relationship model. Subsequently, it calculates displacement values based on the current bearing measurement load and design load [20]. To ensure a fair comparison, each deviation scenario starts from identical initial conditions. The alignment cycle for shafting adjustments is not based on load checking but on a predetermined number of adjustments as the exit condition.

Table 2. Method type.

Serial Number	Method
Method 1	Bearing load and height relationship based on the design shafting
Method 2	Bearing load and displacement relationship modified based on sample transfer learning
Method 3	Bearing load and displacement relationship modified based on model transfer learning

3.3. Construction of Transfer Learning Model

The design shafting numerical model was constructed, setting the height change range of each intermediate bearing at ±2 mm, with a displacement interval defined as a minimum unit of 0.1 mm [35]. For this research, a total of 1900 sets of simulation training samples were generated, with 1800 sets allocated for training samples and an additional 100 sets serving as test samples.

Considering the characteristics of the research subject, the neural network model took as inputs the load values of the four intermediate bearings before and after displacement, while producing the corresponding displacement values for these bearings as outputs. The BP algorithm selects the logarithmic S-type logsig function as the hidden layer node transfer function, with the linear Purelin function serving as the transfer function for the output layer nodes (expressed as y = x). The training times were 10,000, the learning rate was 0.05, and the number of nodes in the intermediate double hidden layers was set to 15. Post-training assessment revealed an average error of 0.40% in the test data.

These results affirm that the neural network model trained extensively on simulation data exhibited superior learning capabilities, effectively portraying the design shafting bearing load and displacement relationship. This model served as the fundamental model for transfer learning in Method 2. When constructing the fundamental model for transfer learning in Method 3, it was crucial to optimize and adjust the number of neurons in proximity to the output layer. This adjustment significantly influenced its efficacy, requiring meticulous optimization. After rigorous calculation, the impact of varying neuron counts in this layer on the calculation error of the migrated model was confirmed and is graphically depicted in Figure 7. Notably, the number of 5 neurons demonstrates the smallest error. Consequently, the number of neurons in the hidden layer adjacent to the output layer is refined from 15 to 5 in this method, while other parameters remain constant.

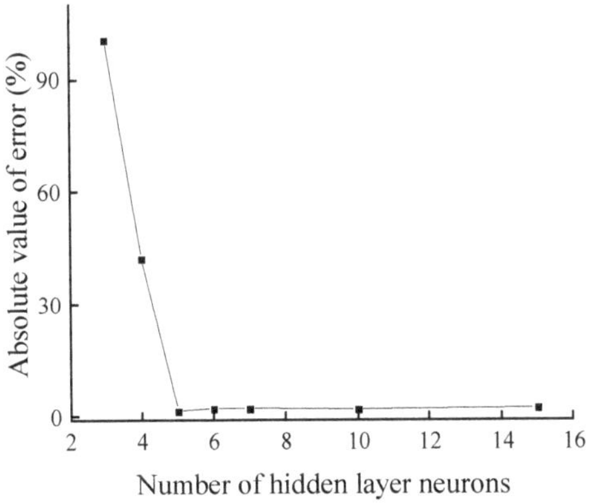

Figure 7. The impact of the number of hidden layer neurons near the output layer on Method 3.

3.4. Application Effect

Figures 8–11 illustrate the effects of the four scenarios on the neural network performance. In each figure, (a), (b), and (c) respectively represent the variation of the measured load error of each intermediate bearing with the number of bearing displacement adjustments. The final error comparisons between the measured load and design load for each scenario are presented in Figure 12; the logarithmic coordinate system is employed for the purpose of comparison.

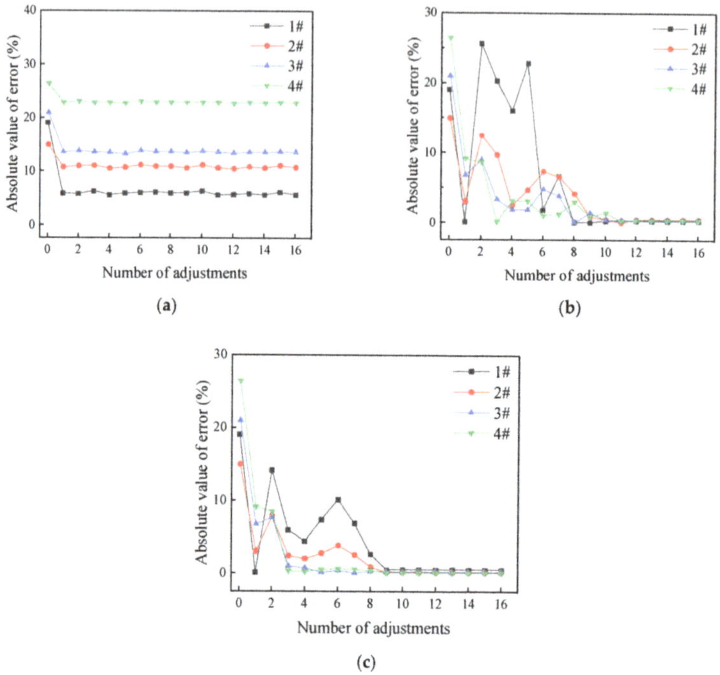

Figure 8. Application effects of three methods in Case 1. (**a**) Method 1; (**b**) Method 2; (**c**) Method 3.

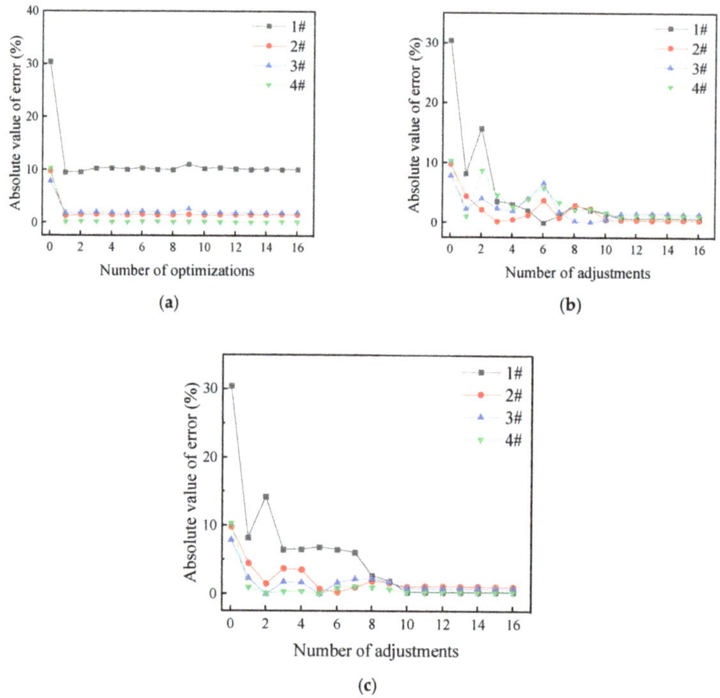

Figure 9. Application effects of three methods in Case 2. (**a**) Method 1; (**b**) Method 2; (**c**) Method 3.

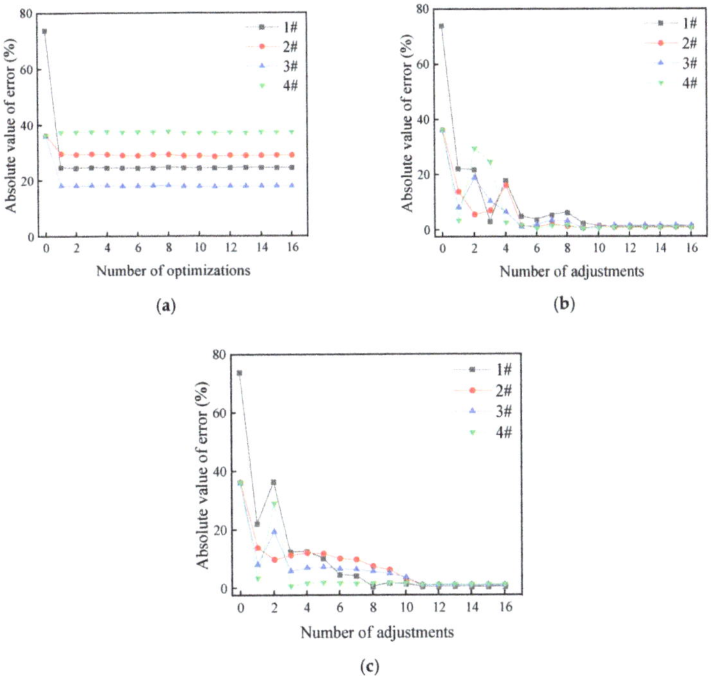

Figure 10. Application effects of three methods in Case 3. (**a**) Method 1; (**b**) Method 2; (**c**) Method 3.

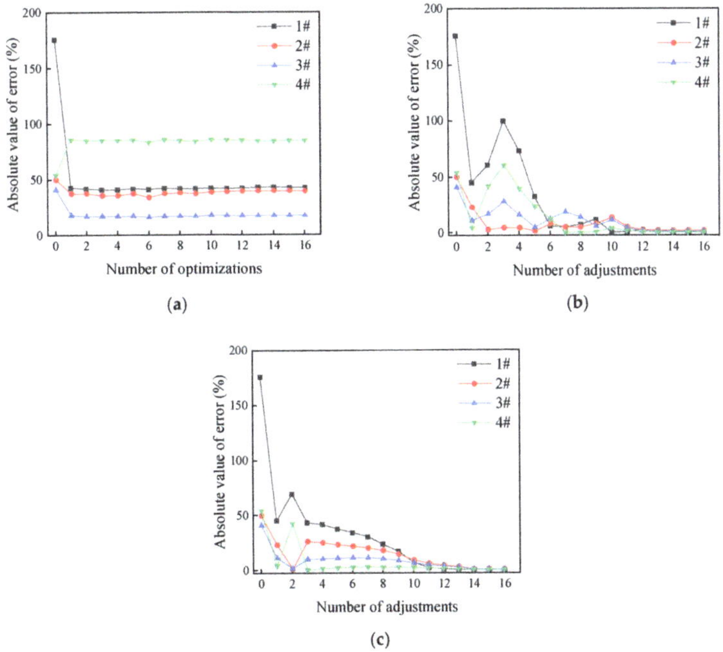

Figure 11. Application effects of three methods in Case 4. (**a**) Method 1; (**b**) Method 2; (**c**) Method 3.

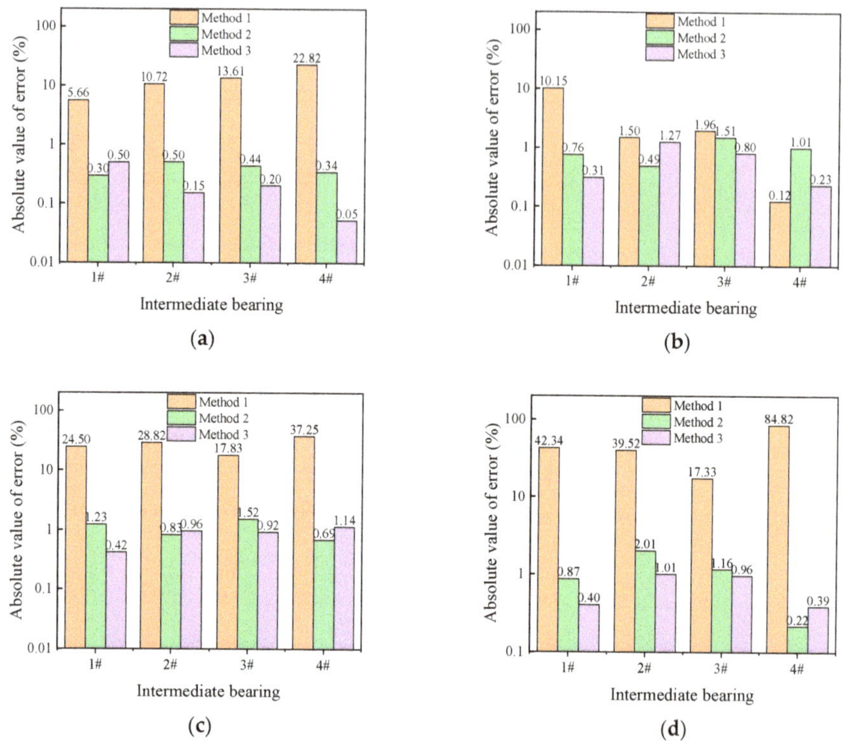

Figure 12. Comparison of final errors in various scenarios. (**a**) Comparison of errors in Case 1; (**b**) Comparison of errors in Case 2; (**c**) Comparison of errors in Case 3; (**d**) Comparison of errors in Case 4.

The effectiveness of Method 1 remained consistent across all scenarios. As the error escalated, so did the deviation between bearing measurements and design loads. Except for in scenario 2, deviations between the design shafting and the actual shafting led to bearing load errors exceeding the allowable limit (±20% of the design load) in all three scenarios. Throughout the shafting adjustment, while neural network errors maintained relative stability, the calculated measured bearing loads displayed minimal fluctuation. Consequently, Method 1 lost its capacity to adjust bearing loads.

In contrast, Method 2 consistently demonstrated effective application across every scenario as bearing measured loads gradually approached design levels during shafting adjustments. Across the four scenarios, the error between each intermediate bearing measured load and the design load remained below 10% at the 6th, 3rd, 5th, and 10th adjustments, respectively, reaching a steady state after the 11th, 11th, 11th, and 14th adjustments. The maximum error after stabilization was 2.01%, the minimum was 0.22%, and the average was 0.87%, significantly lower than those seen with Method 1.

Method 3 exhibited commendable effectiveness across all scenarios as it gradually aligned the measured bearing loads with the design levels during shafting adjustments. Notably, compared to Method 2, Method 3 demonstrated milder fluctuations overall. In the four scenarios, the error between each intermediate bearing measured load and the bearing design load was consistently below 10% at the 5th, 7th, 8th, and 11th adjustments respectively, stabilizing after the 10th, 12th, 12th, and 13th adjustments. Upon stabilization, Method 3 recorded a maximum error of 0.97%, a minimum value of 0.052%, and an average of 0.39%. Method 3 achieved a further reduction in error after stabilization compared to Method 2, positioning the obtained measured bearing load closer to the design bearing load.

The stability, rapidity, and accuracy of all three methods in the shafting adjustment process are summarized in Table 3.

Table 3. Comparison of effects of various methods.

Type	Iterative Stability	Rapidity	Accuracy
Method 1	Less fluctuation	Only one adjustment is required	Major error
Method 2	The error fluctuates is large	Achieving stability is slightly slower but requiring less adjustments to make the deviation less than 10%	High accuracy when the deviation is small; when the deviation is large, the accuracy decreases
Method 3	Relatively stable with slight fluctuations	Achieving stability is slightly faster but requires more adjustment to make the deviation less than 10%	Maintaining high accuracy in various deviations

4. Discussion

(1) Method 1 disregards the discrepancy between the designed shafting and the actual shafting, and can only calculate the displacement values based on the design bearing load and displacement relationship model, according to the bearing measured load and design load. Essentially, it relies on an iterative trial-and-error approach using the alignment path obtained from the design shafting model, without considering data information or experiential knowledge acquired during learning about actual bearing adjustment process. Moreover, when there was a significant initial bearing load error, it lacked the ability to reduce this error in subsequent shafting adjustment processes.

(2) Method 2 and Method 3 are based on the design shafting bearing load and displacement relationship, using measured data obtained from the actual shafting through the bearing alignment process for transfer learning. This enabled them to rectify the actual shafting bearing load and displacement relationship model, gradually enhancing the accuracy of acquired bearing displacement values. Both methods effectively indicated the direction for shafting adjustment and correction, addressing issues commonly encountered in traditional shafting adjustment methods such as difficulty in identifying adjustment patterns, loss of adjustment direction, and increased deviation in bearing load. Consequently, they improved both quality and efficiency in shafting alignment. Compared to Method 2, Method 3 achieved stability more rapidly while further improving accuracy of bearing displacement values after reaching stability. Moreover, it maintained high accuracy under various error conditions. The simulated sample and the measured sample were always combined in the iteration process of Method 2, but the error between them was difficult to completely eliminate through weight, resulting in a slightly lower accuracy of the bearing displacement values after stability. Method 3 solely employs measured samples for training a new fully connected layer, which not only required less time but also yielded a trained model closer to reality, thereby ensuring higher accuracy of displacement values.

(3) Increasing deviation between the design shafting and the actual shafting demanded more adjustments for Method 2 and Method 3 to achieve stability. Consequently, the number of displacement adjustments required for each intermediate bearing load error to reach 10% gradually rose. Moreover, due to a small proportion of measured data during early stages of shafting alignment, Method 2 exhibited fluctuations in bearing measured load.

(4) Neither Method 2 nor Method 3 completely eliminated the discrepancy between the bearing measured load and the design load. This is because the design load is calculated based on the design shafting, while shafting alignment involves adjusting bearing displacements for actual shafting. Consequently, the deviation between the design load and the actual bearing is the reason why the bearing measured load cannot be equal to the design load.

5. Conclusions

In this research, under the conditions of incomplete data of actual shafting during shafting alignment, transfer learning was used to construct the actual shafting bearing load and displacement relationship based on the design shafting model and limited actual shafting measured data. According to the application effects of four scenarios with different degrees of main engine positioning deviation and stern shaft positioning deviation, the following conclusions were obtained:

(1) The multi-support shafting displacement values calculation method based on transfer learning effectively utilized design shafting existing knowledge and the measured data from the actual shafting displacement adjustment process. This progressively improved the accuracy of the obtained displacement values, enhancing shafting alignment quality and efficiency.

(2) While the numerical simulation validates the method's superiority, further research is required to enhance efficiency and minimize the necessary iterations.

Author Contributions: Y.D.: conceptualization, methodology, writing—review and editing. Y.L.: software, formal analysis, writing—original draft. H.Z.: project administration, writing—review and editing. S.F.: project administration, writing—review and editing. All authors have read and agreed to the published version of the manuscript.

Funding: This research was supported by the National Natural Science Foundation of China (NSFC) under grant numbers U2341284 and 51839005.

Institutional Review Board Statement: Not applicable.

Informed Consent Statement: Not applicable.

Data Availability Statement: The data are available upon request from the corresponding author.

Acknowledgments: The authors would like to extend their sincere appreciation to the Researchers Supporting Project Number (U2341284 and 51839005). The authors thank all the participants for their help and friendship.

Conflicts of Interest: The authors declare no conflict of interest.

References

1. Magalhães, D.L.; de SS Martins, D.H.C.; Castro, B.M.; Vaz, L.A.; Monteiro, U.A.; Gutiérrez, R.H. Machine learning performance comparison for main propulsive shafting systems alignment. *Ocean. Eng.* **2023**, *280*, 114556. [CrossRef]
2. Zou, D.; Jiao, C.; Ta, N.; Rao, Z. Forced vibrations of a marine propulsion shafting with geometrical nonlinearity (primary and internal resonances). *Mech. Mach. Theory* **2016**, *105*, 304–319. [CrossRef]
3. Huang, Y.; Chen, T.; Shieh, P. Analytical estimation of the noise due to a rotating shaft. *Appl. Acoust.* **2014**, *76*, 187–196. [CrossRef]
4. Lai, G.; Liu, J.; Liu, S.; Zeng, F.; Zhou, R.; Lei, J. Comprehensive optimization for the alignment quality and whirling vibration damping of a motor drive shafting. *Ocean. Eng.* **2018**, *157*, 26–34. [CrossRef]
5. Guo, J.; Huang, S.; Nikolay, T.; Li, M. Vibration damping of naval ships based on ship shock trials. *Appl. Acoust.* **2018**, *133*, 52–57. [CrossRef]
6. Seo, C.-O.; Jeong, B.; Kim, J.-R.; Song, M.; Noh, J.-H.; Lee, J.-U. Determining the influence of ship hull deformations caused by draught change on shaft alignment application using FE analysis. *Ocean. Eng.* **2020**, *210*, 107488. [CrossRef]
7. Ho, W.-H.; Tsai, J.-T.; Chou, J.-H.; Yue, J.-B. Intelligent hybrid Taguchi-genetic algorithm for multi-criteria optimization of shaft alignment in marine vessels. *IEEE Access* **2016**, *4*, 2304–2313. [CrossRef]
8. The Telegraph. 2023. Available online: https://www.telegraph.co.uk/news/2023/02/04/hms-prince-wales-departure-hit-elementary-blunders/ (accessed on 25 August 2023).
9. American Bureau of Shipping. *Guidance Notes on Propulsion Shafting Alignment*; ABS: Spring, TX, USA, 2019.
10. China Classification Society. *Regulations for the Construction and Classification of Sea-Going Steel Ships*; China Comunications Press: Beijing, China, 2023.
11. Xue, W. The Bearing Load Optimization of 110,000 Tons Oil Tanker. Master's Thesis, Dalian University of Technology, Dalian, China, 2018.
12. Zhou, R.; Yao, S.; Zhang, S.; Li, B. Theoretic studies of the three-moment equation and its application in the vessel's propulsion shafting alignment. *J. Wuhan Univ. Technol.* **2005**, *27*, 77–78. [CrossRef]
13. Li, G.; Wu, W.; Rao, C.; Wu, W. Numerical calculation of whirling vibration of ship shafts based on transfer matrices method. *Chin. J. Ship Res.* **2010**, *5*, 60–63. [CrossRef]

14. Wang, J.; Lou, J.; Li, X.; Chen, R. Research on the shafting alignment calculation based on differential equations for beam deformation and singularity functions. *Chin. J. Ship Res.* **2019**, *41*, 71–75. [CrossRef]
15. Yan, B. Research on Load Adjustment Technology of Intermediate Bearing for Ship Long Shafting System. Master's Thesis, Dalian University of Technology, Dalian, China, 2017.
16. Gong, S. Digital Research on the Shafting Alignment. Master's Thesis, Dalian University of Technology, Dalian, China, 2014.
17. Wang, F. Study on Measuring and Adjusting Pivotal Techniques of Bearing Loads for Propulsion Shafting of Large Ships. Master's Thesis, Dalian University of Technology, Dalian, China, 2016.
18. Cdr Amit Batra, L.; Shankar, K.; Swarnamani, S. Propulsion shaft alignment measurements on warships afloat and alignment solution using multi-objective optimisation. *J. Mar. Eng. Technol.* **2007**, *6*, 39–49. [CrossRef]
19. Wang, F.; Li, R. Optimization of offsets of bearings in propulsion system of large ships and forecast of loads on bearings. *Shipbuild. China* **2015**, *56*, 152–160. [CrossRef]
20. Deng, Y.; Yang, X.; Huang, Y.; Pan, T.; Zhu, H. Calculation method of intermediate bearing displacement value for multisupported shafting based on neural network. *J. Ship Res.* **2021**, *65*, 286–292. [CrossRef]
21. Zhu, Y.; Li, Y.; Sun, Q. Research on manufacture technology of precision spindle system. *Mach. Tool Hydraul.* **2012**, *40*, 3. [CrossRef]
22. Deng, Y.; Yang, X.; Huang, Y.; Zhu, H.; Fan, S. Influence analysis of Installation height deviation of marine main engine on shafting alignment. *J. Wuhan Univ. Technol. Transp. Sci. Eng.* **2021**, *45*, 23–26. [CrossRef]
23. Zhang, X. Research On Efficiency Gains for Long Shafting Installation and Alignment on the Navy Ship. Master's Thesis, Shanghai Jiao Tong University, Shanghai, China, 2017.
24. Gong, T. Process Control for long Shafting Construction Quality and Period on High-Tech New Naval Ship. Master's Thesis, Shanghai Jiao Tong University, Shanghai, China, 2014.
25. Deng, Y.; Yang, X.; Fan, S.; Jin, H.; Su, T.; Zhu, H. Research on the installation and alignment method of ship multi-support bearings based on different confidence-level training samples. *J. Ship Res.* **2022**, *66*, 326–334. [CrossRef]
26. Zhuang, F.; Qi, Z.; Duan, K.; Xi, D.; Zhu, Y.; Zhu, H.; Xiong, H.; He, Q. A comprehensive survey on transfer learning. *Proc. IEEE* **2020**, *109*, 43–76. [CrossRef]
27. Siahpour, S.; Li, X.; Lee, J. A novel transfer learning approach in remaining useful life prediction for incomplete dataset. *IEEE Trans. Instrum. Meas.* **2022**, *71*, 3509411. [CrossRef]
28. Zheng, Y.-J.; Yu, S.-L.; Song, Q.; Huang, Y.-J.; Sheng, W.-G.; Chen, S.-Y. Co-evolutionary fuzzy deep transfer learning for disaster relief demand forecasting. *IEEE Trans. Emerg. Top. Comput.* **2021**, *10*, 1361–1373. [CrossRef]
29. Xiao, J.; Xiao, Z.; Wang, D.; Havyarimana, V.; Liu, C.; Zou, C.; Wu, D. Vehicle trajectory interpolation based on ensemble transfer regression. *IEEE Trans. Intell. Transp. Syst.* **2021**, *23*, 7680–7691. [CrossRef]
30. Covington, P.; Adams, J.; Sargin, E. Deep neural networks for youtube recommendations. In Proceedings of the 10th ACM Conference on Recommender Systems, New York, NY, USA, 15–19 September 2016; pp. 191–198.
31. Zhuang, K.; Ma, C.; Lam, H.-F.; Zou, L.; Hu, J. Pre-trained 1DCNN-BILSTM hybrid network for temperature prediction of wind turbine gearboxes. *Processes* **2023**, *11*, 3324. [CrossRef]
32. Zhu, G.; Sun, T.; Xu, Y.; Zheng, Y.; Zhou, L. Identification of internal short-circuit faults in lithium-ion batteries based on a multi-machine learning fusion. *Batteries* **2023**, *9*, 154. [CrossRef]
33. Ding, J.; Zhang, R.; Wen, X.; Li, X.; Song, X.; Ma, B.; Li, D.; Han, L. Interpretable feature construction and incremental update fine-tuning strategy for prediction of rate of penetration. *Energies* **2023**, *16*, 5670. [CrossRef]
34. Yosinski, J.; Clune, J.; Bengio, Y.; Lipson, H. How transferable are features in deep neural networks? *Adv. Neural Inf. Process. Syst.* **2014**, *2*, 3320–3328. [CrossRef]
35. Yang, X.; Deng, Y.; Zhu, H.; Fan, S. Intelligent assembly and adjustment method of ship multi support shafting based on machine learning. *Ship Build. China* **2021**, *62*, 184–191. [CrossRef]

MDPI AG
Grosspeteranlage 5
4052 Basel
Switzerland
Tel.: +41 61 683 77 34

Journal of Marine Science and Engineering Editorial Office
E-mail: jmse@mdpi.com
www.mdpi.com/journal/jmse

www.ingramcontent.com/pod-product-compliance
Lightning Source LLC
Chambersburg PA
CBHW061620220926
51850CB00005B/552
* 9 7 8 3 7 2 5 8 1 7 3 9 9 *